Science as Public Culture joins a growing number of recent studies examining science as a practical activity in specific social settings. Jan Golinski considers the development of chemistry in Britain from 1760 to 1820, and relates it to the rise and subsequent eclipse of forms of civic life characteristic of the European Enlightenment. Within this framework the careers of prominent chemists like William Cullen, Joseph Black, Joseph Priestley, Thomas Beddoes, and Humphry Davy are interpreted in a new light. The major discoveries of the time, including nitrous oxide (laughing gas) and the electrical decomposition of water, are set against the background of alternative ways of constructing science as a public enterprise. The book makes a significant contribution to our understanding of the relationship between scientific activity and processes of social and political change in a period of great transformations in chemistry and in the conditions of public life.

Science as public culture

SCIENCE AS PUBLIC CULTURE

Chemistry and Enlightenment in Britain, 1760–1820

JAN GOLINSKI

Department of History
University of New Hampshire

CAMBRIDGE
UNIVERSITY PRESS

Published by the Press Syndicate of the University of Cambridge
The Pitt Building, Trumpington Street, Cambridge
40 West 20th Street, New York, NY 10011, USA
10 Stamford Road, Oakleigh, Melbourne 3166, Australia

First published 1992

First paperback edition 1999

Printed in the United States of America

A catalog record for this book is available from the British Library.

Library of Congress Cataloging-in-Publication Data
Golinski, Jan.
Science as public culture : chemistry and enlightenment in Britain,
1760–1820 / Jan Golinski.
p. cm.
Inclues bibliographical references and index.
1. Chemistry – Great Britain – History – 18th century. 2. Chemistry –
Great Britain – History – 19th century. I. Title.
QD18.G7G65 1992
540'.941'09033 – dc20 91-39024
CIP

ISBN 0 521 39414 7 hardback
ISBN 0 521 65952 3 paperback

To my parents

Contents

Acknowledgments   *page* ix

List of illustrations   xi

1   Introduction: Science as public culture   1

2   "The study of a gentleman": Chemistry as a public science
    in the Scottish Enlightenment   11
       Chemistry as an academic discipline   13
       Gentlemanly science in the public realm   25
       The social construction of the Scottish program   37

3   Joseph Priestley and the English Enlightenment   50
       The uses of chemistry in Enlightenment England   52
       Making connections: Priestley's career   63
       The experimenter and the writer   77

4   Airs and their uses   91
       Priestley's chemistry in public education   93
       The birth of pneumatic medicine   105
       The analysis of air   117

5   The coming of the Chemical Revolution   129
       Lavoisier's theory and its reception in Britain   130
       The instruments of persuasion   137
       Demonstration, authority, and community   145

6   "Dr. Beddoes's Breath": Nitrous oxide and the
    culmination of Enlightenment medical chemistry   153
       The Pneumatic Institution   157
       Enthusiastic respirations: The nitrous oxide incident   166
       The end of Enlightenment science?   176

7   Humphry Davy: The public face of genius   188
       Davy's career: The creation of a public audience   190
       The voltaic pile: The making of an instrument   203
       Chlorine and "the lever of experiment"   218

8   Analysis, education, and the chemical community   236
        Specialist careers in the London chemical community   238
        The identity of the discipline and the reception of
        Dalton's atomic theory   255
        Mineralogy and the development of chemical analysis   269
        Conclusion: Discipline-formation and public science   283

Bibliography   289
Index   323

Acknowledgments

Writing this book has occupied me for more years than I care to remember. As a result, I am indebted to a large number of people who, in one way or another, have enabled me to complete it.

First and foremost I must thank the Master and Fellows of Churchill College, Cambridge, who elected me to a Junior Research Fellowship in 1986. Without the confidence they placed in me I should not have had the opportunity even to begin a project such as this. The students and faculty of the Department of History and Philosophy of Science also made my time in Cambridge profitable and enjoyable. The Institute for Research in the Humanities, University of Wisconsin-Madison, granted me a visiting Postdoctoral Fellowship in 1989, and the Huntington Library, San Marino, California, awarded me a V.M. Keck Foundation Fellowship in 1990.

Prior to this, the Royal Society had given a grant-in-aid for research that enabled me to initiate the project while I was at the University of Lancaster in 1985. It was brought to completion in the congenial surroundings of the University of New Hampshire, where the members of the Department of History and the Humanities Program have made me feel thoroughly at home.

Various colleagues have read and commented on whole or part of the manuscript. I greatly appreciate the comments of John Brooke, John Christie, John McEvoy, Roy Porter, Lissa Roberts, Simon Schaffer, Jim Secord, and Steven Shapin. Going considerably beyond the call of duty, Homer LeGrand gave the entire manuscript a very careful reading and then drove through the snows of New Hampshire to discuss it with me.

Other friends in Cambridge and elsewhere have offered general encouragement and advice. For sustaining me in these ways, I am deeply grateful to John Christie, Cathy Crawford, Larry Klein, Javed Majeed, Lissa Roberts, Jim and Anne Secord, Larry Stewart, Mary Terrall, Paul Wood, and especially Simon Schaffer (for his inspiration).

At Cambridge University Press, John Kim, Frank Smith, Helen Wheeler, and Richard Ziemacki have all been remarkably helpful and forbearing.

Audiences at a number of institutions have heard presentations of portions of the work, and I am grateful for their attention and advice: Department of History and Philosophy of Science, University of Cambridge; History of Science and Medicine Seminar, University College, London; Centre for History of Science, Technology and Medicine, University of Manchester; Science Studies Program, University of California, San Diego; Department of History of Science, Johns Hopkins University; Department of History of Science, University of Wisconsin-Madison; and Program in History of Science, Cornell University.

I should also like to thank the staff of the following institutions for their assistance with my researches: Cambridge University Library (Rare Books Room); Whipple Library of the Department of History and Philosophy of Science, University of Cambridge; British Library (Reference Division); Leeds University Library (Special Collections); Glasgow University Library (Special Collections); Royal College of Physicians of Edinburgh; Royal Institution, London; Memorial Library, University of Wisconsin-Madison; Huntington Library, San Marino, California; and Wellcome Institute for the History of Medicine, London.

The British Museum (Department of Prints and Drawings), the Huntington Library, the Cambridge University Library, and the National Portrait Gallery, London, have all helped with the supply of illustrations and have generously granted permission for their reproduction.

I am also grateful to the Glasgow University Library, the Royal College of Physicians of Edinburgh, the Royal Institution, and the Huntington Library for permission to quote from manuscripts in their collections.

Illustrations

1. Portrait of Joseph Black by David Martin — *page* 42
2. Portrait of Joseph Priestley by E. Sharples — 64
3. Priestley's experimental apparatus — 84
4. Anonymous caricature of Priestley as "Docter Phlogiston" — 180
5. "Revolution Anniversary," by W. Dent — 182
6. "M. Francois introduces Mr. Pr***tly," by James Sayers — 183
7. Portrait of Humphry Davy by H. Howard — 192
8. "Scientific Researches," by James Gillray — 202
9. The voltaic pile of Gay-Lussac and Thenard — 215
10. Interior view of the laboratory at the Royal Institution, with the arch leading through to the basement lecture theater at the right — 220
11. Plan of the laboratory of the Royal Institution, with the basement lecture theater at the left — 221
12. The chemical theater at Guy's Hospital — 250
13. Cronstedt's "portable laboratory" for examination of minerals in the field — 280
14. Blowpipe and associated apparatus for mineralogical analysis in the laboratory — 281

1

Introduction: Science as public culture

Appearances in these things are most deceptive: in the theatre experiments are made for illustration, and are generally of a simple kind, and easily comprehended, and the minds of the audience are prepared by the lecturer to follow and understand them. In the laboratory, on the contrary, this aid is wanting when most necessary; and, in consequence, operations . . . of a very accurate kind, and carried on with a perfect design, may appear confused to the uninstructed, or to the uninitiated.

John Davy, *Memoires of the Life of Sir Humphry Davy* (1836)[1]

Certainty, simplicity, vividness originate in popular knowledge. That is where the expert obtains his faith in this triad as the ideal of knowledge. Therein lies the general epistemological significance of popular science.

Ludwik Fleck, *Genesis and Development of a Scientific Fact* (1935)[2]

The career of experimental knowledge is the circulation between private and public spaces.

Steven Shapin, "The House of Experiment" (1988)[3]

Science, it has been said, is "public knowledge." The assertion is an appealing one, but it raises a host of problems.[4] When scientists and philosophers say that scientific knowledge is public, they seem to mean that it is accessible to all. Science has its basis in empirical facts, so anyone with normal senses can come to understand it. It is also thought that everyone can contribute to scientific knowledge, at least in principle. All claims are meant to be judged on their coincidence with the agreed-upon facts, without reference to the circumstances of their origin. Claims about the natural world become accepted scientific knowledge in a process that is supposed to be open and egalitarian. The scientific community is sometimes even taken as a model of an ideal open society.

There are, however, many problems with this view. It might be a de-

---

1 J. Davy (1836), I, pp. 259–260.   2 Fleck (1979), p. 115.
3 Shapin (1988), p. 400.   4 For discussion of these problems, see Ziman (1968).

sirable ideal, but it has evident faults as a model of how science actually works. In particular, it seems not to fit with lay people's experience of the impact of science on their lives. Nonscientists typically do not experience the falsifiability of scientific knowledge or the supposedly democratic character of scientific decision-making. To them, rather, science often appears as a system of authority, the tool of powerful interests in society. Notwithstanding widespread and enthusiastic interest in certain aspects of science, anxiety about unrestrained expertise regularly surfaces in connection with such controversial questions as nuclear power, genetic engineering, and animal experimentation. At such times, the assertion that science is a public activity is called into question.[5]

From the other side of the fence, but frequently in connection with the same controversial issues, scientific bodies have been heard to express concern about the lack of public understanding of science. The gap between expert knowledge and that of the nonexpert population is a matter of anxiety for scientific institutions, perhaps particularly when they feel that their political influence is declining.[6]

The most radical implication of such incidents is that the philosophical image of science that has typically supported assertions of its public character is untenable. Many sociologists and historians of science and some philosophers would now say that a view of science that sees it as entirely open, egalitarian, and consensual is naively unrealistic. A growing body of theoretical and empirical work has considerably modified this image, opening the way to a reexamination of the public nature of scientific activity.[7]

In the first place, this work has shown how the experimental phenomena that lie at the core of science are produced in distinctive local settings. Experimental facts are not simply presented to casual observation like stones picked up on the seashore; they have to be created by active labor with particular kinds of resources. Scientific phenomena are essentially creatures of laboratories, with their particular concentrations of instrumentation and skills. Science, at its point of origin, is not public at all. Nor is this an accidental feature of experimental work; it is arguably quite essential. Sociologists of modern scientific practice have argued that relative privacy is required for successful laboratory work, in order that skills and apparatus can be refined and protected from interference and

5 There have been many studies of controversies of this kind. Among them are some that use a sociological approach to draw out general implications for our understanding of science. These include Wynne (1982), esp. pp. 11–14, 159–176, Collins (1987, 1988), Pinch and Collins (1984).
6 Royal Society (1985).
7 Contributions to this work are surveyed in: Shapin (1982), Whitley (1983), Golinski (1990).

so that confused and chaotic initial perceptions can solidify into clear and distinct facts.[8]

Thus the facts of science emerge initially as a kind of local knowledge, dependent upon the craft skills of the laboratory scientist and his or her specific resources of expertise and equipment. These resources cannot be available to all, so that access to the means by which science is constructed is inevitably restricted. Inequalities of assets between scientists and lay people are not an unfortunate byproduct of current institutional arrangements, but a precondition for the construction of natural knowledge.

In addition to the circumstances of the laboratory, recent studies have also illuminated the ways in which scientific knowledge becomes public, to the extent that it does. The mechanisms by which this occurs have been scrutinized and the resources used, which may be quite different from those employed in the laboratory, have been identified. Science is made public by various kinds of discourse, including conversations and lectures, and the writing of scientific papers, textbooks, and popular works. The results of experiments become defined and gain their significance in the contexts in which they are interpreted. Claims to knowledge become accepted insofar as they are embodied in effective acts of communication. And the persuasiveness of particular claims is not simply a result of what was said, but also of how it was said, where, and by whom. This insight – basically that scientific discourse can be considered as a kind of rhetoric – has led sociologists and literary specialists to examine various genres of scientific writing, with the aim of showing in detail how they are constructed so as to persuade particular audiences.[9]

Rhetoric is one of the requirements for the construction of science in the public domain, but science is not just a product of verbal persuasion. On the contrary, it is intimately involved with manipulation of phenomena and material artifacts. It gains acceptance by mobilizing nature itself, in the form of experimental phenomena that are reproduced by replicating laboratory techniques. Phenomena that have been created by the instruments and skills of the laboratory can be translated to new contexts by extending the practices by which they were originally made. The French philosopher Gaston Bachelard coined the term "phenomeno-technics" to describe the embeddedness of experimental phenomena in bodies of technical practice.[10] Phenomena, the instruments by which they are pro-

---

8 Works on this theme include: Collins (1985), Fleck (1979), Hacking (1983), Rouse (1987), Latour and Woolgar (1979), Latour (1987), Gooding (1985a).

9 Representative studies include: Law and Williams (1982), Myers (1985), Gilbert (1977), and the papers collected in Shinn and Whitley (1985). See also Bazerman (1988), and Cantor (1989), for broader discussion of the field.

10 Bachelard (1980), p. 61.

duced, and the techniques for using those instruments are translated together in the processes by which science is made into public knowledge.

By considering the working of these processes, we may hope to understand more about how the confidence of lay audiences in scientific knowledge may be established and strengthened or (in certain circumstances) undermined. Rather than assuming the public nature of science, this perspective offers a rationale for empirical studies of the ways in which audiences are constructed and transformed in changing historical contexts.

This kind of approach has already begun to be applied in historical studies. It has become apparent that the question of the public status of science has long been a problematic one. Indeed, it is arguable that controversy over whether scientific practice is or is not sufficiently public derives from deeply entrenched features of its social constitution and supporting ideology in Western culture. When experimental natural philosophy emerged in Europe in the seventeenth century, there were already arguments over the degree to which it was open to public view and the degree to which it should be.

These debates arose, for example, in the Royal Society in London in the 1660s and 1670s. The practice in the Society was for experiments to be made initially in relatively secluded laboratories, and then converted into demonstrations before an audience in socially restricted but ostensibly public meetings. They would also be communicated in written descriptions – in correspondence, or in printed texts that were deliberately crafted to place the reader in the position of "virtual witness" to experimental demonstrations.[11]

One area in which tensions occurred was in dealing with rarities (things that had value because of their scarcity) or with commercially valuable processes kept secret by the practitioners of various arts and trades. Francis Bacon had commended such matters to the attention of natural philosophers, and many of them followed his suggestion. Critics of the Society however, such as Henry Stubbe, voiced the anxieties of artists and tradesmen, whose methods were being enquired into and who feared that valuable information would be revealed in too public a forum. For them, intellectual property was best protected by secrecy, whereas publicity was equivalent to theft. On the other hand, by making occasional concessions to commercial demands for confidentiality, the Society invited the charge that it was betraying the Baconian ideal of free communication, and that its knowledge remained the prerogative of an exclusive sect. The degree to which the Royal Society could or should produce public knowledge was thus a hotly contested issue.[12]

11 Shapin (1984, 1988), Dear (1985).
12 For a discussion of these issues in connection with one particular chemical phenomenon, see Golinski (1989).

Historians' understanding of the significance of these arguments, at this critical juncture for the origins of modern experimental science, has been greatly advanced by Steven Shapin and Simon Schaffer's book, *Leviathan and the Air-Pump* (1985).[13] These authors have shown how the formation of the experimental way of life in the early Royal Society involved the constitution of a relatively private space for experimentation and a series of declaredly public settings for communicating its findings. These included the Society's meetings, its publications (especially its journal, the *Philosophical Transactions*), and the extensive correspondence of the Secretary, Henry Oldenburg. In each setting, material, social, and rhetorical techniques were mobilized in order to put the desired message across. Robert Boyle in particular showed how experiments should be displayed and written reports framed in order to convey a persuasive effect. Shapin and Schaffer brilliantly expose the contingent nature of these practices by discussing the alternative views of a perceptive and resolute critic of the Society, Thomas Hobbes. From Hobbes's vantage point, we can see how problematic were the means for making public knowledge bequeathed to modern science by Boyle and his allies.

As methods for converting private opinions into public facts, the Royal Society's procedures involved a reordering of the social setting in which knowledge was pursued. On Shapin and Schaffer's account, the public form taken by the new science was a microcosm of the general pattern of consolidation of the social order after the upheavals of the mid-seventeenth century civil wars. For this reason, controversial issues of metaphysics or religious doctrine were excluded, and a gentlemanly consensus was formed around matters of fact. In addition, as Bruno Latour has pointed out, their study also suggests how experimental natural philosophy contributed to a remodeling of public life as a whole, from the seventeenth century on. Making facts through agreement among the witnesses to an experiment, and then extending them by replicating the experiments in other locations, Boyle showed how social relations should be reorganized by science. The circulation of instruments and the reproduction of knowledge-producing practices were thereafter to become prevalent features of the social landscape. As Latour puts it, "Since Boyle's time, . . . we live in societies built on laboratory-made objects; ideas have been replaced by skills; apodictic reasoning by managed doxa; universal assent by old-boy networks of professional colleagues."[14]

Of course, this situation did not come about overnight. The study that follows focuses on a rather later period in a lengthy and convoluted historical process — that of the extension of the practices of experimental science through society. I shift attention to the period of the eighteenth-

13 Shapin and Schaffer (1985).
14 Latour (1990), esp. pp. 148–155 (quotation on p. 152).

century Enlightenment and its immediate aftermath. By using the term "Enlightenment" I am committing myself to an analysis on a larger scale than a single personality, locality, or institution. Many valuable local studies have appeared in recent years, and my work is heavily indebted to a number of them. But my aim is to illuminate developments in scientific practice and their transmission through society – processes that cannot be described purely at the local level. We need to raise our eyes to a wider horizon in order to grasp how discursive and technical practices may be translated from one local context to others – from the chemist's laboratory, for example, to the bleach fields or the pharmacist's shop. In this connection, Enlightenment seems an appropriate term, provided it is understood as a concrete historical process and not as the diffusion of disembodied ideas. I suggest that the experience of enlightenment, involving certain patterns of communication and social interaction – a certain way of life in the public realm – is of key importance in the extension of scientific knowledge through society at large.[15]

A number of historians have led the way in exploring relations between natural philosophy and Enlightenment public life. The role of public discourse and experimentation in the culture of eighteenth-century Britain has been opened to investigation. Echoing interpretations of science in Enlightenment Scotland, Roy Porter has argued that its manifestations in the English provinces should be viewed as a form of cultural expression by an affluent middle-class elite. Science, like music, literature, or fashion is a cultural form, to be understood historically in relation to social forces such as emulation and consumerism.[16] From a rather different perspective, Schaffer has proposed that public experimental display was integral to the project of natural philosophy at this time. Demonstration of the powers of nature was intimately connected with metaphysical, aesthetic, and ethical principles; but it also gave rise to moral and political dangers that threatened the enterprise of enlightenment itself.[17]

In a substantial new study, Larry Stewart has shown how public suspicion of the experimental philosophy of the early Royal Society was gradually overcome as the settings in which it was practised were multiplied in the early decades of the eighteenth century. Such entrepreneurs as Jean Theophilus Desaguliers and Francis Hauksbee pioneered the presentation of lectures and experimental displays in London coffee houses and inns. Their successors, like Benjamin Martin and John Ferguson, took to the road to tour the burgeoning provincial towns. By attracting

15 My conception of the Enlightenment as a way of life in the public realm is informed by two seminal works in social theory: Sennett (1977) and Habermas (1989).
16 Porter (1980, 1981).   17 Schaffer (1980, 1983).

aristocratic patronage and middle-class subscriptions for their performances and publications, and by linking natural philosophy with a range
of technological activities from navigation to engineering projects, these
men won science its place in British society. They formed an essential
link, though one largely overlooked by traditional historiography, between the mechanical philosophy of Boyle and Newton and the technical
innovations of the Industrial Revolution.[18]

These studies, and others cited in this book, justify my concentration
on the Enlightenment in Britain. To restrict our attention to a single nation is not to narrow the focus, but on the contrary to broaden our coverage of the context in which public science was pursued. Because distinctive national experiences of Enlightenment have been identified in
England and Scotland, research on science in this milieu seems particularly necessary.[19]

Our focus upon chemistry allows us to probe more deeply into the
connections between the practice of a particular science and the circumstances of Enlightenment public life. Neither the continuous identity of
the discipline nor the fixity of its boundaries need be assumed. As we
shall see, the identity of chemistry was frequently redefined in lectures
and textbooks; its disciplinary continuity was sustained through repeated
statements of what the subject was about. Its boundaries were neither
fixed nor impermeable: Chemists profitably recruited phenomeno-technics from many other sciences, including those of heat, pneumatics, and
electricity, to advance their own.

Chemistry began to be constituted in the public realm in Britain toward
the middle of the eighteenth century. In Scotland in particular, efforts
were made to communicate chemistry in the circumstances of Enlightenment public life. The initial focus of this study is on the work of
William Cullen and Joseph Black, who argued for the importance of
chemistry in social circumstances that were themselves the subject of self-
conscious intellectual deliberation. Scottish thinkers felt that the novel
experience of interaction in the civic realm raised deep moral problems
for society and polity. Behavior in the public spaces of the eighteenth-
century city was discussed in relation to its implications for individual
identity, moral responsibility, and social progress.

This debate yielded rhetorical resources for the public presentation of
experimental science, but it also generated possible obstacles to realizing
the ambitions of scientists. On the one hand, chemistry could be shown
as a means of cultural and material improvement, capable of mobilizing
the energies of gentlemen and aristocrats in the pursuit of national prog-

18 Stewart (forthcoming). I should like to thank Dr Stewart for allowing me to read the
   manuscript of this book prior to publication.
19 Porter (1980, 1981); Wilson (1983); Pocock (1980); Gascoigne (1989), esp. pp. 1–3.

ress. This made it appear as a public asset. On the other hand, proposals for technological innovations raised the specter of conflict between private and public good, and the development of specialist skills seemed to threaten ideals of gentility and politeness. Continuing debates about the public status of science were thus an important part of the Scottish experience of enlightenment, and a significant influence on the form that chemical practice took in that context.

In the English provinces in the 1770s and 1780s, Joseph Priestley and his allies set about launching chemistry onto the public stage in a quite different way. Priestley described his discoveries of new "airs" in carefully crafted written narratives, and encouraged lecturers to demonstrate them to public audiences. The rhetoric of both demonstrations and texts was aimed at diffusing factual knowledge among as wide an audience as possible by allowing them to witness, or if possible to replicate, experimental findings. For Priestley, the purpose of this was to provide the population with direct experience of the providential powers of nature in order to liberate them from the ignorance on which corrupt authority was founded. His methods of making his experimental work public were thus subordinated to an overarching moral and political vision of the role of knowledge in spreading enlightenment. The field of pneumatic medicine, with its therapeutic techniques and associated methods of analysis of atmospheric air, was born out of Priestley's experiments and developed by his colleagues and friends in English enlightened circles.

Because of the close relationship between scientific practice and the forms of public life, the development of chemistry was shaped by dramatic changes in the constitution of civic culture at the end of the eighteenth century. This was a crucial element in the radical transformation of the discipline initiated in France by Antoine Laurent Lavoisier which became known as the "Chemical Revolution." We shall see how acceptance or rejection of Lavoisier's new theories was linked with different views as to how scientific knowledge should be established publicly. Arguments about methods of reasoning, about the use of certain instruments, and about the situation of audiences in relation to experiments were all interconnected. In the controversy surrounding the Chemical Revolution, visions of how the chemical community should be structured were at stake just as much as the results of particular experiments.

In Britain, the debate over the new chemistry was heightened by a growing domestic dispute concerning the proper social and political functions of public science. In the 1790s, a period marked by a strongly conservative reaction to the French Revolution, the Enlightenment values that had sustained the public culture of science in Britain were called into question. Political radicals such as Priestley and Thomas Beddoes reasserted the importance of chemistry within a program of social and intellectual progress. Conservatives such as Edmund Burke and John Robison identified such a chemistry as symptomatic of the subversive and pernicious aims of the Enlightenment. The polarized political climate of the end of the Enlightenment shaped reactions to the discovery, by Beddoes and Humphry Davy, of the intoxicating effects of nitrous oxide. The experiments on this "laughing gas" were widely ridiculed as symptomatic of the anarchy and delusion unleashed by supporters of the Revolution. The Priestleian program of pneumatic chemistry in the service of social and moral progress was judged to have degenerated into a fiasco.

This incident had a critical significance in Davy's career, and arguably also in that of chemistry itself as a public science. Davy left Beddoes's employment in Bristol, turning his back on the provincial enlightened milieu in which pneumatic medicine had flourished; he moved to the Royal Institution (RI) in London, where he rapidly gained an outstanding reputation as a lecturer to audiences drawn from the metropolitan social elite. In this setting he articulated a conservative version of the Enlightenment aspirations for chemistry, stressing how its applications could benefit humanity in a stable and stratified society. The personality he projected through his lectures provided an image of a scientific genius that had great popular appeal.

To these audiences, and also to more select specialist groups assembled at the Royal Society and in the RI laboratory, Davy demonstrated the spectacular potency of the voltaic pile, an instrument that conferred unprecedented powers of chemical analysis. He publicly established the efficacy of the pile as an engine of discovery, and on this basis was able to secure acceptance for his isolation of the elements sodium and potassium. Davy showed how command of a public audience enabled experiments to be mobilized with unprecedentedly persuasive effect, as he continued to defeat those who challenged his assertions. Although he had less immediate success in persuading chemists that he was right about the elementary nature of chlorine, his public audience was a considerable asset to him in pressing this claim also.

The form of public science that Davy constructed was in marked contrast to that favored by Priestley. Far from being invited to share in the production of scientific knowledge by replicating experiments, Davy's public audience was expected to remain entirely passive, awed by the power of the philosopher and his instruments, and accepting his interpretation of phenomena. This transformation in the role of the public audience for chemistry was closely connected with the emergence of new instrumentation and a more consolidated social structure for the specialist community. Although he set himself against some of Lavoisier's new doctrines, Davy adopted many of the rhetorical and technical practices that characterized the reformed discipline. He showed how the use of more

concentrated instrumental resources and more refined practices of experimentation required a greater measure of social discipline within the
community of chemists and less direct involvement by a lay public.

To some extent, Davy provided a model for other specialist chemists
in the first two decades of the nineteenth century. His discoveries were
widely hailed and his broadly conservative utilitarian rhetoric was reproduced by other lecturers and writers. Chemists used a variety of new
technical tools to carve out careers for themselves in applied chemistry
and education. But the chemical community achieved only a limited degree of autonomy from the demands of a public audience during this
period. Many chemists continued to practice their science in ways that
acknowledged a duty to make its doctrines accessible and its techniques
widely reproducible. The legacy of Enlightenment public science was a
lasting one, notwithstanding the radical transformation the subject had
undergone.

Looking at early nineteenth-century chemistry in terms of its Enlightenment past, we can get a new outlook on the degree to which radical
change had occurred. I aim to avoid the risk of teleology that lurks in the
use of notions like "specialization" or "professionalization" to charac-
terize developments in this period. Rather than trying to subsume historical change under some supposedly universal process, I shall concentrate
on placing scientific practice in its setting – in the structure of the community of practitioners and their activities in the public sphere. I shall
show how certain techniques, instruments, and modes of discourse continued to be used by the chemical community in its enduring relationship
with a public audience. Davy's efforts to provide the chemist with powerful instrumentation and a pacified audience were respected, but not
universally followed. Most chemists used analytical techniques that were
more accessible (though no less effective) than the voltaic pile, and preserved a more democratic relationship with their public. This dimension
of continuity with eighteenth-century practice was an important aspect
of the changing discipline of chemistry.

By addressing questions of discipline-formation in this slightly oblique
way, I hope to show how a sociologically informed scrutiny of scientific
practice may address issues more traditionally associated with the social
history of science. A focus on techniques, instruments, and discourse, and
their functions in the community of practitioners, can fruitfully complement study of institutions, popularization, and other external aspects of
science. Such a broad-based approach will be necessary if we are to understand in detail how science is constructed as "public knowledge."

2

"The study of a gentleman": Chemistry as a public science in the Scottish Enlightenment

... by communicating their ideas, men improve their knowledge; and, it is only by the enlarged knowledge of the species that the science of individuals is brought to that perfection which does honour to the race ... Thus society is necessary to the human understanding.

James Hutton, *Investigation of Principles of Knowledge* (1794)[1]

The question, therefore, concerning the rise and progress of the arts and sciences is not altogether a question concerning the taste, genius, and spirit of a few, but concerning those of a whole people, and may therefore be accounted for, in some measure, by general causes and principles.

David Hume, *Essays*[2]

When Thomas Thomson, a distinguished British chemist, looked back from the vantage point of the 1820s at the remarkable progress of his science in the previous few decades, he identified the origins of its revival with a specific episode in mid eighteenth-century Scotland. The appointment of William Cullen (1710–1790) to the chair of chemistry at Edinburgh University in 1756 was, in Thomson's view, the event that heralded an upsurge of interest in the discipline. Of course, chemistry had been pursued before by seventeenth-century experimenters like Robert Boyle – but it had languished for many years thereafter. Newton's great achievements in the mathematical sciences had drawn most British men of science into his orbit, to the detriment of chemical studies. Thomson went on:

1 Hutton (1794), III, p. 581.
2 David Hume, "Of the Rise and Progress of the Arts and Sciences," pp. 112–138 in Hume (1903), quotation on p. 115.

But when Dr. Cullen became Professor of Chemistry in Edinburgh in 1756, he kindled a flame of enthusiasm among the students, which was soon spread far and wide by the subsequent discoveries of Black, Cavendish and Priestley; and meeting with the kindred fires which were already burning in France, Germany, Sweden, and Italy, the science of chemistry burst forth at once with unexampled lustre.[3]

Following Thomson's lead, we shall begin our examination of relations between chemistry and Enlightenment public life in Scotland in the middle decades of the eighteenth century. This is not because chemistry was invented at this time. Techniques of chemical operations had been passed down as a craft tradition for many centuries, in association with such practical activities as metallurgy, mineralogy, and pharmacy. The subject had also existed as an academic discipline in universities and other educational institutions for several decades. Cullen's achievement was to give chemistry for the first time in Britain the profile of a public science. He took an existing academic discipline and articulated it in a way which greatly enhanced its acceptability in the culture of the Scottish Enlightenment. In Cullen's hands the subject became an accepted part of general education and came to be looked upon as a potential contributor to technological and economic progress.

Cullen and his great pupil and successor in the Edinburgh chair, Joseph Black (1728–1799), made their careers in unprecedented ways, exploiting the opportunities that their society offered for advancement through chemical skills and shouldering the corresponding burdens. Both seized opportunities to use their knowledge in a variety of fields including medicine, agriculture, and manufacturing, as well as in the university classrooms where they made their reputations. Both exploited the aristocratic patronage prevalent in their social world and participated in the informal societies of Glasgow and Edinburgh where they dedicated themselves to the improvement of their country through civic action.

In this chapter we shall examine how Cullen and Black succeeded in widening the audience for chemistry, extending the range of its perceived applications, and thereby winning it recognition in Scottish society. We shall consider the circumstances in which these developments occurred and the means by which support for the subject was recruited: the experimental practices and theoretical concepts they developed, the specific forms of discourse in which they communicated these, and the social connections they exploited to get their message across. It was by these means that chemistry was made into an enterprise in which civic prestige was invested and in which gentlemen could engage, spurred by the prospect of public honor.

The specific form of activity that chemistry became in Enlightenment

3 Thomson (1802), I, p. 12. Compare Thomson (1830–31), I, pp. 303–304.

Scotland shaped the communication of the subject within the country and beyond. Chemical techniques were diffused among members of improving clubs and societies and in dealings between chemists and their patrons. Generations of students who passed through the Scottish universities, particularly those who studied medicine, took away with them knowledge of the most advanced chemical doctrines, learned at the feet of the leading practitioners of the discipline. But Cullen and Black did not significantly exploit the potential of the printed word. Neither of them published a substantial work on chemistry, so that the Scottish contribution was not widely or authoritatively known outside the country. Nor did they encourage diffusion of their techniques by lecturing outside the universities.

In describing these factors, the discussion that follows will build upon the work already done by Arthur Donovan and John Christie to situate Cullen and his followers in their enlightened context.[4] This was a context in which the public role of the scientist, or more generally the philosopher, was a subject of constant analysis and debate. Cullen and Black participated in the process of career-making and self-fashioning in a public realm that was itself being restructured by the growing influence of commercial activity.[5] The aim of this discussion is to show how the discourse of Scottish chemistry related to the rhetorical underpinnings of social life. We shall see, for example, how prevailing norms of civic responsibility limited the degree to which chemists could assert personal authorship of discoveries or inventions. In this respect, constraints rooted in the public realm significantly hindered the formation of a new discipline.

Chemistry as an academic discipline

The setting in which Cullen and Black worked was that in which the Scottish Enlightenment reached its first pinnacle of achievement. Many of those who contributed to the outstanding intellectual works of the period – in philosophy, social theory, literature, and the arts – made their careers alongside the two pioneers of Scottish chemistry. David Hume and Adam Smith were friends of Cullen. The rhetorician Hugh Blair and the social theorist Adam Ferguson were colleagues on the Edinburgh University faculty. The Scottish Enlightenment owed much to the stimulating local environment of Edinburgh, the city that Tobias Smollett dubbed a "hotbed of genius."[6]

4 See particularly: Donovan (1975a, 1975b, 1976); Christie (1981).
5 Hont and Ignatieff, eds. (1983).   6 Daiches, et al. (1986).

Scotland's flourishing intellectual life was also a part of a wider European movement. This can be seen most clearly in terms of the broad themes identified by the history of ideas. The post-Newtonian natural philosophy that was a central part of the concerns of Scottish intellectuals also found a receptive climate elsewhere at this time.[7] The main concerns of Scottish social theorists – the nature of progress and the tribulations of virtue in a developing commercial society – were shared by enlightened thinkers in other parts of Europe. In addition, the modes of public activity by which intellectual life was sustained show several common features. A similar social mechanics of enlightenment was at work in Scotland and in other European countries.[8]

One aspect of this was the increased publication of printed materials – books, periodicals, and newspapers. Journalism, bookselling, and publishing emerged as profitable occupations and significantly contributed to fermenting intellectual activity. Also important were the numerous clubs and societies, largely self-constituting and independent of government, that dedicated themselves to civic action in pursuit of a broadly conceived process of moral, social, and economic improvement. Experimental science and literature (belles lettres) were frequently pursued in parallel in these societies.[9]

Prominent among the members of these bodies, in all countries that participated in the Enlightenment, were members of what can loosely be called the "professions." These comprised both members of the traditionally recognized professions (clergy, doctors, and lawyers) and such new aspirants to their rank as artists, craftsmen, writers, and academics. These men, more than merchants or industrialists, provided the impetus for intellectual life. One dimension of the Enlightenment as a social movement was the increasing influence and status of professional groups in many European societies.[10]

This process occurred without significant disruption of the established social order. By and large, members of the professions were happy to rise up the social scale rather than to attempt fundamentally to undermine it. Patronage remained key to their advancement in all countries, so that they welcomed aristocrats and landed gentry as coparticipants in enlightened culture. The symbiotic relationship between the established high-status groups and the newly assertive professional classes provided the dynamic of many Enlightenment clubs and societies.

Professional groups advanced steadily in Scottish society, although they

7 Emerson (1986, 1988a).
8 This comes through clearly from the papers in Porter and Teich, eds. (1981).
9 McElroy (1969).
10 Holmes (1982); Chitnis (1976); Emerson (1973a); Roche (1978, 1980); Darnton (1971); Baker (1981).

remained largely dependent on aristocratic patronage.[11] A series of reforms in the Scottish universities in the late seventeenth and early eighteenth centuries opened up new opportunities for those seeking to follow academic careers. Probably the most significant reform was the abolition of regenting (the practice whereby a single teacher would lead a group of pupils through all subjects in the course of their university careers) and its replacement by the appointment of specialist professors. Particularly in medicine and the natural sciences, this development gave new impetus to the refinement of specialist skills.[12]

The universities provided the main location for the growth of the Scottish scientific community in the eighteenth century. Cullen and Black, followed by others such as Francis Home (1719–1813) and William Irvine (1743–1787), made their careers as academic chemists in the reformed universities of Enlightenment Glasgow and Edinburgh. They attached themselves to institutions that were increasingly successful in attracting students, particularly in the growing market for medical education. Chemistry had initially made its entry into universities as an adjunct to medical training, and the heightened profile of the Scottish institutions in that field was a major factor in encouraging its further development.[13] Institutional employment was not however sufficient to make a successful career as an academic scientist. Members of the medical faculties continued to rely on private practice for a significant part of their incomes and still depended on aristocratic patronage for appointments and preferment. They thus remained closely involved in a public intellectual culture that extended well beyond the bounds of the university. This wider culture created both opportunities and constraints in the quest for status. In a gentlemanly world, shaped by humanist notions of civic virtue and social responsibility, and where assertions of personal interest were frowned upon, the building of a professional career could be a worrisome affair.[14]

These factors are well exemplified in Cullen's career. His position on the medical faculty, first at Glasgow University and then at Edinburgh, enabled him to capture a student audience and to claim the fees that were customarily paid directly to professors.[15] He also maintained a private medical practice that he built up energetically after moving to the Scottish capital.[16] Throughout his career he was heavily reliant upon his personal patrons for appointments and promotions and repaid them by

---

11 Emerson (1986, 1988a); Phillipson (1973, 1975, 1981); Sher (1985).
12 Cant (1982) and Shepherd (1982) are very informative on this subject, as is Christie (1974, 1975). See also Grant (1884), I, pp. 258–263, 292–317; and Morrell (1976).
13 Lawrence (1984); Anderson (1978).
14 On the anxieties associated with creating a professional identity, see Camic (1983), esp. pp. 199–223.
15 Morrell (1971a).   16 Risse (1974).

participating in their projects for technological and economic improvements.

Cullen's own medical education was irregular. He studied briefly at Glasgow University and was apprenticed to a physician in that city, but left to sail as a surgeon to the West Indies in 1729. He returned to Scotland in 1731, practising medicine for a while in his home town of Hamilton. During this period he pursued private studies in chemistry and natural philosophy. In 1740 he took an MD degree at Glasgow and moved there four years later to begin lecturing in the city, initially without university recognition. In 1747 he extended his lectures to cover materia medica, botany, and chemistry, and was given an official university appointment and a grant to equip a chemical laboratory. In 1751 he was appointed Professor of Medicine at Glasgow but already had his eyes on a more prestigious position at Edinburgh and on the enhanced opportunities for medical practice that it would bring.[17]

The Edinburgh appointment came to Cullen in 1755. His friend and patron Henry Home (Lord Kames) raised support for him among the members of the Town Council, who had the power of making appointments to medical chairs in the university. The patronage of Archibald Campbell, Earl of Ilay and third Duke of Argyll, was also crucial in securing Cullen's appointment as Professor of Chemistry against the opposition of the ailing incumbent, Andrew Plummer, and many other members of the faculty. Cullen began chemistry lectures in Edinburgh in January 1756. His popularity with students rose steadily until he moved to the chair of Institutes (theory) of Medicine in 1766. Subsequent promotion took him to the chair of Practice of Medicine in 1773, which he held until his death.[18]

The chemistry chair at Edinburgh had been founded in 1713. The first incumbent, James Crawford, had been succeeded by Plummer, who was one of four professors appointed in 1726, in a move that marked the foundation of the Edinburgh University Medical School.[19] The chair was unsalaried but carried with it considerable opportunities to earn student fees and to establish a name as a medical practitioner. As a talented teacher, Cullen gained significantly more than the minimal attendance guaranteed at Edinburgh by the fact that chemistry was a compulsory examination subject for intending medical graduates. His built his classes up from 17

17 For Cullen's Glasgow years, see Thomson (1832–59), I, pp. 1–35; and MacKie (1950). For evidence of his motivations for the move to Edinburgh, see Cullen to William Hunter, August 1751, in Thomson, (1832–59), I, pp. 81–82; Henry Home [Lord Kames] to Cullen, 26 December 1751, early 1753, and 17 September 1754, in ibid., I, pp. 82–84; and Cullen to Kames, 17 Jan 1750, in ibid., I, pp. 593–596, esp. p. 596.

18 On the maneuvers that lay behind Cullen's appointment at Edinburgh, see Thomson (1832–59), I, pp. 81–97.

19 Anderson (1978); Doyle (1982).

in the first year of his appointment to 145 in 1763–64, an achievement he brought to the attention of the Town Council.[20] He also expanded his audience to encompass men from outside the university who had an interest in the subject. Such extramural auditors, described as "gentlemen engaged in any business connected with chemistry," were reported as a significant addition to his student audience in Glasgow, and were presumably also present in Edinburgh.[21] Furthermore, in part to satisfy his patrons, Cullen extended the coverage of his chemical lectures beyond the traditional medical syllabus. For example, in 1749 he told Lord Kames that he was introducing a discussion of agriculture, a subject that had not been given serious academic attention before, "to open young gentlemen's views on the subject."[22]

Cullen's chemistry lectures were thus balanced between satisfying the fairly basic requirements of medical students and cultivating a broader appreciation of the discipline. He acknowledged that his primary role was to teach chemistry "as it may best answer the purpose of students of physic," a purpose that did not necessarily extend to a very wide acquaintance with the subject.[23] On the other hand, he felt obliged to correct erroneous preconceptions about chemistry that medical students might have. The commonly held view of the discipline was, he said, "limited, imperfect and inaccurate," regarding it as subordinate to medicine, or even worse as the art of making gold.[24] A "philosophical" version of chemistry would correct these common prejudices. Thus, while teaching the subject mainly to answer the needs of medical students, Cullen insisted "that it may do so I find it necessary to deliver the philosophical principles."[25] By including some theoretical discussion in his course of lectures, he hoped to reconcile the diverse practical interests of a wide audience. Thus chemistry would be enabled to achieve its destiny as "a considerable part of Natural Philosophy capable of being applied to very important purposes of Society."[26]

Cullen's assertion of the disciplinary autonomy of chemistry, which Arthur Donovan has emphasized, should be seen in this light.[27] He did indeed make repeated claims that chemistry should be regarded as an independent discipline, a philosophical enterprise or science, capable of

20 Thomson (1832–59), I, p. 97; Glasgow University Library, Department of Special Collections, Cullen MSS, box 3, item 9 (hereafter cited in the form: GUL, 3:9).
21 The testimony of a Glasgow surgeon named Wallace, written in 1811, cited in Thomson (1832–59), I, p. 25.
22 Cullen to Kames, [spring] 1749, in Thomson (1832–59), I, pp. 596–597.
23 GUL, 1:8.
24 Royal College of Physicians of Edinburgh, MS no. C.10, p. 15 (hereafter cited in the form RCPE, C.10, p. 15); GUL, 2:44.
25 RCPE, C.11, p. [3], "Introduction."   26 GUL, 1:5.
27 Donovan (1975b), pp. 93–102.

many applications in fields aside from medicine. Subsequently, his pupils and followers among Scottish chemists looked back to him as the author of "The first Dawn of Science in Chemistry" (as George Fordyce put it) or "the true commencer of the study of scientific chemistry in Great Britain" (in the words of Thomas Thomson) – plaudits that confirm Cullen's stature in relation to his successors.[28]

The push toward an independent "philosophical" chemistry was however always qualified by the need to serve the educational needs of medical students. Cullen's own career, with its movement from chemistry into more recognized and rewarding fields of medicine, illustrates the continued dependence of the former discipline on the latter. His pupils mostly followed careers that were more medical than chemical in orientation. Assertions by Cullen or his followers that chemistry in his hands had achieved the status of an independent science must be viewed in relation to this medical context.

In describing chemistry as a putatively autonomous discipline that remained nonetheless bound to a context of medical education, Cullen showed an awareness of the traditions of didactic discourse already established in chemistry.[29] He remained in many respects within the boundaries of the genre developed by his predecessors. In fact he advanced his claims for the disciplinary autonomy of chemistry by critically discussing their views and by differentiating his own version of the subject from theirs.

He repeatedly made the point that his predecessors had overlooked the distinctiveness of chemistry in relation to the mechanical philosophy. As early as the Glasgow lectures of 1748–49, he was attacking the great Dutch chemist Hermann Boerhaave (1668–1738) for failing to specify what distinguished chemical from mechanical operations and properties.[30] Boerhaave was faulted, as were Peter Shaw and G.F. Venel, for not explaining that chemistry was the study of the particular and specific properties of bodies and of the operations by which they were changed.[31] Chemistry should be differentiated, Cullen argued, by reference to terms derived from the German chemist Georg Ernst Stahl (1660–1734). Stahl had distinguished between the physical and the chemical parts of a body – respectively its "integrant" and "constituent" parts. As Cullen put it, the component parts of a body could be either "Physical, otherwise named atoms; [or] chemical, or commonly named Chemical Principles." From this distinction, that between physical and chemical operations followed:

28 George Fordyce, "Lectures on Chemistry" (1786), Royal College of Physicians, London, MS 146, pp. 14–15 fn.; Thomson (1830–31), I, p. 304. See also Playfair (1858).
29 Hannaway (1975); Christie and Golinski (1982); Golinski (1984); Roberts (1991b).
30 Wightman (1955), p. 194.   31 RCPE, C.10, pp. 15–19.

"the resolution [analysis] of the parts implies a chemical, and the division [physical disintegration] a mechanical operation."[32]

Having defined chemistry as the science of the properties specific to certain types of matter, Cullen considered what he regarded as the central problem of chemical didactics, that of "method" or the order in which the subject-matter should be arranged in discourse. Chemistry, he suggested, comprised a huge number of particular and discrete facts, which needed to be welded into a science by a rigorous and thoroughgoing reduction to order. This kind of assertion was far from unprecedented among chemical writers and teachers. In fact, it can be traced fairly continuously back to Andreas Libavius at the beginning of the seventeenth century, but Cullen made it central to his exposition of the subject in a quite novel way. His lectures explicitly discussed the problems of ordering chemical discourse at the same time as they exemplified his proposed solutions.

The urgency of the need for order in chemistry arose from the current situation of the discipline, as Cullen explained in the historical introduction that he inserted into his course when he began to lecture at Edinburgh. Chemistry had made a laudable commitment to free communication, especially through the efforts of Robert Boyle in the seventeenth century. Boyle had "stript Chemistry of all its uncouth and mysterious language"; his writings had cleared the subject of "that jargon & affected mystery which had hitherto deformed it," and "delivered . . . experiments with plainness perspicuity & Candour."[33] But the increased availability of information simply augmented the difficulties of arranging it: the "facts are very numerous and with great difficulty to be collected."[34] To the solution of this quandary, Boyle had contributed nothing.

Another seventeenth-century philosopher, Francis Bacon, had however "planned out the road" by which further progress might be made. Baconian induction was judged by Cullen "the best & safest method of reasoning."[35] Cullen saw the inductive method as an almost mechanical process of arrangement and tabulation of data. He told his students that they could perform the operation in their own commonplace books. As they read chemical works and performed experiments they were to set down facts and observations, initially without any particular order. The facts were then to be indexed with the different processes classified under

---

32 RCPE, C.10, pp. 50–56, esp. pp. 52–53, 54.
33 "Chemical Lectures by William Cullen, MD," Wellcome Institute for the History of Medicine, London (MS 1918), p. 21 (hereafter cited in the form: Wellcome MS, p. 21); GUL, 2:44; Kent (1950b), p. 25.
34 Dobbin (1936), pp. 140–141.
35 Wellcome MS.

headings corresponding to the substances to which they had been applied. By compiling such an index, Cullen told his students, "you will have brought your Knowledge into a Synoptical or Tabular form; and thus ye Business of Induction will be executed almost without your being aware of it."[36]

The preoccupation with the ordering of facts was fundamental to chemistry, as Cullen articulated it. In his Edinburgh lectures, the "objects" of the science were divided into six broad classes: saline, inflammable, watery, earthy, aerial, and metallic. When compared with the distribution proposed by Boerhaave and others, where bodies had been divided into three kingdoms (animal, vegetable, and mineral), Cullen's appears as a distinctively chemical classification, rather than part of a general scheme of natural history. To the classes derived from Aristotle's four elements, he had added two that reflected the growth areas of analytical chemistry in the seventeenth and early eighteenth centuries – metals and salts. The link with natural history was, however, maintained at the level of method – because of their common concern with order, chemistry and natural history were said to remain "necessarily connected together." Chemistry might be said to be aiming for comprehensive classifications of species and properties, along the lines of Linnaeus's system in botany and mineralogy.[37] A similar method was applied in Cullen's work on "nosology," the systematic classification of diseases, to which he turned after 1766. In justification of this approach he wrote, "The distinguishing of things by genus and species is *universally* applicable to every two things in nature that are capable of being distinguished from one another."[38]

In chemistry, as in nosology and natural history, the methodical classification of species served several simultaneous functions. It provided an appropriate means of teaching the subject to students, promised a secure path to scientific or philosophical status, and gave a rationale for practical action. The many potentially useful facts that chemists had discovered were said to "lie scattered in many different books, involved in obscure terms, and joined to a good deal of false philosophy." What was required was a comprehensive survey of chemical literature to extract the facts contained therein and arrange them in a systematic order.[39]

This perspective was undergirded by certain epistemological assumptions. It was taken for granted, for example, that the facts of nature had an external existence independent of human agency, so that they could

36 "Lectures on Chemistry Delivered 1762 and 1763 by Dr. William Cullen," John Rylands University Library of Manchester, Department of Special Collections, MS CH C.121 (4 vols.), I, pp. 10–11 (hereafter cited in the form, Manchester MS, I, pp. 10–11).
37 Dobbin (1936), p. 144; Manchester MS, I, p. 18; RCPE, C.10, p. 34; GUL, 1:11.
38 Quoted in Thomson (1832–59), II, p. 14.
39 Thomson (1832–59), I, pp. 31, 40.

be interpretively disentangled from the distorted language in which they might have been described. Cullen remarked that the chemist should regard himself as subordinate to nature, as a facilitator of natural processes rather than a dominator.[40] The role of the human subject was simply to choose a language of description, and if this were done correctly, then an entirely transparent representation of a natural fact would result. Such representations could be ordered in the synoptic tables through which heterogeneous facts were to be combined and presented to the philosopher's scrutiny.

The preoccupation with ordering what were taken to be constitutively unproblematic facts was shared by many other chemists of the time. A sign of this was the ubiquity of affinity tables in chemical texts of the period. In these tables, displacement reactions were predicted from the relative strengths of attraction between the components of the substances involved. Affinity tables were both emblems of the hoped-for order and demonstrative achievements of a chemistry conceived as the methodization of discrete natural facts. They also symbolized the expectation that future progress in the science would follow the same linear path as developments to date, gradually filling in more of the spaces in a tabular arrangement of which the outlines had already been laid down.[41]

Cullen directed his students to the original affinity table of E.F. Geoffroy, reprinted in Pierre-Joseph Macquer's textbook of theoretical chemistry. When he started lecturing, Cullen's practice was to draw the table on a blackboard, from which the students would transcribe it, but around 1759 he took to having copies printed to be handed out in class. His lectures then made repeated reference to the tables, which served as summaries and *aides-memoires* in rather the same manner as the printed syllabus that he had produced in 1748.[42] Because the value of Geoffroy's table was perceived primarily in relation to its role as didactic aid and encouragement to future research, the fact that he was "chast in theory" was a positive asset. Cullen compared him favorably with Newton and Stahl as contributors to knowledge of elective affinities, and remarked: "It does not derogate from but Honour [him] that later discoveries have found out many Errors in his Tables, and made many additions to them."[43]

As well as using Geoffroy's table, Cullen proposed a new type of diagram to represent double-decomposition reactions. His use of these "arrow-and-letter" diagrams and their derivation from the columns of Geof-

---

40 RCPE, C.10, p. 68.
41 For descriptions of affinity tables and discussion, see: Roberts (1991b); Duncan (1962); Thackray (1970), pp. 90–91, 205–221; Crosland (1962), pp. 240–241.
42 Thomson (1832–59), I, pp. 126–127. A copy of the 1748 syllabus is preserved as GUL, 1:34.
43 GUL, 1:2; Manchester MS, I, p. 317.

froy's table have been described by Crosland.[44] Their significance was primarily didactic rather than theoretical. Cullen told his students that the doctrine of elective attractions "will be best understood by considering an explanation of the following diagrams." As well as providing an aid to explanation the diagrams were said "to exercise y[ou]r memory in ye recollecting ye symbols used in Chem[istr]y." Only after describing in detail the derivation, didactic function, and predictive use of the diagrams did Cullen mention their theoretical implications, concluding rather lamely: "I could wish in this place to give a theory of elective attractions, but the subject is so obscure, that we can only expect to deliver a general view of it."[45]

A "general view" was literally what the diagrams and affinity tables provided. They were taken to comprise knowledge only because knowledge was understood in terms of spatial and visual metaphors. Chemistry, on this model, aspired to the condition of a table of ordered signs, transparently representing discrete natural facts, contemplated from a convenient vantage point by the philosopher's gaze.[46]

Despite the emphasis on the methodical ordering of discrete facts, there was another side to the picture of philosophical chemistry that Cullen presented in his lectures. To portray him as no different in his view of the subject from his French contemporaries, Macquer and Geoffroy, would be a mistake. In fact, Cullen has drawn the attention of historians in substantial part because he set Scottish chemistry on a new path, away from the compilation of data on specific properties and toward a new focus on the experimental manipulation of reactions by changing physical conditions. Specifically, he investigated the role of heat in chemical and physical change and rationalized this in terms of a theory derived from Isaac Newton's notion of the ether.

In Newton's speculations on matter theory, particularly those that found their way into print in the "Queries" to the successive editions of the *Opticks,* the ether was described as a highly tenuous species of matter, composed of mutually repulsive particles that were also repelled by particles of normal ponderous matter.[47] Cullen, who relied particularly on the writings of the Newtonian Bryan Robinson,[48] referred to such an ether as a possible explanation of chemical phenomena such as combustion, reactivity and changes of state. In relation to combustion, the ether took on some of the functions of the traditional element fire, and was

44 Crosland (1959). Cf. Donovan (1975b), p. 131.
45 RCPE, C.10, pp. 72, 83; Manchester MS, I, p. 40.
46 Some of the terminology of Foucault (1970), esp. ch. 5, seems appropriate here to describe the metaphors central to this model of knowledge.
47 Cantor and Hodge, eds. (1981), pp. 19 ff.; Heimann and McGuire (1971), esp. pp. 240–246.
48 RCPE, C.10, p. 87; Manchester MS, I, p. 44. Cullen referred to Robinson (1743).

also linked (more tenuously) with Stahl's principle of combustibility – phlogiston.[49] More novel was Cullen's application of the ether to explain states of aggregation and chemical combination. This theory, which has been discussed by Christie, hinged on microscopic interactions between ether and normal gross matter.[50]

As Donovan and Christie have shown, aspects of the work of the leading Scottish moral philosophers were important to Cullen in this connection. Given the centrality of moral philosophy in the curriculum of the Scottish universities, it is unsurprising that Cullen should have looked to the works of its major practitioners to legitimate his theoretical innovations. He had cemented alliances with both David Hume and Adam Smith at Glasgow, becoming a close colleague of Smith and supporting Hume's candidacy for the Chair of Logic in 1751.[51] From these pace setters of the Scottish Enlightenment, Cullen derived arguments to justify theoretical speculation as an essential part of scientific research. He was also encouraged to develop etherial explanations as a secular alternative to divine action as the cause of physical change.[52] The views of his patron Lord Kames might also have inclined Cullen to take this line. In 1754, Kames drew the fire of theological conservatives for his own suggestions that matter could be self-activating without the need for divine intervention.[53]

Cullen's alliance with this self-consciously enlightened group evidently encouraged him to introduce explicitly speculative material into his lectures. He justified this by noting that, " 'tis natural to mankind to delight in deduc[in]g causes . . . and [this] propensity ought to be encouraged."[54] Edinburgh philosophy suggested that excursions into theory could be a legitimate part of chemistry, and theoretical speculation offered itself as a means of making chemistry truly philosophical. Accordingly, Cullen introduced his ether theory in the course of discussing the specific properties of bodies. He boldly proposed that both the qualities and the state of aggregation of a body could be accounted for by interaction between repulsive etherial matter and attractive normal matter:

The qualities of aggregates, and the forms of aggregation consist in some measure between heat and the particles of matter. It is even probable that all the different

49 Donovan (1975b), pp. 141–143, 148–152.   50 Christie (1981).
51 Donovan (1975b), pp. 46–76; Christie (1981), pp. 87–91; Cullen to William Hunter, 17 September 1776 (on the death of David Hume) in Thomson (1832–59), I, pp. 607–609; Smith to Cullen, 31 August 1773, Glasgow University Library, Cullen Letters, 242; Smith to Robert Cullen (William's son), 9 February 1790, GUL, Cullen Letters, 236.
52 In remarks to his student audience, Cullen dismissed attempts to relate the forces of attraction directly to the agency of God: ". . . this way of reasoning would soon put a stop to philosophical enquiries," he told them. (RCPE, C.10, p. 68).
53 Kames (1754). Cf. Stewart (1754); Ross (1972), pp. 174–175.
54 GUL, 3*:28b; Manchester MS, I, p. 4.

kinds of matter may be reduced to two, viz. the matter of heat, or an elastic matter which seems to have a repulsive power; and that kind of matter which has the power of attraction.[55]

The difficulty that emerged as Cullen developed his theory was that it was much easier to explain the "forms of aggregation" than the "qualities of aggregates." The theory, which explained physical cohesion ("aggregation" in the vocabulary that Cullen took from Stahl) by the relative density of ether to normal matter in bodies, could only gesture toward a similar mechanism to account for chemical attraction.[56] Although experiments performed in the 1750s allowed Cullen to relate the generation of heat or cold to changes in physical state, the problem of the chemical specificity of attractions remained pressing. This continued to be the case through the various changes in his theory that Cullen introduced in the 1760s.[57] He was forced to concede that although his theory was "the most plausible scheme of chemical philosophy, . . . the difficulty is to find the cause of *Elective* attraction, or why aether does not admit of an equal union with all bodies."[58] It was a difficulty he never resolved. In the face of intractable theoretical problems he retreated, as we have seen, to the general view offered by the affinity table.

This withdrawal from the heights of theoretical speculation to the more mundane classification of phenomena could also find endorsement from Hume and Smith. Cullen recommended a Humean skepticism about theoretical doctrine to his students as an appropriate form of reasoning in chemistry.[59] Smith had given more precise approval of the method of classification in his essay of the early 1750s, "The Principles which Lead and Direct Philosophical Enquiries; Illustrated by the History of Astronomy." Smith wrote: "It is evident that the mind takes pleasure in observing the resemblances that are discoverable betwixt different objects. It is by means of such observations that it endeavours to arrange and methodise all its ideas, and to reduce them into proper classes and assortments." This procedure was the origin of "those assortments of objects and ideas which in the schools are called Genera and Species," and was applicable to chemistry as much as to any philosophical subject.[60]

Cullen's indulgence in his lectures in a measure of theoretical speculation could thus be reconciled with the doctrines of the leading intellectuals of his enlightened milieu. But so could his careful demarcation of such speculation from solid empirical facts, and his insistence on meticulous compilation and classification of data as the basis for future prog-

---

55 RCPE, C.10, p. 65.   56 RCPE, C.10, pp. 65–68.
57 Thomson (1832–59), I, pp. 50–54, 580–583; Christie (1981), pp. 99–101.
58 RCPE, C.10, p. 87.   59 RCPE, C.10, p. 27.
60 Smith (1980), quotations on pp. 34, 37–38. (The essay was originally composed c.1750–53, and is likely to have been known to Cullen in manuscript.)

ress. To impose a theoretical structure upon the subject would have been to follow the mistaken path of seventeenth-century mechanistic chemists such as Boyle and Nicholas Lemery, in Cullen's view. His own lectures, in contrast, sought legitimacy within the intellectual culture of his time by a judicious deployment of speculative notions in an overall structure that remained firmly anchored on the bedrock of recorded facts.

By declining to organize his lectures as if all of chemistry could be deduced from a single definition (the form Boerhaave had followed), Cullen avoided the appearance of dogmatism or what Bacon had called the "magistral" mode of discourse. He thereby placed himself on a level of equality with his students, as if accompanying them on the gradual progress from factual to theoretical knowledge, rather than delivering a completed doctrine from on high. This was a wise choice for a situation in which Cullen was attempting to gain acceptance for chemistry from an audience he envisaged as independent gentlemen. Theory appeared in his lectures in a way that was consistent with the gentlemanly educational ideal then prevalent at the Scottish universities.[61] Far from imposing dogmatically upon his students, he was consistently courteous toward them, even taking occasion to compliment them on "the decent behaviour which is to be expected from Men of good breeding."[62] By insinuating sophisticated theory into a nondogmatic discursive structure, Cullen earned the respect of his successors such as John Robison (1739–1805), chemistry lecturer at Glasgow in the late 1760s and subsequently Professor of Natural Philosophy at Edinburgh. It was Robison's judgment that Cullen had succeeded in making chemistry "a liberal science, a study of a gentleman."[63]

Gentlemanly science in the public realm

This gentlemanly science was one with great appeal to the students and others who attended Cullen's course of chemical lectures. With this style of discourse he laid claim to a philosophical status for the discipline, without imposing upon his audience with dogmatic expositions of theory. Cullen read well his auditors' preference for a mode of lecturing that emphasized the steady accumulation of phenomenal facts and their methodical classification. He offered them the vantage point of a general view over chemical doctrine, to be attained by Baconian induction and symbolized by the tables of affinities that he displayed. Thereby he fostered their sense of gentlemanly autonomy, rather than stifling it by an overassertive or magisterial style of discourse.

61 Jones (1983).   62 GUL, 1:8.   63 Quoted in Thomson (1832–59), I, p. 46.

At the same time, Cullen's chemistry also took its place in a public realm that extended beyond the walls of the lecture theater to settings outside the university. In the scientific societies with which he was involved, and in his dealings with his patrons, Cullen also articulated his chemical doctrine. Here again he had to strike a balance between deployment of theoretical perspectives and the compilation of experimental facts. In these settings, confronting the particular problems that they posed, he further developed the forms of discourse that would allow chemistry to assume the functions of a public science.

One setting in which Cullen worked, and in which he was soon made aware of the hazards of going beyond the level of a polite exchange of facts, was the Philosophical Society of Edinburgh. His involvement with the society began in 1753, with the submission of a paper on the chemical history of salts, written at the urging of his patron Lord Kames and addressed to the Edinburgh physician John Clerk, a longstanding supporter and friend. Here Cullen displayed the method he was using to organize this topic in his Glasgow lectures.[64] While ostensibly requesting the society's assistance to find an appropriate didactic method, Cullen was perhaps also aiming to gain their support in his hoped-for move to the Edinburgh chemistry chair.

He began by pointing to the low esteem in which chemistry was currently held, and remarked (as he did in his lectures) that the problem was not the chemists' use of obscure jargon, but their failure to find a comprehensive system to order all the known facts. Chemistry was a science of wide potential utility but its philosophical and general uses were being ignored due to lack of method.[65] Accordingly, Cullen went on to present a classification scheme for organizing all of the known salts, listing four recognized acids (vitriolic, nitrous, muriatic, and vegetable), and three corresponding alkalis (vegetable, fossil, and volatile). The innovation here consisted in identifying a new fixed mineral alkali (soda), distinct from the traditionally recognized fixed vegetable alkali (potash) and volatile alkali (ammonia). This was to prove a contentious point with at least some members of his audience. By combinations of these acid and alkali components, twelve neutral salts could be formed. Cullen proceeded to list them and commented on their identifications in the chemical literature. Although he did not actually show a summary table of salts in the paper itself, such a scheme offered itself very naturally to tabulation, as he subsequently demonstrated in his lectures.[66] Looking ahead to this didactic innovation in his paper, he described the classification of salts as

---

64 For background, see: Thomson (1832–59), I, pp. 12, 57–59, 64, 83–84, 525–536. For the text: Dobbin (1936).
65 Dobbin (1936), pp. 140–142.
66 Wightman (1956), p. 202; Manchester MS, I, p. 84v; RCPE, C.10, p. 145.

a "view" of chemical doctrine with potential utility in research and teaching. At the same time, he strove to maintain a properly modest stance with regard to theoretical speculation, insisting that his classification was the simple outcome of induction applied to chemical facts, and noting that chemistry "hardly admits of any theory but what immediately depends upon an induction of facts."[67]

The reaction to Cullen's paper must have disappointed him. In a letter to his former teacher in Glasgow, Black reported that the audience in Edinburgh had been unreceptive and the incumbent chemistry professor, Plummer, downright hostile.[68] Plummer had taken Cullen to be contradicting one of his discoveries (a new volatile alkali produced from coal) by identifying a new species of fixed alkali.[69] Lord Kames attempted to defend Cullen by offering to make corrections to the paper before it was printed. David Clerk (son of the addressee of the paper) also tried to mediate between the two chemists. At the meeting, Clerk expressed the view that the introduction to the paper had been "too bold," but he wrote to Cullen that he felt the members lacked the chemical knowledge to understand it fully.[70]

Black's assessment — that few of the members of the Philosophical Society had any appreciation of what Cullen was trying to do — implies that the disagreement over details was a sign of a more fundamental difference of outlook. There seems to have been no comprehension among members of the extent of Cullen's ambition to reform chemical method. Instead, he was taken to have violated the ethics of public discourse in the society by disputing the factual claims of a member. This was the main obstacle to their perceiving the wider implications of his paper as a model of philosophical chemical discourse.

From its foundation, the Philosophical Society had been constituted as a forum for factual communication. It was created in 1739, following the publication of a pamphlet that specified that reports submitted to the Society should comprise narratives of experiments, with "great Caution and Accuracy" observed in their composition. "Metaphysical Subtleties" were "not to be insisted upon."[71] A similar stipulation was made in the preface to the first volume of the Society's *Essays and Observations, Physical and Literary*, published in 1754, a piece that may have been written (at least in part) by David Hume. The author of the preface explained that the Society's aim was the same as that of other learned acad-

67 Dobbin (1936), p. 144.
68 Black to Cullen, January 1754, in Thomson (1832–59), I, pp. 58–59.
69 For Cullen's comments on Plummer, see Dobbin (1936), p. 151.
70 David Clerk to William Cullen, 23 January 1753, GUL, Cullen letters (quoted in Emerson (1981), p. 149).
71 Royal Society of Edinburgh [1739], quotations on p. 26.

emies in Europe – namely, to communicate observations of facts to the public. These facts might eventually contribute to the establishment of general truths, if "bold and happy geniuses" arose to reduce observations to "method and simplicity," but the Society would not preempt that process by judging the significance of contributions. The public alone had the right of making judgments of merit.[72] The founding rhetoric of the Society, with its emphasis on politeness and formal equality, required the ascription of credit to reporters of factual observations, and hence underwrote the enthroning of facts as public knowledge. On the basis of agreement upon facts, mutual respect among gentlemen could be maintained.

Apparently having learned this lesson, Cullen had more success with his second contribution to the Society, "Of the cold produced by evaporating fluids," which was read in 1755 and published in the *Essays and Observations* the same year. This was a much more empirical piece, a record of the results of experiments on the cooling effect observed when liquids evaporated. Cullen noted that he was confining himself to "facts already sufficiently verified," and explicitly declined to enter into speculative comments in advance of further experimentation. He encouraged other researchers to join him in further enquiry into this new realm of phenomena.[73] This approach seems to have found more favor with the members of the Society. David Clerk reported to Cullen that the paper was judged "ingenious & good"; and he went on, "There were many remarks made by way of Additions not Objections; nothing however, but what either has or naturally will occur to yourself. . . ."[74] The Society was evidently more impressed by factual reports of experimental enquiries, particularly if they opened up new fields of research, than by attempts at reorganizing a whole discipline. In communicating in this setting, Cullen learned that theoretical speculation or even synoptic tabulation had little value. The economy of discourse in the Society was driven by the communication of facts, not by attempts at general overviews of a field.

The balance between reported facts and interpretive generalizations was also at issue in discussion of the relationship between chemistry and the practical arts. Cullen had a clear interest in asserting that knowledge of chemistry gave the educated man an advantage over the unreflective practitioner. Such technical fields as pharmacy, bleaching, agriculture,

72 "Preface" to *Essays and Observations, Physical and Literary,* 1 (1754), iii–viii; quotations on pp. vi, vii. On the authorship, see Emerson (1981), pp. 146–147.
73 Cullen (1755). The paper was dated 1 May 1755, and the author identified as Professor of Medicine at Glasgow.
74 David Clerk to Cullen, 19 July 1755, GUL, Cullen letters (quoted by Emerson (1981), p. 149).

and metallurgy were said to be illuminated by the principles yielded by philosophical chemistry. Hence the deployment of chemical theory in connection with technical activities was of critical importance; but it was also fraught with risk. The relationship had to be worked through in Cullen's practical dealings with his patrons, and in participation in their projects, as well as being articulated in his lectures. Practical achievements were highly problematic, and a maladroit insistence on the relevance of chemical theory to practice could be taken for arrogant projecting or self-serving puffery. In a climate that was broadly, but by no means totally, favorable to entrepreneurial scientific improvement, Cullen had to step carefully.

In the lecture theater he could be relatively unrestrained in the scope of his claims. There he asserted how indispensable philosophical chemistry was to improvements in the arts. He did this in terms derived from Bacon: The "principles" of chemistry were both the causal laws discovered by induction and the precepts that would guide applications to practice. Chemical theory would yield the principles that would direct innovations in the practical arts. Predictably, this declaration remained more in the nature of a promissory note than a cashed-in achievement. Cullen commented upon various arts in his lectures under the headings of the chemical substances they exploited. "Salts" covered the making of saltpeter and potashes and the refining of sugars, "oils" the making of soap and wine, and so on. But he never claimed to have prosecuted the inductive "chemical history" to its conclusion or to have produced the fundamental principles of these practices. Instead, using the visual metaphors for knowledge that we have noted already, he proposed that chemistry offered a general view of the arts, one that supposedly held the advantage over the "various & discordant" views of their practitioners. The benefit of this metaphorical high ground permitted him to omit detailed descriptions of practices "in the way of Trade & Business," a decision that he acknowledged might disappoint some among his audience, but conveniently avoided trespassing on the protectively guarded territories of particular occupational groups.[75]

Even in his lectures, however, Cullen acknowledged the existence of a gap between the rules produced by philosophical chemistry and their actual implementation. He allowed that there would be problems in putting philosophers' insights into practice. By this stipulation he distanced himself from "extravagant projects," which he acknowledged had disgraced many previous chemists. Chemists were still notorious for projecting what they could not deliver, he noted, notwithstanding that they had abandoned their predecessors' promises of turning base metals into gold. Cul-

75 RCPE, C.11, p. 3.

len had to avoid being tarred with the brush of projector, although he also had to insist on the value of chemistry to practical improvement.[76]

In this predicament, Adam Smith's "Essay on the History of Astronomy" offered some encouragement. Smith provided a rationale for the philosopher's interest in the arts, while admitting the problematic aspects of the enterprise. In general, he proposed that the philosopher's aim should be to find connections between perceived phenomena, so that the passage of the mind from idea to idea would be rendered "smooth, and natural, and easy." In this way the anxiety experienced whenever one cannot connect novel impressions with existing ideas would be overcome. In the case of the arts, the philosopher should not expect his task to be understood by artists themselves. They, being accustomed to the experiences of their trade, would not comprehend the importance for others of having ideas connected in a rational scheme:

When we enter the work-houses of most common artizans; such as dyers, brewers, distillers; we observe a number of appearances, which present themselves in an order that seems to us very strange and wonderful. Our thought cannot easily follow it, we feel an interval betwixt every two of them, and require some chain of intermediate events, to fill it up, and link them together. But the artizan himself, who has been for many years familiar with the consequences of all the operations of his art, feels no such interval. They fall in with what custom has made the natural movement of his imagination: they no longer excite his Wonder, and if he is not a genius superior to his profession, so as to be capable of making the very easy reflection, that those things, though familiar to him, may seem strange to us, he will be disposed rather to laugh at, than sympathize with our Wonder.[77]

Philosophers had the duty to brave such ridicule, Smith proposed, to expose the connecting principles that bind together the artisan's unreflective and habitual actions. Entering the artist's workshop, the philosopher "endeavours to introduce order into this chaos of jarring and discordant appearances, to allay this tumult of the imagination, and to restore it . . . [to] tranquillity and composure." This activity was identified by Smith with a translation into the public realm of hitherto private artisanal craft. In this respect, the philosopher was a delegate of the public, revealing to their view the principles of the arts and thereby making them available to improvement. With reference to chemistry, Smith asked why it had, "in all ages crept along in obscurity, and been so disregarded by the generality of mankind." His answer was that:

The connecting principles of the chemical philosophy are such as the generality of mankind know nothing about, have rarely seen, and have never been acquainted with; and which to them, therefore, are incapable of smoothing the passage of the imagination betwixt any two seemingly disjointed objects. Salts, sulphurs, and mercuries, acids, and alkalis, are principles which can smooth things

76 RCPE, C.10, pp. 21–22.   77 Smith (1980), pp. 42, 44.

to those only who live about the furnace; but whose most common operations seem, to the bulk of mankind, as disjointed as any two events which the chemists would connect together by them.[78]

The remedy for stagnation rooted in obscurity was public scrutiny of the artisan's thoughtless practice. The philosopher, because of his skill in classifying and explaining phenomena, was specially qualified to act on behalf of his fellow citizens in bringing the arts into public view.

Cullen's lectures show him helping himself to this kind of legitimating argument, and applying it to the relations between the philosophical chemist and the chemical arts. He announced that he would bring these arts into the public realm and show their connections in a general view that was denied to the practitioners themselves. In this way he attempted to insinuate himself as an expert mediator between the patron of improvement and the artisan, scrutinizing with a keen eye the activities of the artist, while providing strategic knowledge to the patron.[79]

Such claims were relatively easy to make in lectures. When Cullen came to deal with the practical arts in the course of the projects in which he was involved, he had to grapple with a more complex reality. The sheer intractability of the material difficulties involved in arts such as agriculture and bleaching severely limited his freedom of maneuver. He also had to negotiate such social and cultural constraints as the unwillingness of workers to alter their habitual practices and the need to avoid imputations of projecting. Cullen had to tread a difficult path, insinuating the applicability of philosophical chemistry to public improvement of the arts, while avoiding allegations that he was seeking to satisfy private avarice.

Cullen became involved in agriculture through shared enthusiasm with his patron Lord Kames. It was the subject of a correspondence that the two began in 1748, and of a book that they apparently considered writing about that time. In the early 1750s, Cullen was also himself managing a farm, which he had purchased for his brother Robert at Parkhead near Glasgow; by the late 1770s he had acquired his own.[80] The fruits of his attempts to apply chemistry to agriculture were however limited. Some small-scale programs of experimentation were pursued with Kames's encouragement, such as one on the effectiveness of niter as a fertilizer, but only tentative efforts were made to articulate the relevant theory. After trying to deal with the subject in his lecture course, Cullen decided that its uncertainties made it inappropriate for discussion in that setting. He

78 Smith (1980), pp. 45–46, 46–47.   79 GUL, 1:7.
80 Thomson (1832–59), I, pp. 60–64, 67–69. Cullen to William Hunter, 21 August 1778; Henry Home [Kames] to Cullen, 23 December 1749, and 5 January 1750; Cullen to H. Home, 17 January 1750, and [spring 1749]; in Thomson (1832–59), I, pp. 564–565, 592–597.

had admitted to his students in Glasgow that "there are many who would smile at any attempt to treat of Agriculture or other such practical Art in a College."[81] In 1768 he therefore covered the subject in a separate course, given to an invited audience of trusted friends. In this private setting he explicated theories of plant nutrition and the role of manures, while stressing the "extreme difficulty of attempting to establish any principles of this art." Repeatedly he emphasized how doubtful was the chemical theory he was articulating, and how tenuous its connection with agricultural practice.[82]

This cautious attitude was entirely consistent with the view taken by Kames in his own *Gentleman Farmer*, published in 1776. Kames allowed that theoretical speculation might be indulged in for amusement, but denied that any systematic or demonstrative reasoning would be possible in agriculture. Though farmers might usefully learn a few basic chemical facts, "to be an expert farmer, it is not necessary that a gentleman be a profound chemist." Indeed, "the life of an antideluvian" would be required to make all the experiments necessary to establish theoretical principles of agriculture. Fortunately then, "agriculture depends not much on theory."

The means of improvement that Kames proposed for Scottish agriculture relied less upon sophisticated theory than upon a gentlemanlike communication of factual reports. The model of the Edinburgh Philosophical Society was to be extended throughout Scotland by the inducement of emulation and limited financial rewards. Spurred largely by patriotic spirit, the proposed Board for Improving Agriculture would embark on a comprehensive survey of practice in all regions of the country, followed by distribution of premiums to farmers who communicated notable innovations. The Board would act as a clearinghouse for information and would issue instructions on advanced methods of farming. By his knowledge above all of a more comprehensive range of facts, the gentleman farmer would distinguish himself from the habit-bound and idle simpleton who was the typical Scottish husbandman.[83]

Like Kames, Cullen was evidently wary of exaggerated claims for the applicability of chemical theory to agriculture. He was happy to indulge his patron in the private entertainment of speculative conversations and letters, but was well aware of the ridicule that might greet ambitious attempts to extend scientific progress into the field. This was presumably the reason why he decided to give his agricultural lectures only to a select audience and declined to publish them. As Kames argued, the gentlemanly path to improvement was through encouragement of experiment

81 GUL, 1:16.   82 Cullen (1796), pp. 1, 27.
83 Kames (1776), pp. 290–292, 357, 359–378.

and the methodical compilation of factual results, not through excessive trust in unfounded theories.

Cullen's public reticence in this connection is highlighted by contrast with the rather bolder stance of Francis Home, his rival for the Edinburgh chemistry chair in the mid 1750s. In his *Principles of Agriculture and Vegetation* (1756), Home was notably more sanguine about the possibility of reducing agriculture to chemical principles. Without a knowledge of chemistry, he noted, the principles of the art could never be fixed, and without scientific principles, improvement would scarcely be possible. Home went on to suggest an account of the mechanism of plant nutrition, the "fixed point from which we may have a full view of this extensive art." By the end of the book, however, he was downplaying these rather grandiose assertions of the importance of theory and calling instead for further experimental research. He suggested that a committee of the Philosophical Society should be set up to collect and publish experimental narratives, to be written "in the plainest and most distinct manner," without interpretive comments. This blend of ostentatious theoretical claims with an endorsement of cautious experimental research won Home a prize from the Society for Improvement of Arts and Manufactures.[84]

Home was also more willing than Cullen to commit himself publicly on the value of chemical theory in relation to bleaching, a crucial case for attempts at improvement of the arts at the time. In his *Experiments on Bleaching*, which he presented to the Board of Trustees for the Encouragement of Fisheries, Arts and Manufactures, and published in 1756, Home insisted that chemists were qualified to direct improvements in many of the arts. Apparently unconcerned about allegations of projecting, Home endorsed Peter Shaw's conception of a universal chemistry. "If chymistry was once too wild and extravagant," he wrote, "it has been for many years too tame and confined." In bleaching, chemistry was capable of providing principles that were directly applicable to practice and preferable to the purely experiential knowledge of practitioners. Home did not hesitate to criticize habits of secrecy and a lack of ambition for material progress among workers in the bleaching industry. Chemistry could solve problems of stagnation and slovenliness, he proposed, for it

... looks upon the operators, as entirely under its guidance; the operations to be accounted for only on its principles; and the hopes of further perfection in the art, so far as human judgment is concerned, to rest on it alone.[85]

Cullen, in his work on bleaching, avoided such vociferous advocacy of chemistry as the motive force of improvement. Instead, he insinuated himself more modestly, but arguably more successfully, into the structure

84 Home (1756a), pp. iii, 4–6, 174–176.   85 Home (1756b), pp. 8–9, 16–18, 21, 328.

of social relations through which improvements in the industry were being explored. His involvement in projects sponsored by the Board of Trustees has been recounted by Thomson and Donovan.[86] In the early 1750s, Cullen gathered information for the Board on all aspects of practice in the industry, and initiated a number of experimental trials of the effectiveness of alkali extracted from wood ashes as a bleaching agent. His efforts consolidated his connection with the Duke of Argyll, who had recently gained control of the Board through his clients, and were rewarded by a premium to the value of 21 pounds.[87] Throughout, he carefully presented himself as an observer and experimenter, rather than a projector. In the essay on bleaching, which Kames forwarded on his behalf to the Board in 1752, he made the point that "a person who proceeds on general principles, without being exactly acquainted with the practice & detail of the Art, is in danger of proposing refinements, that are either impracticable in the way of trade or of little use when practised."[88] And when the Board commissioned comments on his essay from the bleaching entrepreneur John Christie of Ormiston, Cullen modestly deferred to Christie's greater practical experience:

I am glad to find a man of judgment and experience agreeing with me in so many particulars. The most part of the instruction I had received on the subject, I had got from his directions; and from his notes on my remarks I have learned a good deal more. In some particulars he seems to differ from me, and I am ready to believe he is in the right.[89]

This kind of deference to the experience of practitioners, and Cullen's unwillingness to promote chemistry as an instant solution to all the problems of agriculture or bleaching, gained respect. His subtle and tactical deployment of chemical skills earned him continuing patronage from the powerful individuals and organizations driving the Scottish program of improvement. Francis Home, though he won some recognition for his efforts in applied chemistry, also excited suspicion – Lord Deskford, a leading member of the Board of Trustees, thought Home had stolen his basic ideas about bleaching from Cullen.[90] Home had to wait until 1768 for the Edinburgh chair (in materia medica), which he coveted. Cullen secured the patronage that was so vital to his success in moving to Edinburgh in 1755, and continued to consolidate his place in the public cul-

86 Thomson (1832–59), I, pp. 74–79; Donovan (1975b), pp. 78–83. See also Durie (1979), pp. 55–59, 84–85.
87 In addition to the sources mentioned in note 86, see Shaw (1983), pp. 60, 67–77.
88 William Cullen, "Remarks on the Art of Bleaching or Whitening Linen," RCPE, Cullen MSS, no. 35, f. 1r. (Another draft, GUL, 40b, is quoted by Donovan (1975b), pp. 80–81.)
89 Cullen to Kames, February 1753, in Thomson (1832–59), I, pp. 76–68 (quotation on p. 77).
90 Joseph Black to Cullen, 22 November 1755, in Thomson (1832–59), I, p. 93.

ture of gentlemanly improvement. In doing so he negotiated constraints that were simultaneously practical and social, involving both the material difficulties that bleachers and farmers confronted and the social conventions that disparaged projectors seeking private gain.

Kames's *Gentleman Farmer,* in its discussion of how agricultural improvement should be structured as a public activity, throws light on the social constraints that Cullen faced. According to Kames, "no other occupation rivals agriculture, in connecting private interest with that of the public." Improving farmers were natural patriots, because in seeking to raise their own productivity they were simultaneously serving national and civic needs. To ensure that this continued to be the case, a particular form of organization would be necessary to sustain patriotism and public spirit and to curb avarice and factionalism. The model of a "Board for Improving Agriculture" was proposed as a way of creating "A society of gentlemen, who serve for honour not for profit."[91]

Such reassurances reveal the anxiety they were intended to allay: The fear of seeming to follow a purely private interest was a prevalent one, and a consistent factor shaping attempts to constitute improvement as a public activity. The imputation of personal ambition was to be avoided if possible, in a culture that still tended to regard private interest as incompatible with social cohesion and moral propriety. This was the fundamental reason why accusations of projecting were shunned by Cullen and his contemporaries.[92]

Cullen experienced this problem as a pressing one in his dealings with his patrons. To Kames, he felt able to reveal his anxiety that a public avowal of his personal interest in a project would bring him discredit. Asking Kames to relay to the Board of Trustees a request to support experiments on new methods of alkali manufacture, he wrote:

If the Trustees are not likely to be very prompt in this affair, I would not have it mentioned. I shall not pretend [to you] to be disinterested. On the contrary, I should be very glad to have my interest connected with schemes of this kind; but at present I would be hurt by suspicions of interested views.[93]

Although Cullen did not conceal from Kames his eagerness for the rewards to be gained by using his skills, he was aware that he had to be cautious in admitting his interest openly because the overt expression of an expectation of gain would be considered dishonorable. With Argyll, a much more exalted patron than Kames, Cullen was wary of appearing too eager for material rewards even in private correspondence. Drafting a letter to the duke in 1751 to report his work on salt purification, Cullen

91 Kames (1776), pp. xvii–xviii, 373–374.
92 On this general problem, see: Hont and Ignatieff, eds. (1983); Pocock (1985).
93 Cullen to Kames, February 1753, in Thomson (1832–59), I, p. 77.

struggled to reconcile his anticipation of profit with the need to keep up an appearance of disinterestedness. In one version of the document, he turned the issue of projecting into a joke, perhaps relying on the fact that his patron had a virtuoso devotion to chemistry himself. He wrote:

My Chemical Labours have as yet yielded me no profit but I begin to wish they may. . . . How far they may be useful to the public, I subject [to] your Grace's opinion. How they may be beneficial to my self I humbly beg to have your Grace's advice. In the meantime till I shall have the honour of your Grace's opinion, I presume to beg that no part of these observations may be communicated to any body else. I have not hitherto dealt in Secrets & never shall have any reserve with your Grace but in this the projecting spirit has seized me & I hope your Grace will for one indulge my request.[94]

The problem of private and public interest, and the propriety of expressing them in particular circumstances, was one that was central to Cullen's career as a proponent of philosophical chemistry. His aim was to show that chemistry was not simply a concern of the practitioners of the chemical arts, but neither need it be confined to the service of medicine. It could be a philosophical enterprise, assuming responsibility for stimulating improvements in many arts, and forming part of gentlemanly education. In other words, chemistry could take its place in the public culture of scientific and moral improvement that comprised the core of the Scottish Enlightenment.

This aim was clearly incompatible with explicit avowal of an interest in personal gain. Cullen had to distance himself from suspicions that he was a projector, driven by private ambition or greed. Far from constructing an identity for chemistry as a specialist profession, he had to show that the subject was simply part of the liberal education of any gentleman, and hence a natural contributor to the civic endeavor of improvement. This balancing act was a difficult one for someone in Cullen's ill-defined and potentially insecure social position. Dependent as he was on the fickleness of patrons, and necessarily anxious to gain their favor, he had nonetheless to mask his personal ambitions to operate in a culture supposedly driven by devotion to public duty. The situation was easier for aristocrats like Lord Milton, Argyll's leading agent in the management of Scottish society, who could afford to be cynical about the private interests underlying the prevalent rhetoric of public good. In a letter to a friend, Milton wrote:

94 GUL, Cullen letters, no. 61. Compare drafts of other letters to Argyll in GUL, Cullen letters no. 60, and printed in Thomson (1832–59), I, p. 75. The similar form of words used in Cullen to Kames, July 1752 (printed in Thomson (1832–59), I, p. 599) suggests that the letters to Argyll might never have been sent. (I am grateful to Michael Barfoot for advice on this matter.) On Argyll's scientific interests, see Murdoch (1980), p. 33; and Argyll to Cullen, 7 March 1751, in Thomson (1832–59), I, p. 71.

I have lived long enough to see that in private as well as public treaties mankind will be no further bound than they find their interest leads them ... After all thats commonly said of publick spirit we in fact see that tis private interest that makes every person active and diligent, opens his eyes and his ears, raises his invention, supports him under fatigue and makes his business an entertainment.[95]

At least in his public discourse, Cullen could afford no such relaxed cynicism. His advocacy of philosophical chemistry sought unreservedly to assimilate it to the prevailing culture of civic improvement, even if the tensions between public values and private ambitions and interests were on occasion unavoidable. Difficulties of this kind stemming from the unique cultural conditions of Enlightenment Scotland added to the numerous practical problems that Cullen faced in his attempt to establish chemistry as a public science.

The social construction of the Scottish program

Through his lectures at Glasgow and Edinburgh, through scientific societies and participation in his patrons' projects for improvement, Cullen won for chemistry an unprecedented measure of acceptance and esteem in Scottish society. He succeeded in presenting the discipline as relevant to the interests of his students and of sections of the population outside the university. By his efforts, chemistry was constituted in a particular nexus of public spaces: lecture theaters, informal associations for civic improvement, the farms of aristocratic patrons, bleach fields and alkali factories. It became a public science in a quite specific sense, imbued with the civic values and aspirations prevalent in Enlightenment Scotland.

Cullen gave Scottish chemistry the social profile that it was to retain for the remainder of the eighteenth century. The subsequent program of research by Joseph Black and his associates on heat transfer in relation to physical and chemical change was significantly shaped by this legacy. A highly innovative program in terms of its conceptual and instrumental resources, it was nonetheless formed on Cullen's pattern for a philosophical discipline. The university lecture theater remained the prime site for rendering chemical research public. Despite the evident limitations of this means of communication, Black devoted himself, even more single-mindedly than Cullen, to teaching and demonstrating experiments in lectures. He declined to publish more than a tiny fraction of his researches, a decision that has normally been ascribed to personal idiosyncrasy, but was in some respects quite in keeping with the national tradition.

95 Lord Milton to John Goodchild, 14 August 1744, quoted in Shaw (1983), p. 157. Shaw gives more information about Milton's role as Argyll's "leading henchman" on pp. 68, 147–191.

Cullen had set the example in this respect. He had used lectures to articulate theoretical doctrines that had not found an outlet in publications. He communicated his view of chemistry primarily by teaching and diffused his ideas more widely by fostering lasting personal relations with his students. Edinburgh graduates, many of whom had studied under Cullen, comprised an identifiable network within the medical profession in late eighteenth-century Britain. They proved disproportionately influential in the development of new experimental approaches to medicine.[96] Among the former students with whom Cullen maintained contact after they left Edinburgh were George Fordyce and William Saunders, both of whom gave lectures on chemistry and other medical subjects in London.[97] Thomas Percival, William Withering, Matthew Dobson, John Haygarth, and William Falconer were to be among the pioneers in applying pneumatic chemistry to medical treatment in the decades after they passed through Cullen's tutelage.[98] Percival later sent his own son to study medicine in Edinburgh, while Withering and Haygarth both recalled their mentor's instruction with deep gratitude. Withering told Cullen that it was due to him that chemistry "for the first time could . . . claim the title of a science."[99]

Cullen provided an excellent model for Black of the practice of communicating science through cultivating lasting relationships with students. Indeed, Cullen's position as father-figure to Scottish chemistry as a whole probably owes a lot to his carefully nurtured friendships with a large number of his pupils. Fordyce was almost certainly not alone in regarding him as "a father who lives in the hearts of a thousand sons."[100] Though his charisma was evidently less than Cullen's, Black followed his example in earning admiration and devotion from some of his pupils.

As early as his Glasgow days, Cullen had shown the warmth of his affection for favored students and his willingness to extend their instruction beyond the confines of the lecture room. He employed several students (including Black himself) as research assistants, and offered private tuition to clarify problems arising from his lectures.[101] He also took the quite novel step of inviting students to attend sessions in his laboratory, where they could try experiments themselves and where he could help

96 See Chapter 5 (below); and (for a slightly later period) Chitnis (1986), pp. 36–37, 58–59, 133–181.
97 Cullen's correspondence with Fordyce is in Thomson (1832–59), I, pp. 123–129, 627–628. Other students are mentioned on pp. 460–462.
98 Correspondence with them is in Thomson (1832–59), I, pp. 625–627, 635–642, 648.
99 Percival to Cullen, 17 October 1785; Haygarth to Cullen, June 1772; Withering to Cullen, 19 October 1786; in Thomson (1832–59), I, pp. 635–638.
100 Fordyce to Cullen, 22 October 1774, GUL, Cullen Letters no. 180.
101 Thomson (1830–31), I, pp. 308–310; Thomson (1832–59), I, p. 134; GUL, 1:13.

them acquire practical knowledge of chemical techniques.[102] This extension of instruction from lecture to laboratory was a move Black was to replicate during his own period as lecturer at Glasgow from 1756 to 1766, when he worked with his students John Robison and William Irvine, and with the then instrument maker to the university, James Watt.[103]

Cullen had also shown Black the way to extend his students' education by encouraging them to participate in the flourishing student societies at Edinburgh, where the doctrines of the professors were debated and applied. Cullen himself had been a founder member of the Medical Society, while pursuing his own education in the city in the mid-1730s. The Society continued to flourish thereafter, so that when he returned to the Scottish capital as a professor he encouraged his own pupils to join. Among those who did so were Saunders, Fordyce, Fothergill, Withering, and Black himself.[104] In a similar way, Black as a professor encouraged the formation of the student Chemical Society in 1785. The Society was composed entirely of his pupils, who contributed papers discussing and extending his chemical theories.[105]

Black's social world was no more bounded by the university walls than Cullen's. He had important associates who shared his interests in chemical manufacture and the arts. For example, he took a close interest in the manufacturing projects of his brothers James and Alexander in Belfast. As with Cullen, Lord Kames also provided a source of extramural encouragement, sounding out Black's knowledge of bleaching techniques in the 1750s and later seeking his imprimatur for *The Gentleman Farmer*. While Kames gained legitimacy for the chemical content of his work, Black used the correspondence between them to try out speculative ideas that he could not articulate more publicly. He deployed notions of specific attractions between different substances to try to account for the effect of moisture on the consistency of clay.[106]

Black's relationship with Kames was less crucial than Cullen's in determining the course of his career; the two men seem to have regarded one another as fellow laborers in the fields of applied chemistry. A more important association of this kind was the longstanding one with James Watt, who was originally Black's laboratory assistant at Glasgow and subsequently became a business partner and friend. Although they were

102 Anderson (1978), p. 11; Wightman (1956), p. 196; GUL, 1:9.
103 Donovan (1975b), pp. 250–271.
104 Thomson (1832–59), I, pp. 9–11; Cullen's list of his students, annotated with dates at which they joined the Medical Society is GUL, Cullen Letters, no. 191.
105 Ramsay (1918), pp. 110–111; Kendall (1949–52); Perrin (1982).
106 Kames (1776), p. xiii, fn.; Black to Kames, 23 May, 7 and 15 September 1775, in Woodhouselee (1807), II, pp. 75–81; Kames to Black, 30 May 1775, in Lehmann (1971), pp. 299–300.

geographically separated after Black's move to Edinburgh in 1766, their correspondence shows the two becoming increasingly close in personal terms. In his letters to Watt, Black felt able to discuss sensitive issues of secrecy and property rights – for example, in the late 1760s, when he was concerned that their work on methods of alkali manufacture might be appropriated by other chemists. Anxiously, Black urged Watt to secure a patent for the process for making alkali from lime and salt, despite continuing difficulties with putting it into practice.[107] In a similarly confidential moment, he warned one of his brothers to beware of the Austrian visitor Dr. Schwediauer, who was "a great Projector & pryer into Secrets."[108]

These questions of privacy and property were difficult but keenly felt ones for Black, apparently reflecting a considerable degree of self-doubt about his public role. Even more than Cullen, he was anxious not to appear as a projector. In 1789 he wrote to Ronald Crawford, a Glasgow chemist, that he was "habitually very diffident of all new projects, on account of the time, expense & trouble they generally require."[109] To Alexander Black he conveyed the following advice for "brother James":

I have little faith for the Success of . . . [his] Projects for the Publick and still less would I trust to the Gratitude of the Public for him even supposing they should succeed. . . . I always fear this hunting after Schemes & Projects will get him the Character of a Projector . . . when they find his Projects are unsuccessful or impracticable.[110]

The anxiety about getting "the character of a projector" was clearly one that Black felt for himself. As we have seen, the stigma of appearing motivated by personal interest was shunned by those who wanted to assimilate chemistry into the public culture of the Scottish Enlightenment. With trusted friends or relations, the issue could be addressed directly, or alternatively its potency neutralized by humor. An incident reported by John Hope, professor of botany at Edinburgh, reveals how groups of friends might reassure one another by joking about the matter. In 1784, Hope wrote to the English iron master Matthew Boulton, describing a meeting of an informal subgroup of the Philosophical Society:

Last night we had a full meeting at the Oyster Cellar, Mr. Cort, Lord Dundonald, Hutton, Black, MacGowan, etc. Dr. Hutton whispered to me, what a number of

107 Robinson and McKie, eds. (1970), esp. pp. 16–23; Musson and Robinson (1969b).
108 Joseph Black to his "brother" [Alexander or James Black], 18 July 1789, MS HM 49188 in Huntington Library, San Marino, California. On Franz Xavier Schwediauer, an Austrian physician who lived in Edinburgh during the late 1780s, see Linder and Smeaton (1968).
109 Black to Crawford, September 1789, in Ramsay (1918), p. 104.
110 Joseph Black to Alexander Black, [undated], in McKie and Kennedy (1960), p. 137.

projectors, and Black said I was a fool of one myself. We had as usual a great deal of pleasantry. . . .[111]

Black's notorious reluctance to publish his research may be related to this; it seems to manifest an extreme anxiety about his public reputation. Despite repeated urging by his friends, he never brought to press his seminal research on latent and specific heats, originally carried out in the late 1750s. At least twice he promised Watt that the work would shortly appear in printed form, but when John Robison took over the task of literary executor on his friend's death he found that no prepared manuscript existed.[112] Watt and Robison in particular were frustrated by Black's failure to assert his priority as discoverer of specific and latent heats against the competing claims of Jean André DeLuc in France and Johan Carl Wilcke in Sweden. Robison's rather rueful conclusion was that ". . . there never was a man less assuming, or less eager to obtrude himself on the public attention"; notwithstanding the appeals of his supporters, ". . . nothing could prevail upon him (so much did he dislike authorship) to vindicate his claims."[113]

The tendency of most commentators has been to accept the excuses Black offered at the time as explanations of this. He pleaded excessive commitments to his teaching as the reason why he could not assume the responsibilities of an author. McKie and Heathcote identify his devotion to his students and medical patients, "his heavy duties, his ill-health and his fastidiousness in composition," as the reasons why he could go no further than making an outline of the work.[114]

It seems likely, however, that more than just shortage of time was involved. Although the teaching schedule at Edinburgh could indeed be punishing, like all the other professors Black enjoyed the luxury of a six-month vacation during which there was plenty of time for writing (Figure 1).[115] A possible justification for declining this task, which was employed on Black's behalf by his followers, was that lecturing to students was publication enough. Watt told the Portuguese scientific writer John Hyacinthe Magellan (1722–1790) in 1780 that Black ". . . is very far from concealing his discoveries[;] he has always taught them publickly, and I fancy has had some thousands of scholars." When Robison produced his posthumous edition of Black's lectures in 1803, he emphasized that the

111 John Hope to Matthew Boulton, 22 May 1784, quoted in Clow and Clow (1952), p. 415. Cort was a steel maker, Dundonald later acquired a reputation as an agricultural improver and chemical manufacturer, and James Hutton was an agricultural improver and distinguished geologist.
112 Black to Watt, 15 March 1780 and 30 January 1783, and Robison to Watt, 23 July 1800, in Robinson and McKie, eds. (1970), pp. 83–84, 119–120, 342–347; Robison to James Black, 21 April 1802, in McKie and Kennedy (1960), pp. 161–170.
113 Black (1803), I, p. 523.   114 McKie and Heathcote (1935), pp. 34–35.
115 Morrell (1971a).

*Figure 1. Portrait of Joseph Black by David Martin. [Reproduced
by permission of the Royal Medical Society, University of
Edinburgh.]*

doctrine of heat they contained had already been rendered fully public
by Black's teaching and by the circulation of students' notes.[116]

This claim had a certain plausibility in the context in which Black operated. Many Edinburgh professors, including Black's successor in the

116 Watt to Magellan, 29 March 1780, in Robinson and McKie (1970), pp. 86–88 (quotation on p. 87); Black (1803), I, pp. 505, 531; II, pp. 216–217.

chemistry chair Thomas Charles Hope, who was also renowned as a lecturer, published no textbooks. Nor had Cullen produced a text to accompany his chemistry course. Credit and reputation in the Scottish academic world rested much more on teaching performance than on success as an author. In terms of the conventions of scientific communication in that milieu, we might therefore agree with Donovan that "Ideas and discoveries which have been presented in lectures have in fact been made public."[117] Such an indulgent view was not universally accepted, however. Magellan rejected the arguments made by Black's supporters, and told Watt that communicators deserved more credit than discoverers who refused to publish. "It is to those who publish their discoveries, and also those of others, that the public is indebted for them," he wrote elsewhere.[118]

If the positive side of Black's failure to publish was a belief that he had made his doctrine public enough, then the negative side was an almost paranoid fear of plagiarism. In letters to his friends he expressed the apprehension that his discoveries would be stolen from him if he communicated them too openly. As early as 1754, he wrote to Cullen concerning the proposed publication of his thesis on magnesia alba and fixed air: "If the experiments are worth any thing they will be stolen by others; but yet I would rather have it so than make them public, unless you think it would be of some sort of service to me."[119] In this instance, Cullen did persuade him that publication would be personally advantageous, and portions of the renowned thesis appeared in print.[120] When Watt tried to get him to publish on heat in the 1780s, however, Black responded by making a comparison with the situation if Watt himself were to write on the steam engine. As an author Watt would be obliged to show a conventional modesty, and to describe his invention "in such a Cold and modest manner that Blockheads would conclude there was nothing in it, and Rogues would afterwards by making trifleing Variations vamp off the greater Part of it as their own and assume the whole merit to themselves."[121]

The comparison was one that Watt could well appreciate. In partnership with Boulton he had worked strenuously to protect his patent rights over the separate-condenser steam engine in the face of constant challenges from other manufacturers with similar machines. In his correspondence with Magellan, Watt himself used the same analogy, though in

117 Donovan (1975b), p. 271.
118 Watt to Magellan, 29 March 1780, in Robinson and McKie (1970), p. 87; quotation (in French) in McKie and Heathcote (1935), p. 41.
119 Black to Cullen, 18 June 1754, in Thomson (1832–59), I, pp. 50–51.
120 Black (1756).
121 Black to Watt, 13 February 1783, in Robinson and McKie (1970), pp. 123–124.

doing so he accounted for Black's behavior in a way that the latter might not have found acceptable:

Each man who is obliged to live by his profession ought to keep the secrets of it to himself so far as is consistent with the use of them: It is only People of Independ[en]t fortunes who have a right to give away their Inventions without attempting to turn them to their advantage.[122]

Turning his discoveries to his advantage was, however, far more difficult for Black than for Watt. The analogy between the steam engine and the doctrine of heat foundered because Black could not assert his property rights over knowledge as explicitly as his friend. Although he pressed privately for Watt to secure patent protection for the method of making alkali, Black had to evade the public label of projector. Watt, an inventor and businessman working in the entrepreneurial climate of the English Midlands, could affirm his rights over intellectual property quite overtly. For Black, the son of a mercantile family, slightly insecurely inhabiting aristocratic and genteel Edinburgh, avowels of private interest were altogether less acceptable. The professional prerogative articulated by Watt was not one that Black would have been able to acclaim.

Black's intense anxieties about publication of his work seem to have derived from the conjunction of his highly sensitive personality and the particular public culture of Enlightenment Scotland. Undoubtedly there was a measure of personal idiosyncrasy in his repeated refusal to heed his friends' urgings and bring his work to press. On the other hand, the centrality of university teaching to the transmission of Scottish science made available the justification that lecturing was publication enough. At the same time, assertions of personal property in knowledge ran the risk of being seen as projecting. A man as sensitive as Black to the possibility of plagiarism could therefore feel deprived of resources for protecting his rights as an author.[123]

Given the absence of a published text, the problems faced by Robison when he came to prepare his edition of Black's lectures were considerable. He complained to Watt about the fragmentary nature of the surviving notes, and their lack of literary style. He seems to have found more helpful one of the sets of handwritten notes normally sold to students,

122 Watt to Magellan, 9 March 1780, in Robinson and McKie (1970), pp. 80–81.
123 An incident that throws some light on this problem was that of the unauthorized publication of Cullen's lectures on materia medica in 1771 (Thomson 1832–59, I, pp. 142–144, 611–618). The London publishers accepted that Cullen had a legitimate interest in the quality of the published text, but William Hunter warned him that he would risk public opprobrium if he avowed his interest to the degree of trying to profit financially from a deal.

which Black himself had purchased and annotated.[124] From such documents Black's authorial intentions were posthumously reconstructed. The primary aim of the edition was, as Robison put it, "to support the Claims of [D]r Black to his various chemical discoveries," and retrospectively to secure his reputation as a philosophical chemist in the wake of the radical changes in theory at the end of the eighteenth century.[125] Robison's editorial procedures were subordinated to this aim; his was an explicitly reconstructive (not in any sense original) edition of the lectures. It is scarcely surprising that they continue to pose problems for historians who seek to read through them to find out what Black actually said.[126]

One respect in which this difficulty arises concerns the demonstrations incorporated in the lectures. We know from the standard notes sold to students, and more tellingly from those taken verbatim in 1767–1768 by Thomas Cochrane, that Black used demonstration experiments extensively to make his points.[127] Robison appears to have found this rather embarrassing. He mentioned the fact that Black's own notes were "full of references to processes going forwards, or which have been gone thro' in the Class, and pointings to things on the table," as an aspect of their artless style.[128] In his own edition he replaced these references by narrative records (in Black's voice) of particular experiments performed on particular occasions. This was one way by which Black's priority in his great discoveries could be secured.

In the original lectures, such insistent claims to originality and precedence over competitors do not seem to have occurred. Black did make it clear to his students that the doctrine of heat he was presenting was his own, but he did so without giving circumstantial narratives of particular experiments. Demonstrations were simply presented in the classroom to establish the doctrine as a public fact. For example, the absorption of heat in certain processes was shown to prove that the fluidity of a body depends upon a latent heat content that is not apparent to the thermometer. "These Exp[erimen]ts prove . . . all I alledge," Black told his students, ". . . These are the facts upon which my opinion is founded."[129]

Observers at the lectures remarked upon the skill Black showed in these demonstrations and on the simplicity of the apparatus used. Henry Brougham recalled the "perfect philosophical calmness" Black displayed in dextrously pouring liquids from vessel to vessel without spilling a sin-

---

124 Robison to Watt, [25 February] and 23 July 1800, in Robinson and McKie (1970), pp. 335–347; McKie and Heathcote (1935), p. 12.
125 Robison to James Black, 21 April 1802, in McKie and Kennedy (1960), pp. 161–170 (quotation on p. 165).
126 Christie (1982).   127 Black (1966).
128 Robison to Watt, 23 July 1800, in Robinson and McKie (1970), pp. 342–347 (quotation on p. 343).
129 Black (1966), pp. 12–15 (quotation on p. 14).

gle drop.[130] But Robison gave voice to the suspicion that these experimental demonstrations were just for show, that they did not substantially contribute to advancing chemical theory. He wrote to Watt, "Dr. Black seems to have turned his whole attention to rendering his Lectures as popular and profitable as possible, by a neat exhibition of Experiments – he multiplied these, without any new Views."[131] Writing after the transformation in chemical discourse wrought by Lavoisier, Robison could not help but see Black's lectures as distressingly unsystematic. He became increasingly despondent as he continued his editorial work, believing that this verdict would be shared by other readers and would permanently damage Black's established reputation. He therefore accepted with relief an alternative way of regarding the lectures suggested by Watt. Watt proposed that they should be seen as a "history," without theoretical pretensions. In October 1800, Robison wrote:

> Mr. Watt has given me another Notion of the subject from what I had allowed myself gradually to form. He considers it as a *History of Chemistry,* for forty years, by one of its greatest Masters. [D]r. Black never attempted to give a *System* of Chemistry. He was unfriendly to all systems of an experimental Science, and with respect to chemistry, he thought a system was an Absurdity, because, altho' we think ourselves very wise, and have made many discoveries in it, the science is still but beginning, in its Infancy . . . His Lectures therefore *profess no system,* and, for this reason, may chance to be little thought of by refined Theorists. But they will contain complete Instruction for those who are not acquainted with Chemistry, and wish to learn it.[132]

By regarding Black's lectures as history, Robison could find an excuse for their unsystematic character – for the disordered accumulation of experimental processes without general conclusions. Brougham echoed this interpretation, recording how Black had appeared in his lectures as "the historian of his own discoveries," lengthily reiterating the experiments by which he had originally made his name.[133] From a post-Lavoisian perspective, a cumulative style of discourse like this could only be glossed as history, and was inevitably seen as less effective than the integration of a few precise results into a systematic theoretical structure. The contrast of styles between Black and Lavoisier indicates much about the unprecedented novelty of the latter's mode of chemical discourse, a topic to which we shall return in Chapter 5.

Such retrospective views do little justice to the effectiveness of Black's

130 Brougham (1872), pp. 19–20.
131 Robison to Watt, [25 February 1800], in Robinson and McKie (1970), pp. 335–341 (quotation on p. 339).
132 Robison to George Black, Jr., 18 October 1800, in Robinson and McKie (1970), pp. 360–363 (quotation on p. 361).
133 Brougham (1872), pp. 20–21.

method of teaching in its own context. Although his wider reputation undoubtedly suffered from his failure to publish his own work, he did succeed in inspiring a local tradition of experimental and theoretical research on the phenomena of heat in relation to physical and chemical change. As Donovan has shown, William Irvine extended Black's work by assimilating the discovery of latent heats of fusion and vaporization into an expanded theory of heat capacities. Irvine proposed that bodies release or absorb heat when they change their physical state due to a dramatic change in their heat capacities.[134] Although Black rejected Irvine's theory, he noted the degree to which it had been derived from his own while Irvine was his student at Glasgow.[135] When Adair Crawford in turn extended Irvine's theory to give an explanation of animal heat in relation to respiration, he recorded that he had been led to the subject by the lectures Irvine had given after he succeeded to Black's Glasgow chair.[136]

Black appreciated that such a program of experimental work could not be sustained by lectures alone. Hence his induction of Irvine and Watt into methods of laboratory research and his work in support of standardization of thermometer scales. In his lectures, Black pointed out the crucial importance of standard instruments for any experimental study of heat.[137] He backed the efforts of George Martine, whose *Essays . . . on the Construction and Graduation of Thermometers* (first published in 1740) proposed the universal use of mercury instruments calibrated at the two fixed points of freezing and boiling water.[138] Black himself performed experiments on the uniformity of expansion of mercury, which he reported to an experimental society in Glasgow. In his lectures he circulated copies of Martine's table, on which fifteen different thermometer scales were compared and reduced to a single standard, greeting the achievement as the establishment of a "universal language" for measuring temperature.[139]

Work of this kind to create standardized instruments was essential to the success of experimental studies of heat. We shall see examples in subsequent chapters of experimental programs that depended on new instruments and on the communication of rigorous procedures for their use. In comparison with the refined apparatus and disciplined practices introduced by Lavoisier and Davy, Black's efforts seem unsystematic indeed. He introduced some new apparatus and exploited connections with some instrument makers, particularly Watt who manufactured and calibrated thermometers for him, but these innovations were not pursued

134 Donovan (1975b), pp. 265–277.   135 Black (1803), I, pp. 137–142.
136 Crawford (1779), pp. 2, 12fn, 17fn.   137 Black (1803), I, pp. 50–68.
138 I have used Martine (1787), an edition dedicated to Black.
139 Black (1803), I, pp. 57, 66–68.

with the energy or rigor to be shown by later researchers.[140] The lecture hall remained the prime site at which the public profile of Black's science was created. This was one reason why Black's own reaction to Lavoisier's new chemistry was mixed. He expressed admiration for the forcefulness of the French chemist's experimental demonstrations, but doubted that they could be effective in persuading a nonexpert audience. He thus undertook to teach the new system, but to do so in keeping with the Scottish didactic tradition of methodical, inductive exposition. By this means, he claimed, Lavoisier's theory could be taught "with an impression of truth which his method can never make."[141]

Assessments of the successes and failures of Scottish chemistry are perhaps inevitably shaped by hindsight, even if we could evade Robison's continuing influence on our perceptions of Black. Having emphasized the achievements of Cullen and his successors, I shall therefore mention some limitations of the program as they appear in the light of subsequent developments. This will help prepare the way for the chapters that follow.

The first point to be made is that the disciplinary identity of chemistry remained highly vulnerable. Cullen strenuously proclaimed the autonomy of the subject, and carefully demarcated a realm of distinctively chemical phenomena concerning combination and analysis. He also loosened the ties between chemistry and medicine by forging links with the other practical arts. But this was all done without making a radical break with the discursive forms through which the subject was taught. Black in his turn suggested a novel definition of the discipline – as the science of phenomena concerned with heat and mixture – but he failed to follow this through with a comprehensive revision of the structure of his course of lectures. In its rhetorical conformation, chemistry was not to be radically transformed until Lavoisier put into effect his Chemical Revolution, with its cardinal innovations in textbook structure and nomenclature.

At that time and in the decades following, study of the phenomena of heat, pneumatics, and electricity was to remold chemistry in the most fundamental manner. This occurred when the phenomena could be reliably reproduced by the use of refined instrumentation and associated technical practices. Such innovations required new rhetorical and insti-

140 Black (1966), pp. 17, 21; Black to Watt, 20 December 1769, in Robinson and McKie (1970), pp. 22–23.
141 Black (1806), I, pp. 548–549. The remarks have clearly been reconstructed by Robison, who admitted to having "been obliged to manufacture this part of the lecture considerably" (Robison to James Black, 21 April 1802, in McKie and Kennedy (1960), pp. 161–170 (on p. 168)). It nonetheless seems likely that the remarks reflect Black's views, at least approximately. (See Robison to Watt, 23 July 1800, and [October 1800], in Robinson and McKie (1970), pp. 342–347, 355–357; and Christie (1982), p. 51 and fn. 38. For more on Black's reaction to Lavoisier, see Kendall (1949–52); Perrin (1982); Christie (1979).

tutional forms – a more consolidated structure of discipline – to sustain and diffuse them. The failure of the Scots to achieve this is the second and most important point that must be made in assessing their program. Scottish chemistry remained most at home in the peculiar institutional environment of the universities, although Cullen and Black did succeed to some degree in extending it also to other areas of public life. They exerted most influence through their pupils, and did not show themselves willing to publish textbooks on chemistry or to lecture outside the universities. Similarly their relatively restrained initiatives in laboratory training and development of new instrumentation set a limit on the diffusion of experimental techniques.

When we turn to the career of Joseph Priestley, we shall find numerous points of comparison. Priestley was not concerned with the disciplinary identity of chemistry, but he did contribute the discovery of a whole new class of chemical phenomena concerning gases. By the use of certain rhetorical techniques of authorship, he communicated knowledge of these phenomena and the means to reproduce them. Disavowing theory, he explored literary and social methods to diffuse the techniques of pneumatic chemistry throughout society. He worked to extend the perception of his new "airs" in provincial England, as part of his own program to make chemistry a public science. Rather than making chemistry the servant of gentlemanly improvers, however, Priestley saw it as a way to assert the citizenship rights of a disenfranchised population.

3

Joseph Priestley and the English Enlightenment

It may be my fate to be a kind of comet, or flaming meteor in science, in the regions of which (like enough to a meteor) I made my appearance very lately, and very unexpectedly; and therefore, like a meteor, it may be my destiny to move very swiftly, burn away with great heat and violence, and become as suddenly extinct.

Joseph Priestley, *Philosophical Empiricism* (1775)[1]

In this passage, as in much of his scientific career, Joseph Priestley (1733–1804) displayed a peculiar blend of diffidence and bravado. On the one hand, the image of the meteor modestly understates the degree of his commitment to science and the permanence of his achievement. His work in chemistry and natural philosophy was no nine-days' wonder; his discovery of a series of previously unknown gases or "airs" remains the basis of a considerable reputation among scientists to this day. On the other hand, the cometary metaphor accurately captures the dramatic impact of much of his experimental work and its widespread appeal in his own day as public spectacle. This was a matter of some importance to Priestley, and one on which he frequently deliberated. Though he liked to appear diffident, he was far from indifferent to the enthusiastic public interest that his discoveries provoked.

This curious passage provokes a series of questions concerning the constitution and context of Priestley's accomplishment in science. How did he present himself and his work to a public audience? Why was so much interest shown in his discoveries? How did they become more than just short-lived phenomena? These questions form the focus of this and the next chapter. We shall see that Priestley played an important and unique role in the development of chemistry as a public science in the era of the Enlightenment.

Unlike Cullen and Black, Priestley never identified himself as a chemist, even a philosophical one. Although he lectured on chemistry for a short

1 Priestley (1775), p. 67.

while toward the end of his life, he did not seek to present it as a comprehensive system or an autonomous discipline, and he sometimes admitted that he lacked qualifications in the subject.[2] Rather than being seen as a chemist, Priestley is better seen as an experimental natural philosopher, who also worked as a metaphysician, theologian, political writer, and historian. His experimental practice was informed by deeply held moral, political, and theological convictions, but the question of the disciplinary identity of the subject in which he was engaged was not his concern. Priestley's new airs nonetheless came to be widely perceived as chemical phenomena and absorbed into the repertoire of lecturers and writers. His work strikingly directed chemists toward the phenomenotechnics of gases, which had already begun to claim their attention. Subsequently, chemists such as Thomas Beddoes and Humphry Davy looked back to Priestley as a pioneer of their science.

His reputation rested not just on what he discovered but also on the effectiveness of the way he communicated it. He presented his work in a literary form designed to gain its acceptance in the public culture of science that he saw developing around him in the English provinces in the 1770s and 1780s. Descriptions of discoveries in factual narratives with exhaustive details of apparatus and techniques allowed them to be replicated easily. Priestley articulated a vision of a scientific community that would unite experimental investigators with a large general audience, and would advance in step with religious and political emancipation and social progress. His form of discourse was designed to serve the overarching purpose of encouraging such an enlightenment.

In this chapter we shall first consider the milieu in which Priestley worked. By comparing the social context of the Enlightenment in England with that in Scotland, we can learn something of the climate in which chemistry was developed and applied at the time. We shall discuss the structure of higher education, the growth of scientific societies, the continuing importance of personal patronage, and the establishment of a market for cultural goods and services. These were the elements of the context in which Priestley and many of his followers built their careers. In the next section we look at the pattern of Priestley's career and consider the interpretive stance that is appropriate for understanding it. Priestley will be presented here as a social actor and communicator; his experimental work will be shown to have been given meaning in the course of communication with his audiences, including patrons, colleagues, and the readers of his books.

The contours of his career provide essential background for the interpretation of Priestley's discourse in the final section of the chapter. Against

2 Priestley (1794); Schofield (1966), pp. 164, 180; Priestley (1806), pp. 61–62.

this backdrop, we can understand his adoption of a particular narrative style – the historical or "analytical" style – to convey descriptions of his experiments. This mode of writing served his purpose of enrolling a public audience in the pursuit of natural philosophy and hence spreading the benefits of enlightenment. Its appropriateness was evidenced by the diffusion of knowledge of his discoveries to an extensive and heterogeneous audience, spread over a wide geographical area. Some problems did however arise from the choice of this mode of discourse, and these will be examined by focusing on a dispute between Priestley and the chemical lecturer Bryan Higgins in 1774–75.

The uses of chemistry in Enlightenment England

In the last chapter we saw how Cullen and his followers strengthened the position of chemistry in Scottish university education and convinced their patrons of its potential utility in relation to a range of economically important arts. In England also, in the decades following the middle of the eighteenth century, people were persuaded that chemistry could be of use to them. Through the work of Priestley and others, an audience was attracted to the subject and patrons convinced of its utility. To understand how this happened we need to grasp something of the social context of the Enlightenment in England, and how it differed from that in Scotland. Several factors that distinguish the situations in the two countries are apparent: We shall focus on the situation in the English universities, the role of personal patronage and the institutions through which it was channelled, and the growth of a market in education and culture.

The reforms of the Scottish universities had been vital in establishing an academic community anxious to improve professional education in subjects such as medicine, and hence attracted by the Dutch model of medical education in which chemistry had a central role. In England, however, the two ancient universities remained conservative and relatively insulated from competitive pressures. Although students continued to graduate in medicine from Oxford and Cambridge, and thereby gained the privilege of recognition by the Royal College of Physicians in London, they studied a traditional syllabus that emphasized the classical sources of medical doctrine and ignored clinical teaching and the sciences. The relative neglect of chemistry teaching at the English universities was one aspect of their backwardness in medical education as a whole. The institutions were also closed to those who were not orthodox Anglicans and hence lost potential students to the Scottish and Continental universities.[3]

3 On medical education at Oxford, see Webster (1986). More generally, see Poynter (1966). The role of chemistry teaching in medical education is discussed in Crellin (1974).

Although Cambridge had had an established professorship in chemistry since 1703, teaching in the subject was not continuously available. The first professor, John Francis Vigani, was recruited on an ad hoc basis, having established himself privately as a lecturer in the city. He appears to have taught with some success to an audience of university members and local medical practitioners from the 1680s until about 1708. His successor, John Waller, is not known to have lectured at all. In 1718, John Mickleburgh, a Fellow of Corpus Christi College, was appointed to the Chair of Chemistry and delivered five courses of lectures in the subsequent three decades. On his death in 1756, Mickleburgh was succeeded by John Hadley, who delivered only two series of lectures before departing for London where he assumed a more lucrative post at St. Thomas's Hospital. Hadley did not relinquish his Cambridge position until his death in 1764, when he was succeeded by Richard Watson. The appointment of Watson, who notoriously confessed when he took the job that he knew not one word of chemistry, may be taken to symbolize the unsupportive climate for the subject in Cambridge.[4]

The situation in Oxford was even worse. There was no established chair in chemistry in the eighteenth century and lectures were given only occasionally. John Freind, who delivered a highly theoretical course in the Ashmolean Museum in 1704, does not appear to have had a successor for the following three decades. About 1739, Nathan Alcock began to lecture privately on chemistry and anatomy in the city, and overcame opposition from conservative factions in the university to get his courses established on the curriculum for medical students. When Alcock left to take up medical practice in Bath in 1757 he was followed by John Smith, who appears to have given a few lectures until he moved on to a chair in geometry in 1766. After that, the teaching of chemistry in Oxford lapsed for a further fifteen years.[5]

Only in the 1780s did the poor record of chemical instruction in the English universities begin to improve, and even then the improvement was not sustained for long. Watson's widely read *Chemical Essays* (1781–87), based on his Cambridge lectures, were outdated by the time they were published. They did not even mention Priestley's pneumatic chemistry or Scottish work on the chemical effects of heat.[6] Isaac Milner, on the other hand, who took over chemistry teaching from Watson in 1782, did present what appears to have been a moderately well-informed course of lectures during his decade-long tenure of the Jacksonian chair of natural philosophy.[7] Concurrently, Martin Wall, who was appointed to a new readership in Oxford endowed by the Earl of Lichfield in 1781,

4 The careers of the Cambridge chemistry professors are described in a series of articles by
   Coleby (1952a, 1952b, 1952c, 1953).
5 Turner (1986), esp. pp. 663–668.
6 Watson (1783), p. [iii].   7 Milner ([c.1784]); and *D.N.B.*

initiated some improvement in the situation there. He informed his audiences about Black's work on heat, and that of Priestley, Henry Cavendish, and others on gases.[8] Wall's successor from 1788, Thomas Beddoes
(formerly a medical student at Edinburgh), brought Oxford audiences up
to date with the latest Continental developments but was forced out of
his post in 1793 for his political radicalism. Beddoes had written in 1789
that "The spirit of Chemistry has almost evaporated at Oxford, as indeed
I always expected it would."[9] The expectation was a well-founded one,
given the history of the subject at Oxford.

It is clear that the English universities did not provide a context in
which vigorous advocacy of the utility of chemistry was well rewarded.
The ancient institutions only grudgingly accorded a place to the subject and responded to developments in it in a tardy and haphazard
manner. Such teaching as occurred tended to be narrowly directed at medical students, without indicating the potential of chemistry for other
practical arts. Unsurprisingly, the English universities played only a
small part in recognizing and extending Priestley's experimental researches.

In contrast, chemistry had a more secure and prominent position in
some of the dozens of Dissenting Academies that were established in the
eighteenth century to provide higher education for those debarred from
the universities on sectarian grounds. The academy at Warrington, where
Priestley was introduced to chemistry by Matthew Turner, had been offering the subject as part of the natural philosophy course for third-year
students since its foundation in 1760. Turner was succeeded as chemistry
lecturer there by Dr. John Aikin, who published the syllabus of his lectures and translated the French textbook of Antoine Baumé for his students. The Manchester Academy, founded in 1786, came to enjoy the
teaching services of the distinguished chemists Thomas Henry and John
Dalton. Priestley himself taught chemistry to students at the Hackney
Academy in the mid-1790s.[10] Through his connections with the Dissenting Academies, Priestley gained a larger and more appreciative audience
for his chemical work than he would have had teaching at Oxford or
Cambridge.

Another factor that encouraged the development of chemistry in Scotland was the commitment of the aristocracy and gentry to a paternalistic
role in economic and social progress. Scotland's ruling classes formed
close connections with the professional intelligentsia in numerous clubs
and societies devoted to improvement. English scientific societies were no

8 Wall (1783); D.N.B.
9 Thomas Beddoes to William Withering, [1789?], quoted in Levere (1981a), p. 62.
10 McLachlan (1931), pp. 210–211, 219, 250–251, 256–259, 308; McLachlan (1943),
     pp. 51–59; Hans (1951), pp. 54–62.

less variegated than these, and the role of some of them in encouraging interest in chemistry was similar to that of the Scottish bodies. At the apex of national scientific endeavor was the Royal Society, which since its foundation in 1660 had become the most prestigious such organization in Britain and the model for imitation by others throughout Europe. Election to a Fellowship of the Royal Society was a recognized honor among experimental philosophers. To publish a paper in its journal, the *Philosophical Transactions,* was seen as a sign of merit; and its most prestigious awards such as the Copley Medal (founded in 1736) were universally esteemed. Through these honors and awards, the Society defined standards of work in experimental science. Science was not yet a profession in the eighteenth century, but through the operations of the Royal Society and the other national academies it gained structure and definition as a "vocation."[11]

At a time when the boundaries of the scientific community were quite permeable, the Royal Society consolidated the structures of credit and prestige among experimental philosophers by mobilizing the accepted hierarchies of status in society at large. Aristocrats were recruited on preferential terms to the Society and continued throughout the century to comprise a disproportionately large number of members of its council. Links with the government were also maintained. Sir Joseph Banks, President from 1778 to 1820, came to exercise considerable powers of patronage through his influence on the civil service, the colonial administration, and the court. Building upon the patronage practices of the previous long-serving presidents, Sir Isaac Newton (president, 1703–27) and Sir Hans Sloane (president, 1727–41), Banks steadily acquired greater powers within a growing "learned empire."[12]

Banks's attempt to take over the Royal Society, apparently to satisfy personal ambitions, was particularly resented by those who were dissatisfied with the political order in eighteenth-century Britain. Radicals came to view the Society as part of a corrupt political administration. Priestley had flourished under the patronage of Banks's predecessor, Sir John Pringle (1707–1782), who encouraged his work and that of other pneumatic chemists. But after Pringle's retirement, Priestley distanced himself from the Society, perceiving that Banks was acting against him in a partisan way. Relations between Banks and Priestley became even more strained in the 1790s, after the outbreak of the French Revolution. Such stresses

11 On the Royal Society in the eighteenth century, see McClellan (1985); Weld (1848); Heilbron (1983); Miller (1989). McClellan makes some interesting claims for the role of societies in consolidating "preprofessional" scientific careers, especially on pp. xxiv–xxvi. But these should be compared with the perspectives on careers and vocations in eighteenth-century science suggested by Porter (1978); and Outram (1980).

12 McClellan (1985), pp. 21–22, 32–34, 61–62, 150–151, 243–244; Miller (1981), pp. 9–74.

were inevitable, given the Royal Society's scientific and socio-political functions.

A society that had a similarly close relationship with the centers of social and political power, and played a more direct role in encouraging the development of chemistry, was the Society of Arts, founded in London in 1754. This institution gave paternalistic encouragement to applied science by offering a series of premiums to individuals who had made significant innovations in production methods and were prepared to reveal their methods to the public.[13] The close dependence of the Society on court and ministerial patronage enabled its more humble members to attach themselves to the coat tails of the government machine. Such chemists as Nicholas Crisp (an apothecary) and Robert Dossie (a former apothecary's apprentice) gained preferment and perks through the Society.[14]

The chemical initiatives of the Society of Arts did not prove particularly successful, however, and they will not help us to understand the contexts in which Priestley's work found its audience in the second half of the century. Priestley kept himself entirely aloof from the activities of this institution, perhaps because he regarded it as an arm of the court establishment from which he always kept his distance. His associates in Birmingham also maintained their independence: James Watt and Josiah Wedgwood both seem to have regarded the award of premiums as a threat to the property rights of inventors. Perhaps they would have agreed with William Cullen that Dossie was to be "noted for prying into Arts, yet not for his penetration and Judgement."[15] These entrepreneurs preferred to trust themselves to the market economy rather than to a paternalistic reward system.

This suggests that the influence of paternalistic institutions was offset by the availability of other sources of reward for scientific and technological accomplishments. Although aristocratic patronage was important to many individuals in building their careers, aristocratic cultural values had little purchase on upwardly mobile social groups. This is especially the case with provincial groups interested in science. In the societies that sprang up in English provincial towns from the 1780s onward there was frequent involvement of the local gentry, but most of the membership was composed of professional men such as doctors, lawyers, clergy, teachers, and artists. Although they presented themselves as emulating the Royal Society, and although they arguably reflected a gentlemanly

13 Allan (1979a, 1979b).
14 Allan (1973–74); idem. (1979a), pp. 44, 70, 297 (on Crisp), 71, 84, 184 (on Dossie). For Dossie's career, see also Gibbs (1951a).
15 Cullen, Manchester MS, I, p. 96; Schofield (1959), esp. pp. 668–670.

ethos by treating science as a branch of polite literature, these groups must be seen primarily as the creations of aspiring local elites and as expressions of an independent provincial cultural identity.[16]

The Lunar Society of Birmingham, with which Priestley was closely associated from 1780 to 1791, was the harbinger of this dawn of provincial intellectual culture. The society was an informally constituted group of friends with shared interests, which took its name from its monthly meetings at the time of the full moon. Its most important members were James Watt (the steam engine manufacturer), Matthew Boulton (his partner who owned a large metalware factory at Soho outside Birmingham), Josiah Wedgwood (the Staffordshire pottery manufacturer), James Keir (a physician, writer, and chemical manufacturer) and Erasmus Darwin (a physician and natural philosopher).[17]

The Manchester Literary and Philosophical Society, founded in 1781, pointed the way toward the establishment of more formal literary and philosophical societies in the major provincial towns, and initiated a movement that continued to grow through the succeeding four decades. In the Manchester Society an alliance was forged between local professional groups, manufacturers and gentry, all of whom were united around a program for the pursuit of knowledge that would be both polite and useful.[18] Priestley's discoveries were of great interest to the leading members of the Society and were applied by them in a number of different fields of research. His credit was high also in the Derby Philosophical Society, a smaller circle created by Erasmus Darwin in 1784 when he left his Lunar Society friends to settle in Derby. The members of this group (which included medical men, manufacturers, gentry, and clergy) dedicated themselves to the pursuit of "gentlemanlike facts," and gave particular attention to chemistry.[19] The following decade, the Newcastle Literary and Philosophical Society was brought into being by the initiative of William Turner. Here again, the membership was composed of clergy,

16 In saying this, I am joining those who have taken issue with Morris Berman's argument that eighteenth-century English provincial science labored under the "hegemonic" influence of the aristocracy over national cultural life. See Berman (1975); and compare with Thackray (1974). The issue is also excellently discussed in Inkster (1983). Berman's claim parallels the analysis of eighteenth-century English social structure by E.P. Thompson, especially in Thompson (1978). An alternative view of social structure, which shows that pervasive patronage was not inconsistent with social mobility, is given by Perkin (1969), esp. ch. 2; and Porter (1982a), esp. ch. 2.

17 Schofield (1963). Biographical sources on members of the Lunar Society include McKendrick (1973); Smith and Moilliet (1967); Robinson and McKie (1970); King-Hele (1977); McNeil (1987).

18 The Manchester "Lit. and Phil." is discussed in Thackray (1974); and Musson and Robinson (1969b), pp. 89–118.

19 On the Derby society, see Musson and Robinson (1969b), pp. 190–199; and King-Hele (1977), ch. 8.

medical men, and a smattering of local gentry, and again chemistry was given considerable attention and Priestley personally honored.[20]

One notable feature of these societies was the extent to which they represented Scottish influences on English intellectual life – a factor that predisposed them toward an interest in chemistry. Among the members of the Lunar Society, William Withering and Darwin both had MD degrees from Edinburgh, and Keir had also studied there. Watt kept up a lengthy correspondence with his Glasgow mentor Joseph Black. In the subsequent growth of English provincial science, Edinburgh-educated medical men were notably prominent.[21] Members of the Manchester Lit. and Phil. who had studied there included Thomas Percival, John Ferriar, and Matthew Dobson.

None of these societies were in any sense vehicles for paternalistic activity by the aristocracy. They are more plausibly interpreted as cultural outlets for upwardly mobile, largely professional and commercial, groups. Dramatic economic growth in the English provincial towns had swept along those engaged in commerce and manufacturing and the doctors, lawyers, and clergy who served them. As they became more prosperous, members of these groups asserted their rights to religious emancipation and political reform. Their scientific and cultural preoccupations were closely connected with these social aspirations. Many of the leading figures in the movement were medical practitioners who sought to improve public health and at the same time to deploy their professional expertise. Others were attracted by the broader aims of advancing learning, which they associated with progress and spreading enlightenment.

Being excluded in many cases from an influence on government and the established church, the provincial intelligentsia found another way to participate in the development of their society. They organized themselves in line with a very different model of the public realm from that underlying the activities of the Society of Arts. Rather than paternalistic rewards for revealing inventions, they prized their own independence from political or sectarian domination. Members of these societies valued social contact with other members of the emergent local elite in a setting that permitted both formal discourse and unstructured discussion. Relaxed social mixing, conversation, and the opportunity to present one's thoughts for friendly criticism and appreciation conveyed a feeling of participation in the progressive enterprise of enlightenment.

Another result of the steadily increasing prosperity of the middle classes was the growth of a market for cultural goods and services. This development complemented the formation of an institutional basis for provincial culture. Writers, including authors of scientific texts, could now

20 Watson (1897).   21 Chitnis (1986), esp. pp. 36–37, 56–64, 133–181.

support themselves (at least partially) by the sales of their books. Public lecturers could drum up paying audiences to hear news of the latest scientific discoveries. The purveyors of Enlightenment culture acquired new skills of self-promotion and tapped new springs of financial support. The extent of the consumer market was a major factor in shaping the culture of the English Enlightenment, and clearly distinguishes it from that in Scotland.[22]

The market for popular texts and lectures in natural philosophy had been expanding since the early decades of the century.[23] Alongside an extension of the market for scientific books went a significant expansion in the activities of public lecturers.[24] Itinerant public lecturing in the provinces was first practised in a sustained way by Benjamin Martin (1704–1782). His arrival in a town would be preceded by advertisements in the local press and the collection of subscriptions by a bookseller or publican. If the reception looked promising, the lecturer would give one or more courses, each of a dozen or so lectures, generally charging one or two guineas per course. Martin first took his philosophical show on the road in the 1740s and visited Bath, Birmingham, Chester, and Shrewsbury in his first few years. For a decade and a half thereafter he toured regularly, supplementing his income from subscriptions by writing and making scientific instruments.[25] The itinerant lecturers who incorporated Priestley's discoveries into their performances in the 1770s and 1780s were following a career path first beaten by Martin.

Before Priestley's day, however, chemistry had no place in the repertoire of most public scientific lecturers. The majority concentrated on a syllabus of standard topics in experimental natural philosophy: statics, mechanics, hydrodynamics, physical pneumatics, and perhaps also heat, light, and electricity. There were some who taught only chemistry, but they tended to pitch their appeal at medical practitioners and apothecaries rather than the general public. This pattern had been established in the late seventeenth century. George Wilson, who taught several courses of chemistry in London in the 1690s, had advertised to "Doctors of Physick, Apothecaries, Chirurgeons, and others, Studious of Physick, or curious in Chymical Operations." Wilson charged his audiences three guineas for a course – an indication that they probably expected to derive commercial or professional benefit from its contents.[26]

In the 1730s, a successful attempt was made to introduce general pub-

---

22 Interpretations of the English Enlightenment which emphasize this factor include Porter (1980, 1981); Borsay (1977); McKendrick, Brewer and Plumb (1982).
23 Rousseau (1982); Meyer (1955); Secord (1985).
24 Hans (1951); Rowbottom (1968); Stewart (1986a, 1986b, forthcoming); Millburn (1983, 1985); Inkster (1980).
25 Millburn (1976).   26 Gibbs (1953) (quotation on p. 183); Wilson (1709).

lic audiences to lectures on chemistry by Peter Shaw (1694–1763). Shaw taught in London in 1731 and 1732 on chemistry and its applications to fields such as brewing, wine making, and mining. The following year he took the course to the resort town of Scarborough, where he drew a fashionable audience of visiting aristocracy and gentry. Alongside his lectures he offered for sale a "portable laboratory," a mobile furnace he had manufactured with the assistance of the instrument maker Francis Hauksbee. This apparatus was said to be suitable for recreational use by leisured gentlemen or for a small-scale chemical business. Shaw's entrepreneurial initiatives laid the foundation for a successful medical career; he gained sufficient patronage and reputation to break into medical practice and subsequently became a prominent London physician.[27]

Apparently following Shaw's lead, William Lewis (1708–1781) began offering public chemistry courses in London in the mid-1730s. By the end of the following decade, Lewis had become a very successful lecturer and had attracted royal and aristocratic patrons. He had also achieved a respected position as a research chemist in his private laboratory at Kingston-upon-Thames. But his economic position remained rather precarious; he continued to be dependent on sporadic patronage, occasional rewards from the Royal Society and the Society of Arts, and the modest revenue from his books.[28]

In this regard, Shaw's achievement had another lesson to teach: He had also shown that chemists would have to continue to build their careers in close association with the medical community. His chemistry lectures were successful, but they comprised only one step on a career path aimed toward medical practice. Before he began lecturing, he had published editions of works by John Quincy, Hermann Boerhaave, and G.E. Stahl, which related closely to the concerns of leading members of the London medical establishment. He seems always to have been aiming to become a physician, and his eventual success may have owed more to the patronage of the prominent practitioners to whom his works were dedicated than to his success as a lecturer.[29]

Others followed Shaw's lead in using chemical lecturing as a resource to build a medical career. Three decades later, two of Cullen's students, George Fordyce (1736–1802) and William Saunders (1743–1817), began offering lectures on chemistry in London. As possessors of Edinburgh MDs they had to work hard to achieve standing in the London medical establishment, but both succeeded in doing so. Both men started to build their reputations by lecturing on chemistry and other medical specialties in the capital's growing market for medical education.[30]

27 Shaw [1734a]; Gibbs (1951b); Golinski (1983).
28 Gibbs (1952).   29 Golinski (1983), esp. pp. 20–21.
30 Crellin (1974), p. 11. On the market for medical education more generally, see: Poynter (1966); Gelfand (1985).

Fordyce began teaching chemistry and materia medica in 1759, and added classes on the practice of physic five years later. For the following thirty years he taught all three subjects on six mornings a week throughout the year. His sustained popularity as a lecturer compensated for an initially slow start in gaining repute as a practitioner. He was elected to a physician's post at St. Thomas's Hospital in 1770, became a Fellow of the Royal Society six years later, and in 1787 finally achieved the fellowship of the Royal College of Physicians, despite having previously antagonized the College authorities by lobbying for recognition of practitioners without Oxbridge degrees. Saunders's career followed a similar pattern. He lectured on chemistry and pharmacy from 1766, switching to theory and practice of medicine after being elected physician to Guy's Hospital in 1770. He gained his fellowship of the Royal College of Physicians in 1790 and was subsequently honored by the College as Gulstonian Lecturer and Harveian Orator.[31]

The careers of Fordyce and Saunders established a model that was widely imitated in the last two decades of the century. In that period, a substantial population of practitioners, including many with Scottish degrees, gravitated toward the major London hospitals to pursue opportunities in teaching and research. Chemistry gradually became a standard part of regular courses of medical education. But in the middle decades of the century, when the two Scots arrived in the capital, the subject was less widely recognized as an essential part of doctors' training. Their decision to float the subject on the medical market, inspired no doubt by Cullen's example, was an entrepreneurial initiative akin to William Hunter's promotion of innovative courses in anatomy about a decade earlier.[32] Fordyce and Saunders showed that there was a market for chemical lectures in London, and that individuals with a knowledge of the subject could use it to gain status in the metropolitan medical community.

Chemical skills could also be used in other ways in the medical profession. In a situation dominated by competition between practitioners, scientific expertise was sought for professional gain. A display of chemical knowledge could yield rewards in attracting clients. Because medical practice was still shaped significantly by the demands and prejudices of patients, however, practitioners had to establish their credentials in public view.[33] A reputation gained solely in the scientific community conferred no particular competitive advantage. The point is illustrated by the analyses of mineral waters published at the time. This was an important field for the development of chemical expertise in relation to medicine, and one to which a number of Priestley's medical disciples made

31 George Fordyce, "Lectures on Chemistry" (3 vols., MSS, Library of the Royal College of Physicians, London, 1786); Saunders (1766); Howell (1930).
32 Porter (1985b).   33 Jewson (1974); Porter (1982b, 1985a).

significant contributions. The growing popularity and commercial development of the English spa towns in this period provided plentiful opportunities for doctors to develop a therapeutics based on mineral waters, and to profit by putting it into practice. By publishing analyses of the waters, practitioners could display their scientific skills and hope to enhance their reputation with prospective patients.

At least one hundred books describing mineral waters from one or more locations were published in Britain in the course of the eighteenth century.[34] The literature included simple descriptions of spas, comprehensive surveys of their contents and medical virtues, and attempts at chemical analysis of their constituents. This large body of writings has generally been discussed in terms of progressive improvements in methods of analysis and certainly some of the leading experimenters involved, notably Henry Cavendish and Stephen Hales, were motivated more by the technical problems posed than by medical ambitions.[35] Nonetheless, the medical context of much of this work should not be ignored. Throughout the eighteenth century, practitioners flocked to such spas as Tunbridge Wells, Epsom, Harrogate, and Scarborough, all of which underwent dramatic expansion as leisure and health resorts. The ancient hot springs at Bath presented a particularly alluring promise of prosperity for doctors who succeeded in practice there. Physicians' fees were higher in Bath than anywhere else, and their numbers greater than average in proportion to the size of the population.[36] The Bath waters were repeatedly examined for the chemical sources of their reputed efficacy in treating palsy, gout, wounds, urinary disorders, diabetes, skin diseases, stomach problems, sciatica, and many other complaints.

Many practising (and aspiring) doctors wrote works on mineral waters to display their scientific abilities to the public on whom their career success depended. Authors mobilized the authority of experimental science in an opportunistic and entrepreneurial way, as required by the competitive medical market in which they found themselves. In this respect they were acting like the lecturers who carved out niches in the market for medical education. In both situations, knowledge of chemistry was a resource for individual self-promotion.

Priestley (through his work as a writer), and those of his followers who were public lecturers, had considerable experience of cultural markets. But, as Priestley's case will confirm, markets were not the only or even the dominant feature of the social landscape of the English Enlighten-

34 This estimate is based on the very useful bibliography in Mullett (1946). For background on the popularity of the English spas, see: Turner (1967); Lennard (1931).
35 The most useful discussions of methods are: Coley (1982); and Eklund (1976).
36 Lane (1984), esp. pp. 359–360. Compare the contemporary account by Adair (1786), pp. 115–124.

ment. Independent entrepreneurialism was cut across by relations of patronage and clientage, which were still pervasive and fundamental features of this society. Individualism was offset by the solidarity of the Dissenting community and its institutional alternatives to the educational and scientific establishment. Priestley escaped the intellectual torpor of the ancient English universities and largely shunned the paternalistic overtures of the Society of Arts and Sir Joseph Banks's "learned empire." He was able to do so partly because of his success as a writer but largely because of the support of provincial intellectuals and Dissenters. He studied and taught at the Dissenting Academies, associated with his colleagues in the Birmingham Lunar Society, and cultivated a network of followers and friends in other provincial circles. He also maintained relations with a number of prominent patrons who provided financial and moral support. In all of these respects, Priestley's work was firmly rooted in the social conditions of the English Enlightenment (Figure 2).

Making connections: Priestley's career

Much recent discussion of Priestley's career has been dominated by the problems of distinguishing and connecting the wide range of interests to which he devoted time in the course of an outstandingly busy life.[37] The barest outline of his biography gives evidence of a succession of different occupations and a constant engagement in multifarious intellectual activity. He began his adult life as a Dissenting preacher in Suffolk, moving to Cheshire shortly before being appointed in 1761 to a post as tutor in languages and belles lettres at the Warrington Academy. While there, he began his lifelong occupation as a writer, initially producing historical and literary works. In 1767 he moved to Leeds where he became minister at the Mill Hill Chapel, and shortly thereafter emerged as a writer of works on experimental science. Among his first scientific books were histories of electricity and optics, which combined chronological resumes of previous experiments with records of new facts that Priestley himself had discovered. By making trials of disputed points in electrical science, Priestley had been led into a program of experimental investigation that he continued to record in subsequent writings.

In 1773, Priestley changed his occupation again when he became librarian and intellectual companion to the Earl of Shelburne. He re-

37 The standard edition of Priestley's works is Priestley [1817–31]. Important collections of letters are in Schofield (1966); and Bolton (1892). A useful general biography is Gibbs (1965), which is authoritative though it lacks full documentation. There are also some biographical remarks in Lindsay (1970). The standard bibliography is Crook (1966). Other interpretive work includes: Hiebert. et. al. (1974); and Schofield (1983).

Figure 2. Portrait of Joseph Priestley by E. Sharples. [Reproduced by permission of the National Portrait Gallery, London.]

mained in Shelburne's employment for the following seven years. During this period, which he spent largely at his patron's country seat at Calne in Wiltshire, Priestley's writings extended into the fields of theology, metaphysics, and politics. He also initiated what was to become a series of books describing his experimental work on the new airs. This work continued after 1780 when he left Shelburne's employment and moved to Birmingham.

It was while in Birmingham, working as a minister at the New Meeting House, that Priestley became associated with the other members of the Lunar Society, many of whom shared his interests in chemistry, natural philosophy, and the practical arts. The time in Birmingham was perhaps the most public period of Priestley's life: He acquired a reputation as a preacher and polemical writer on behalf of political reform and the rights of Dissenters. He also became closely associated with several scientific lecturers operating in the West Midlands area. This high profile was in part responsible for the tragic end of Priestley's contented life in Birmingham in 1791, when his house and laboratory were ransacked by a loyalist mob. Urged on, or at least tacitly licensed, by several local landowners and magistrates, the mob was apparently incensed by Priestley's explicit support for the French Revolution. Following the destruction of his house, he withdrew to London where he took refuge in the Dissenting community in Hackney. Three years later he emigrated to the United States and passed his final years in retirement in rural Pennsylvania.

Faced with an intellectual biography of such polymathic breadth, some recent writers have attempted to unify Priestley's interests within an overarching conceptual structure, which promises to connect his theological, metaphysical, scientific, and political views. For Robert Schofield, the core of this conceptual structure is to be found in a commitment to a particular metaphysical theory of matter. Priestley's variety of "mechanism" is supposed to have derived from the Newtonian tradition and to have determined the orientation of much of his scientific work.[38] John McEvoy has quite persuasively argued against this interpretation and has identified the core of Priestley's beliefs with a cluster of theological and philosophical commitments focused around the theme of "rational dissent." Once this cluster of basic ideas is grasped, McEvoy suggests, everything in Priestley's thought can be related to everything else within "the totality of his intellectual vision."[39]

My own interpretation is much indebted to McEvoy, particularly for the connections he has exposed between Priestley's way of communicating his experiments and his theological, epistemological, and political commitments. But I shall not try to encompass Priestley's work within a synoptic conceptual or metaphysical structure. Rather, I shall follow a line of interpretation delineated by Simon Schaffer and John Money.[40] I shall analyze Priestley's work in relation to the communities in which he practised and the audiences to which he addressed his writings. He will

38 Schofield (1961, 1967, 1970 (esp. pp. 261–273), 1974).
39 McEvoy and McGuire (1975); McEvoy (1978–79, 1979, 1983, 1987, 1988).
40 Schaffer (1983, 1984, 1987); Money (1988–89).

be shown as a dedicated communicator, motivated by the determination to establish experimental facts in the public realm. Priestley's prime aim was to encourage the widest possible participation in natural knowledge in order to advance enlightenment. He used the resources provided by the cultural market to serve his distinctive religious and political goals. The widespread demonstration of the facts of natural providence was to lead to the emancipation of reason and the displacement of illegitimate and corrupt authority in Church and State.

My approach is thus sociological rather than conceptual, rhetorical rather than philosophical. Buttressing this approach is the claim that to discuss Priestley as a writer and communicator is not merely to isolate another theme from the melange of his ideas and beliefs, but to define his identity as a social actor. By understanding Priestley's activity in this way, we can avoid the problems that arise if we confine him to a single disciplinary identity such as "chemist" and can begin to connect the various enterprises in which he was engaged with the broader social context. For, like the comet with which he compared himself, Priestley was an unusual figure, following an unfamiliar path. He had to forge his own career, pursuing his particular politico-religious aims, with the resources available to him in the English provincial Enlightenment. He made personal links with numerous people who could help him, mostly those whom he could repay in some respect with his work. He also made connections of a different kind with the more diffuse and distant audience that he reached through his published works. Priestley's vision of a public culture of science was extrapolated from the social and political circumstances in which he worked from the 1760s to the 1790s.

In this context, academic disciplines were simply not important to Priestley, either as audiences toward which he could direct his writings or as sources of support. Much more significant were small and informal scientific groups and ostensibly disinterested private patronage. These were the mechanisms for sustaining science as a public enterprise that he praised in the preface to his *History and Present State of Electricity* in 1767. Small self-supporting societies that pursued their own program of experimentation and publication were highly valued, though Priestley also recognized at this point that there remained a role for incorporated national societies capable of supporting such large-scale projects as voyages of exploration. Continuing patronage from private individuals was also seen as a necessity: "Natural philosophy is a science which more especially requires the aid of wealth . . . ," he wrote. "From the great and opulent, therefore, these sciences have a natural claim for protection."[41]

The Lunar Society was highly valued by Priestley as an example of how

41 Priestley (1767), p. xviii.

science ought to be constituted in the public realm. It helped sustain his researches in a number of respects, including the straightforwardly financial. Priestley was the beneficiary of a subscription fund, organized after he lost Shelburne's patronage, to which Boulton, Wedgwood, Darwin and Samuel Galton contributed an annual total of 200 pounds.[42] As well as money, members of the Society supplied equipment and material to aid Priestley's researches. Wedgwood provided a seemingly endless supply of ceramic retorts and tubes, Boulton sent samples of air and minerals for analysis, and all the members rallied around to reequip his laboratory after it was destroyed in 1791.[43] To Priestley, this last supportive action symbolized the values of a true philosophical community, in poignant contrast with his subsequent isolation in London and America. He wrote from London: "Such assistance from philosophical friends I should in vain look for here, and as long as I live I shall look back with pleasure and regret to our Lunar meetings."[44]

As this remark seems to imply, Priestley derived support from the Lunar Society in psychological as well as purely material senses. Maurice Crosland has pointed out how, at certain points in his life, Priestley saw scientific intercourse as "likely to provide the principal basis of friendship."[45] The regular meetings of the Society provided an opportunity for joint experiments and for witty and speculative conversation. Priestley showed experiments to the assembled members, as he had previously at Lord Shelburne's Bowood House and among philosophical friends in London and Bath.[46] His correspondence with other members of the Society provided occasions for explicitly speculative comments by a man who rarely allowed himself such freedom in print. To Priestley, such an opportunity to try out his ideas was a further encouragement to continue his experiments: "For the pleasure of communicating our discoveries is one great means of engaging us to enter upon and pursue such laborious investigations."[47]

As well as being a source of support, the Lunar Society thus provided part of the overall audience for Priestley's experimental work. They were a proximate audience (as opposed to the distant audience reached through publication) whom Priestley addressed in person or by letter and with

---

42 Priestley (1806), pp. 91–93; Bolton (1892), p. 25 fn.
43 Priestley's correspondence with Wedgwood is in Bolton (1892), pp. 23–27, 49, 68, etc.; and that with Boulton is in Schofield (1966), pp. 151–153, 161. On the reequipping of Priestley's laboratory, see Bolton (1892), p. 118; and Badash (1964).
44 Priestley to William Withering, 5 November 1791, in Bolton (1892), pp. 118–119.
45 Crosland, (1983), p. 227.
46 For records of these events see: Priestley (1775), pp. 4, 78; idem., (1775–77), I, pp. 90, 219 (in London); II, pp. 34 (at Calne), 222 (in Bath); idem., (1779–86), II, p. 398 (to the Lunar Society); and Robinson and McKie (1970), pp. 118, 121–122, 166–167.
47 Priestley to Wedgwood, 30 November 1780, in Bolton (1892), p. 21.

whom communication was a particular pleasure. He sometimes made it clear that such discourse was intended as repayment for the support he had received. In one instance, he wrote to Wedgwood: "Such is the interest you take in philosophical discoveries, and such are my numerous obligations to you with respect to those that I had in this business, that I cannot help giving you an early account of everything that I do that I think will give you any pleasure."[48] News of his philosophical discoveries was one of the few resources Priestley had available with which to repay favors from others. He resorted to this kind of communication even when he was not certain how it would be received, for example when he wrote to his brother-in-law William Wilkinson: "Though I do not know that such experiments as mine are interesting to you, yet as I am desirous of giving you any satisfaction that may be in my power, I wish to communicate the result of some that I have lately made, and that my friends think to be of importance."[49]

As simultaneously audience and support network, the Lunar Society presented Priestley with a model of a genuine philosophical community. Simon Schaffer has rightly emphasized the value that Priestley ascribed to such communities in providing a sustaining context for experimental work.[50] Perhaps because his expectations of social relations among philosophers had been shaped by his experiences among Dissenters, Priestley set great store by the cultivation of networks of acquaintances. His autobiographical *Memoirs* catalogued a large number of friendships and connections based on shared religious, political, or philosophical interests.[51] As well as yielding support, such networks provided a frequently used means of communicating the results of his work.

The Lunar Society was a successor in this respect to the first informal network of natural philosophers with whom he had come into contact – the "electricians" to whom he was introduced in the course of compiling his *History of Electricity* in the mid 1760s. This group overlapped with the philosophical and political club dubbed by Benjamin Franklin "the Club of Honest Whigs," which met at St. Paul's Coffee House and with which Priestley had also become associated at this time.[52] Members of the London group such as Franklin, Richard Price, John Canton, and William Watson provided him with information, electrical equipment, and books. They also supplied less material, but no less essential, moral support in the form of comments and encouragement in response to

48 Priestley to Wedgwood, 18 August 1788, in Bolton (1892), p. 92.
49 Priestley to William Wilkinson, 16 June 1784, in Bolton (1892), p. 71.
50 Schaffer (1984), pp. 157–162.
51 For examples, see Priestley (1806), pp. 22, 36, 50, 54, 63–65, 70, 88, 97.
52 Crane (1966); Griffith (1983).

Priestley's reports of his experiments. Shortly after returning from London to his home in Warrington in 1766, Priestley wrote to Canton, "The time I had the happiness to spend in your company appears upon review like a pleasing dream."[53]

If the community of electricians and the "honest Whigs" seemed in retrospect like a "pleasing dream," the Royal Society was to prove something of a nightmare. Priestley had joined in 1766, at the suggestion and with the support of his London friends. He published on electricity and airs in the *Philosophical Transactions* in the 1760s and 1770s and was well regarded by Pringle (President of the Royal Society from 1772 to 1778), who strongly supported the award to him of the Society's Copley Medal in 1773.[54] But by 1790 he was complaining to the succeeding President, Sir Joseph Banks, that the Society was displaying factionalism and "party spirit," which could only be "injurious to the interests of philosophy."[55] Particularly in the climate of political antagonism toward him that prevailed after 1791, Priestley found himself shunned by other members of the Royal Society and soon ceased to contribute to its journal.

In contrast, one of the reasons why the Lunar Society worked as a philosophical community, in Priestley's view, was that it deliberately eschewed such factionalism. He wrote that "we had nothing to do with the religious or political principles of each other."[56] The fact, for example, that he and Boulton had taken opposite views on the rights of the American colonists had not been allowed to disrupt philosophical intercourse between them. For Priestley, the failure of the Royal Society to live up to this ideal showed that it had become corrupted by submitting to the overbearing influence of the ministry and the court. Against such corruption it was necessary to make a stand for "independence" – a central term in Priestley's definition of his own social identity.[57]

Priestley's ideal of political independence from central government institutions was shared by other provincial intellectuals, and underwrote the rhetoric of many small clubs and societies. Members of these groups cultivated sociability on an egalitarian basis, the ideal of independence facilitating social mixing and consolidating solidarity.[58] The pattern was a common one among intellectual groups in the European Enlighten-

53 Priestley to John Canton, 14 February 1766, in Schofield (1966), p. 15.
54 Priestley (1806), p. 52; Singer (1948–50), pp. 145–146; McKie (1961).
55 Bolton (1892), pp. 100–101.   56 Bolton (1892), p. 195.
57 Schofield (1966), pp. 136, 138–139. On the vocabulary of "corruption" and "independence" in the English opposition politics, see: Jacob and Jacob (1984), esp. chh. 2, 6, 8, 9; Dickinson (1977), chh. 5, 6; Pocock (1975), chh. 13, 14; idem. (1985), ch. 11; Goodwin (1979), ch. 2.
58 Brewer (1982).

ment.[59] But an insistence on independence was not inconsistent with the acceptance of aristocratic patronage. Priestley's readiness to accept patronage of experimental research was entirely in line with the tradition of eighteenth-century "country" politics.[60] Although, when originally approached by his friend Richard Price to become Lord Shelburne's companion, he expressed apprehensions that "the state I should be brought into would be too dependent and humiliating, and not leave me sufficiently master of my own conduct," he was won over by Shelburne's generosity and sympathy for opposition political causes.[61] Even after the break with Shelburne, he remained attached to the aim of finding a disinterested aristocrat or gentleman who could serve as a worthy and high-minded patron of his science. Indeed, his egalitarian picture of the Lunar Society understates the degree to which powerful individuals within that group, particularly Boulton, were able to dispense patronage to other members and to claim favors and respect in return.[62]

What Priestley firmly and repeatedly rejected was "any pension from the court." When he was offered allowances by the Rockingham and Pitt administrations, he refused them, wishing (in his own words) "to preserve myself independent of any thing connected with the court."[63] It was not patronage as such that Priestley shunned, but the tentacles of the government patronage machine, which he identified with corruption of the original purity of the constitution. It is in these terms – of a striving to maintain independence against the encroachments of corruption – that his attitude to patronage can most authentically be portrayed. Already a generation later this political vocabulary was becoming obscured. A reforming Whig of the early nineteenth century, Henry Brougham, contemplated Priestley's dependence on the financial support of his patrons while he boasted of his independence from the court and concluded rather

59 See the remarks on masonic societies in provincial France, in Roche (1978), I, pp. 257–280; and compare Baker (1981); and Emerson (1973a).
60 On Priestley's politics in general, see: Fitzpatrick (1987); Thomas (1987); Fruchtman (1983); Kramnick (1986); Graham (1989).
61 Priestley to Richard Price, 27 September 1772, in Schofield (1966), pp. 108–109. A more complete collection of the correspondence surrounding Priestley's employment by Shelburne is given in Peach and Thomas (1983), pp. 124–128, 132–138, 145–148.
62 As Schofield has pointed out, "the meetings [of the Lunar Society] were dominated by the wishes and determined by the presence in Birmingham of Matthew Boulton" (Schofield (1963), p. 145). Boulton was the founder of the Society, usually hosted meetings at his house and (according to one witness) "took the lead in conversations, and . . . had a grandiose manner like that arising from position, wealth, and habitual command" (Samuel Galton's daughter, Mary Anne, quoted in Bolton (1892), p. 203). Boulton was the employer of several members of the Society: Dr William Small and Dr William Withering were successively his personal physicians, John Whitehurst was his instrument maker, and James Keir was employed for a while to reside at Soho to advise on production and management problems (Schofield (1963), pp. 35, 124, 128, 152–159).
63 Priestley (1806), pp. 92, 94–95.

hopelessly, "We must on this be content to remark, that different men entertain different notions of independence."[64]

Priestley's notions of independence might also have been relevant to his decision to appeal to a wider audience through publication. For eighteenth-century writers, resort to the literary marketplace could be a means of preserving one's intellectual independence.[65] Priestley was an author before he was a natural philosopher, and it can plausibly be argued that his identity as a writer for commercial publication was a crucial long-term factor in shaping the course of his career. From the moment he embarked on literary work to support his modest income as a teacher and minister, he showed himself willing to engage in a measure of self-promotion to publicize his writings. Readers found that several of his early works contained advertisements for his other products – for example, electrical machines "made by the direction of Dr. Priestley" were offered at five guineas apiece in the *Familiar Introduction to the Study of Electricity* (1768). His histories of electricity and optics included advertisements for the "Chart of Biography" and "Chart of Universal History," pedagogic aids that subsequently proved very successful sellers.[66] Like many other writers, Priestley formed a close association with his publisher, Joseph Johnson of St. Paul's Churchyard, who held the rights to almost all of his works. Johnson was known as a successful businessman, whose list extended from scientific and medical works to politics and religion, and who maintained close links with Dissenters and political reformers.[67] Even the dedications of Priestley's works to prominent and popular figures, such as the Prince of Wales, the Duke of Northumberland, the Earls of Morton and Sandwich, and the opposition politician Sir George Savile, can be read as promotional ploys.[68] Alert both to expanding commercial opportunities and to traditional consumer deference to the aristocracy, Priestley promoted his books in a way well suited to the social conditions of the English Enlightenment.

Notwithstanding his participation in the consumer market for printed scientific works, Priestley's entrepreneurialism was never more than a means to a basically religious and political end. He may have been more willing than Cullen and Black to engage in commercial behavior because

64 Brougham (1872), p. 82.
65 David Hume's career provides a good example (Mossner (1970), pp. 611–615). On the general issue, see Korshin (1973–74); and Rousseau (1982).
66 Priestley (1769), p. 85. According to Crook (1966), the descriptive booklet for the *Chart of Biography* (originally published in London in 1765) had at least fourteen editions until 1820, that for the *New Chart of History* (originally published in London in 1769) had at least fifteen editions until 1816.
67 On Johnson, see: Chard (1975, 1977); and for background Belanger (1982).
68 These dedications appear in (respectively) Priestley (1790, 1772a, 1767, 1772b, 1779–86).

it was approved by his peers among provincial intellectuals and in the Dissenting community. It must be emphasized that Priestley saw his commercial activities as serving the fundamental aims of enlightenment.[69] His moral goals were deeply implicated in his career choices and were very evident in his texts. As we shall see, the result of this was that the character he presented in his writings became exposed to moral criticism.

It should also be noted that Priestley's books enjoyed only limited commercial success. His triumphs were *The History of Electricity* (which reached a fifth edition in 1794) and the less technical *Familiar Introduction* (which attained a fourth edition in 1786). In other cases the readers were harder to find. *The History and Present State of Discoveries Relating to Vision, Light and Colours* (1772) was an expensive production for which Priestley had to raise funds by subscription. However, the sales were too poor to recompense him for the costs of printing and buying the books he had needed for the research, a fact that helped persuade him to base his later works on his own experimental investigations.[70] The success of the *Experiments and Observations on Different Kinds of Air* (originally put out in three volumes in 1774–1777 and republished in five editions up to 1790) suggests that Priestley's efforts at reaching an audience became at least moderately successful. But, by his own admission, he never again matched the popularity of his early books on electricity.[71]

There are three sources of evidence to which we can look to assess the size and composition of Priestley's readership: contemporary reviews of his works, records of borrowing from libraries, and the list of subscribers published in the *History of Vision*. Reviews of scientific works were relatively rare in England before the foundation of the major review journals and the expansion in the range of scientific periodicals at the end of the eighteenth century. Regular reviews of Priestley's works, addressed to an audience without detailed knowledge of natural philosophy, did appear in the *Monthly Review* (founded in 1749). Most of these were written by William Bewley (1726–1783), a surgeon and apothecary who lived at Great Massingham in Norfolk, and with whom Priestley began a lengthy correspondence after Bewley wrote a notice of the *History of Electricity* in 1767.[72]

69 In this respect, I take issue with the generally helpful discussion in Crosland (1983).
70 Priestley (1806), p. 64; Crosland (1983), pp. 231–232.   71 Priestley (1803), p. 52.
72 Bewley's reviews include: *Monthly Review*, 1st ser., 37 (1767), 93–105, 241–254, 449–464; 47 (1772), 227–230; 51 (1774), 136–147, 361–368; 54 (1776), 107–114, 425–435; 57 (1777), 5–8; 58 (1778), 60–68; 60 (1779), 441–449; 61 (1779), 161–171; 65 (1781), 337–346. Letters from him to Priestley appear in the appendices to all three volumes of Priestley (1775–77) and in the first two volumes of Priestley (1779–86). On the friendship between the two men, see: Priestley (1806), p. 66. On the *Monthly Review* and the identity of its reviewers, see: Nangle (1934–55).

Bewley's reviews generally spoke of Priestley as addressing a homogeneous audience, called simply "the public." Within this public, however, there were recognized to be some readers with specialist knowledge who could benefit in a particular way from his writings. Bewley described electricity and the new airs that Priestley had discovered as capable of generating widespread curiosity and enthusiasm. The astonishing nature of discoveries like the Leyden Jar or dephlogisticated air would capture the attention of the most superficial observer, because they were described with accuracy and dramatic verisimilitude in Priestley's texts. On the other hand, Priestley's exhaustive experimental descriptions were able to provide detailed factual information to readers who already possessed knowledge of the relevant phenomena. Thus philosophical readers would be particularly able to discern the significance of the experimental facts that Priestley had "thrown into the public stock." Or, as Bewley put it in connection with the *History of Electricity*, "insulated electricians in the country . . . will be able to rejoice in receiving sparks of electrical intelligence from so excellent a prime conductor."[73] Priestley was seen as capable of holding the attention of a large public audience and simultaneously conveying factual information to specialists, by the skill and accuracy with which he had compiled his texts.

Further information on Priestley's success in conveying facts to the reading public can be gleaned from records of books borrowed from libraries. Paul Kaufman has published a helpful analysis of borrowing for the years 1773–1784 from the Bristol public library, the only English city library from which a record is known to have survived from the eighteenth century.[74] The popularity of the library during this period was one sign of a flowering of the local literary culture, following the city's sustained economic expansion over the previous decades.[75] The patrons of the Bristol Enlightenment were mainly affluent bourgeois men and women, who had benefited from the city's outstanding commercial success and were generally well integrated within the Anglican Church and the prevalent political order. The middle and upper-middle class status of the customers of the library is indicated by their ability to pay a fairly substantial subscription fee: one guinea to join and one guinea annually thereafter.

To this audience, Priestley's books appealed rather more than others in the realm of natural philosophy, though inevitably they excited less interest than works in the more popular genres of travel literature, moral advice, and political history. Priestley was the author with the most titles in the library (10 being listed during this period), but the total number of

---

73 *Monthly Review*, 1st ser., 37 (1767), p. 464.   74 Kaufman (1960).
75 On the Enlightenment in Bristol, see Neve (1983).

borrowings of his works (89) comprises less than half the number for the most popular single works, such as John Hawkesworth's *Voyages* (201 borrowings), Lord Chesterfield's *Letters to his Son* (185 borrowings) and David Hume's *History of England* (180 borrowings).

On the other hand, Priestley fared relatively well by comparison with other scientific authors: His *Experiments and Observations* was the most popular scientific work, borrowed forty times after the appearance of the first volume in 1774. The *History of Electricity* (twelve borrowings from 1773 to 1784) and the *History of Vision* (nine borrowings from 1773 to 1784) also did moderately well. Priestley's work on electricity clearly outpaced competing works, such as James Ferguson's *Introduction to Electricity* (four borrowings from 1777 to 1782) and Tiberius Cavallo's *Complete Treatise on Electricity* (two borrowings from 1779 to 1784). Similarly, his work in the currently more fashionable field of airs out-shone Cavallo's *Treatise on Air* (eight borrowings from 1782 to 1784) and David Macbride's *Experimental Essays* (fourteen borrowings from 1773 to 1782). Priestley's works on airs were largely responsible for creating the widespread public interest in these phenomena, an interest succeeding that which had been shown in electricity a few decades earlier. The figures for borrowings from the Bristol library bear out the extent of this public interest, though they also show that natural philosophy failed to attain the level of popularity of certain other genres of reading.[76]

A more precise specification of Priestley's audience is possible in the case of one of his works, the *History of Vision*, which contained a list of subscribers who had supported its publication. Publication by subscription, though used only for a minority of books, was not uncommon in the eighteenth century.[77] A prospective author and his publisher would solicit promises to purchase copies of the book, and those who subscribed would be recorded in the work when it was printed. This method of support for literary production was less risky than relying completely on post-publication sales and provided an alternative to dependence on a single person.[78] Looking back in his *Memoirs*, Priestley recalled the difficulties of finding enough subscribers and other purchasers to recompense him for the expense of producing this work – a comment that may

76 All statistics extracted from Kaufman (1960).

77 The subscription list is bound in the first volume of Priestley (1772a). Issues of the text after the first contained addenda to the original list. On the circumstances of publication of the work, see Priestley (1806), p. 64. For general information on publication by subscription in the eighteenth century, see: Robinson and Wallis (1975), pp. i–xxi; Alston, et.al. (1983).

78 See Korshin (1973–74); and (for a very jaundiced contemporary view of the process) Malton (1777).

indicate that the range of its readership was narrower than for some of his other books. However, the 505 subscribers he secured (some of whom agreed to take multiple copies) represents a healthy number. Priestley's total was less than the approximately 670 that subscribed for Thomas Malton's *Compleat Treatise on Perspective* (1778), but was about twice the number who signed up for Robert Smith's *Elementary Parts of Optics* (Cambridge, 1778).[79]

Lists of subscribers can only give a first approximation to the breakdown of a book's total readership. They are likely to show an over-representation of certain categories of persons – friends of the author, for example, and members of the aristocracy, who were strenuously pursued by those seeking subscriptions and prominently displayed in lists.[80] Priestley's friends were indeed well represented, including the London electricians Canton, Watson, and Franklin, along with other friends such as the Leeds surgeon William Hey, the Warrington professor John Aikin, and the experimental philosopher Henry Cavendish. A few members of Priestley's family were also included. Nevertheless, there was no overwhelming geographical bias among subscribers, apparently because the publisher, Johnson, collected subscriptions in London. Hence, although there was a strong contingent (fifty-two names) from Leeds and the surrounding area (Priestley's home town, where he was living at the time), there were also many subscribers from other parts of England.

Priestley's list included a high proportion of members of the traditional professions, whose role as significant contributors to Enlightenment culture has been mentioned. Listed were: sixty-seven clergymen (mainly Anglicans), twenty teachers at educational institutions (mostly at Oxford and Cambridge), thirty-eight medical practitioners (mainly physicians), and ten lawyers. Among institutional subscribers were a few of the intellectual clubs and societies whose contribution to spreading the Enlightenment in the English provinces has also been noted by historians.[81] Book clubs in Manchester, Taunton, and Yarmouth subscribed, along with the Educational Society of Bristol and the Manchester Circulating Library. The aristocracy was also well represented, with thirty-eight subscribers, and men of commerce were not absent, twenty-one being designated "merchants."

The occupations of about 200 men were recorded in Priestley's list,

---

79 Malton (1778) attracted a large number of subscriptions from architects, builders and artists. Smith (1778) was an abbreviated edition of a textbook originally published in 1738, and was mainly subscribed to by students and teachers at Cambridge University. Neither work is precisely comparable with Priestley's in terms of content or presumed readership.

80 Speck (1982).   81 Porter (1982a), ch. 6; Kaufman (1969); Hamlyn (1947).

and their breakdown suggests an audience for his work comparable with those identified for other aspects of Enlightenment culture.[82] The prominence of "professional" groups, the persistent dominance of the aristocracy and gentry, and the lesser but not insignificant presence of the commercial classes, all fit this broad picture. On the other hand, it is notable that Priestley's list included no women, despite considerable evidence that they attended lectures and read texts on natural philosophy.[83] Their omission may simply reflect their lack of financial independence, though it is possible that Priestley's text was not as directly aimed at attracting women readers as were certain other works of the period. In contrast, Abraham Bennet's *New Experiments on Electricity* (Derby, 1798) numbered 46 women among approximately 400 subscribers.[84]

Taken together, the evidence of reviews, library borrowings, and subscriptions suggests that Priestley's works mostly succeeded in reaching a broad readership among the enlightened public. His audience was geographically well dispersed, at least within England. His works achieved a popularity among middle-class readers that was greater than that of other scientific texts of the time, though less than that of the most popular literary genres. His readership comprised a mixture of educated professionals, gentry, aristocrats, and the commercial classes. Remarkably, at least in the early 1770s, there were no clear exclusions along sectarian or political lines.

This was the "public" to which Priestley's writings were addressed. It was a notably heterogeneous audience in geographical, social, and ideological terms. It embraced a wide range of degrees of technical knowledge of the scientific fields described. Twenty-six of the subscribers to the *History of Vision* were fellows of the Royal Society, and hence could be said to have some degree of special interest in the sciences, even if not all of them were active experimentalists. Priestley clearly saw no difficulty in addressing them as part of his overall audience, though (as we have seen from Bewley's reviews) they could be differentiated from the others for particular purposes. The question that arises is how did Priestley face up to the rhetorical task of communicating with this diverse audience? How did he frame his texts to convey factual knowledge to his readership as a whole?

---

82 Compare the data on the diffusion of the later editions of the *Encyclopédie*, in Darnton (1979), pp. 246–323; and on memberships of French provincial academies, in Roche (1978), I, pp. 185–256.

83 On the participation of women in Enlightenment culture in general, and in science in particular, see Meyer (1955); Schiebinger (1989), chh. 1, 2; and note the pertinent remarks of David Hume in "Of essay writing," in Hume (1903), pp. 568–572.

84 Bennet (1789) was published in a provincial milieu very similar to Priestley's. Among the subscribers were Priestley himself and other members of the Lunar Society such as Darwin, Boulton, Watt, Keir, and Wedgwood.

The experimenter and the writer

There are no truths which more readily gain the assent of mankind, or are more firmly retained by them, than those of an historical nature, depending upon the testimony of others. It is a kind of evidence to which all men are most accustomed, so that it is quite familiar to them; and it is peculiarly adapted to the great bulk of mankind, who are unused to abstract speculation.

Priestley, *An Answer to Mr. Paine's Age of Reason*[85]

As to his talents as a writer, we have only to open our eyes to be convinced that they are far below mediocrity. His style is uncouth and superlatively diffuse. Always involved in minutiae, every sentence is a string of parentheses, in finding the end of which the reader is lucky if he does not lose the proposition they were meant to illustrate.

William Cobbett, "Observations on Priestley's Emigration"[86]

The construction of scientific discourse in textbooks and research reports has been the focus of much recent research. Discussions of the different forms of scientific rhetoric have highlighted the importance of written discourse in enabling experimental effects to become reproducible outside the settings where they originate. Priestley's writings clearly offer themselves to this kind of analysis. His laboratory work – for example that on different kinds of air – was reported in a specific style of literary exposition. Carefully framed descriptive narratives were composed with the explicit aim of making experiments reproducible by relatively unskilled practitioners working with minimal equipment. Priestley's writings were designed to involve a wide and diverse public in the expanding scientific culture of his time.

Priestley established his name as an author of important discoveries concerning airs in 1772, when his pamphlet *Directions for Impregnating Water with Fixed Air* showed how "fixed air," already identified by Stephen Hales, Joseph Black, and others, could be dissolved under pressure in ordinary water to produce an artificial equivalent of certain well-known mineral waters. For Priestley, this manufactured mineral water promised to have "all the medicinal virtues of genuine Pyrmont or Seltzer water; since these depend upon the fixed air they contain."[87] In the *Philosophical Transactions* for the same year he also published a paper, "Observations on Different Kinds of Air," in which he described how the newly discovered "nitrous air" could be applied as a test of the relative "goodness" of different types of air – in other words, the degree to which they were able to support respiration and combustion.[88] In the same paper he showed how air that had been "vitiated" by combustion or respiration

85 Priestley (1795), p. 9.   86 Cobbett (1801), p. 191.
87 Priestley (1772b); idem. (1775–77), I, p. 33.   88 Priestley (1772c).

could be "restored" by vegetation or by agitation over water. As he explained in the first volume of *Experiments and Observations on Different Kinds of Air* (1774), Priestley was convinced that this process for restoring the goodness of the atmosphere had been created by God as a providential means of maintaining the balance of nature.[89]

In the second volume of *Experiments and Observations* (1775), Priestley described the discovery for which he remains most celebrated, "dephlogisticated air," subsequently named "oxygen." This air was prepared in August 1774 by heating mercury calx, and was mentioned by Priestley to Lavoisier and other French chemists in the course of a visit to Paris in Autumn of that year.[90] In March 1775, Priestley tried the nitrous air test on this new air and (to his great surprise) found it superior in "goodness" to ordinary air. It also supported combustion and respiration better than normal atmospheric air. Priestley respired the new air and found it gave him a light and easy feeling in the chest; he speculated, "Who can tell but that, in time, this pure air may become a fashionable article in luxury."[91]

These important discoveries have been well recognized in historical treatments of Priestley's work. Less well considered have been the textual means by which they became established and accepted among his audience. It is important to recall that Priestley was always both an experimenter and a writer, and that his readers knew him as an experimenter only through his writings. Significant continuities underlay his apparent switch from historian to experimentalist in the course of his career. Even his later works, in which he reported mainly his own experimental results, comprised narrative descriptions in a basically historical form. Texts like this had a crucial role in Priestley's model of the public culture of science. They also presented an image of Priestley himself as an experimenter, in which he was shown as representative of certain favored moral values. The keynotes of his character were openness to accidental discovery, economy in the use of resources, and modesty in theoretical speculation.

Priestley's first scientific works, the *History of Electricity* and the *History of Vision*, could easily have been assimilated to his previous historical writings, such as the "Chart of Universal History" and the Warrington lectures on "History and General Policy." This may explain why his election to the Royal Society was opposed by some fellows on the grounds that he had established a reputation only as an historian, not as an ex-

89 For Priestley's work on this process, and the importance he ascribed to it in relation to the balance of nature, see McEvoy (1978–79), pp. 94–103, 158–164.
90 Perrin (1969) and Kohler (1975) give some details of this episode.
91 Priestley (1775–77), II, p. 102.

perimenter.[92] But these texts, though historical in their overall construction, actually contained much of Priestley's own experimental work. Having come across factual discrepancies in the literature, he had been led to make his own investigations of subjects like the electrical conductivities of various materials and the refraction of light. He incorporated the results in the appropriate places in the overall historical scheme.

Priestley began his preface to the *History of Electricity* by situating his work in the genre of philosophical history. "Philosophical" as opposed to "civil" history would be composed of certified natural facts arranged in chronological order of their discovery. It would catalogue the progressive revelation of the wonders of nature, and hold out the prospect of continued progress in the future. Contemplating this prospect of indefinite future progress, readers would experience feelings of the "sublime."[93] In the lectures on oratory he delivered at Warrington, Priestley had defined sublime emotions as those provoked by objects or ideas that "relate to great objects, suppose extensive views of things, require a great effort of mind to conceive them, and produce great effects."[94] By calling forth feelings of this magnitude, a historical exposition of the development of science would teach the moral lesson that human progress depended on a properly humble and empirical approach to nature.

The writing of philosophical histories was also to play a part in the formation of a community of experimental philosophers. Texts of this kind could recruit an audience for experiments by showing how entertaining they could be. They could also inspire readers to become experimenters themselves. Priestley promised that his narratives of discovery would demonstrate how even unqualified practitioners with few resources had made significant contributions to natural philosophy; and he undertook to recognize future amateur contributions by mentioning them in subsequent editions of the work.[95]

In the preface to the *History of Electricity* and that to the subsequent *History of Vision,* Priestley drew upon his own experience of beginning experimental philosophy to show how historical reviews of discoveries could make it more accessible. He recalled his own difficulties in obtaining texts which contained the information he sought. The historical method of composition would be the appropriate one "to engage the attention, and to communicate knowledge with the greatest ease, certainty and pleasure." Texts written in this style would recruit investigators and provide them with essential basic knowledge. They would thus support the activities of self-organizing philosophical societies, which Priestley strongly

---

92 Schofield (1966), p. 62.   93 Priestley (1767), pp. i–ii, iv–v.
94 Priestley (1777), p. 154.   95 Priestley (1767), p. xi.

favored, and would attract patronage for natural philosophy. This was the social significance assigned to the historical or analytical mode of composition:

... the recital of the labours of philosophers in an historical method gives a writer a better opportunity than a systematical method would do, of transmitting them to posterity in such a manner as will operate most powerfully on the minds of readers, and be a motive with them to exert and distinguish themselves in philosophical pursuits.[96]

In order to "operate most powerfully" on readers, philosophical histories would comprise an accumulation of narratives of experiments, if possible written by the experimenters themselves. Primary sources were valued by Priestley over "lame and superficial" secondary sources, because only the former would be sufficiently detailed to engage the readers' interest, convince them of the factuality of what was related, and permit them to repeat the experiments at will. Excessive prolixity in descriptive prose would, however, risk boring apprentice philosophers, who were always likely to prefer doing to reading.[97]

The fact that experimental discoveries were, to a large extent, accidental occurrences, was a further encouragement to amateurs. Although Priestley omitted "mistakes" and "opinions" as distractions from a narrative of the progressive accumulation of facts, he included details that showed that discoveries were largely accidental, because these would teach philosophers that their actions were caused by divine providence, not by their own will. Adulation of human genius would thus give way to worship of the divinity in nature: ". . . electrical discoveries have been made so much by accident, that it is more the powers of nature, than of human genius, that excite our wonder with respect to them."[98] The accidental nature of the discoveries related would therefore increase the sublime feelings provoked by contemplation of the progress of science, and would demonstrate that the experimental study of nature was at the same time an act of religious devotion. It was this emotional and religious conviction that would bind together Priestley's projected community of practising experimental philosophers.

In substantial respects, Priestley's later works, the *Experiments and Observations on Different Kinds of Air* and the *Experiments and Observations Relating to Various Branches of Natural Philosophy*, continued to use the mode of textual exposition he had already established. They carried forward the project of concurrent experimentation and writing that the previous histories had announced. Although the texts that appeared were not precisely the *"history of discoveries relating to AIR"*

96 Priestley (1772a), I, p. vii.   97 Priestley (1772a), I, pp. vii–ix.
98 Priestley (1767), p. xiii.

that Priestley had originally promised, because they mainly recounted his own work, their use of the historical method and their projected function in the philosophical community were very much in line with the program he had earlier outlined.[99] The volumes of *Experiments and Observations* presented the materials, in the form of narratives of experiments, from which future works of philosophical history could be compiled. But by becoming author of the experimental phenomena themselves, as well as author of their descriptions, Priestley created a new textual persona for himself, that of discoverer as well as historian. At this point, as one reviewer put it, "he steps forth in a new character."[100]

The preface to the first volume of *Experiments* (published in 1774) explained Priestley's emergence in this new character by reference to a contemporary climate of intellectual and moral enlightenment. Aware that he was courting the label "enthusiast," he did not hesitate to identify "a very particular providence" underlying the rapid extension of scientific knowledge recently experienced. More widespread experimental knowledge of natural phenomena would soon bring about other changes, such as progressive moral improvement and an ending of the power of illegitimate and corrupt political and religious authorities:

The rapid process of knowledge, which like the progress of a wave of the sea, or of light from the sun, extends itself not in this way or that way only, but in all directions, will, I doubt not, be the means, under God, of extirpating all error and prejudice, and of putting an end to all undue and usurped authority in the business of religion as well as of science; and all the efforts of the interested friends of corrupt establishments of all kinds, will be ineffectual for their support in this enlightened age. . . . And the English hierarchy (if there be anything unsound in its constitution) has equal reason to tremble even at an air pump, or an electrical machine.[101]

In the subsequent volumes of *Experiments,* Priestley reiterated the vision of progressive enlightenment spearheaded by a widening awareness of the natural wonders that experimental science revealed. The preface to the second volume asserted that the subject had now "gained almost universal attention among the philosophers, in every part of Europe." Only the study of electricity had generated a similar degree of excitement, and now that philosophers had connected the two realms it was clear that they were witnessing the approach of "one of those great epocha's . . . [distinguished] by some great discovery, for the benefit of mankind."[102] Again in the preface to the third volume, and in those to the subsequent three volumes of *Experiments and Observations Relating to Various Branches of Natural Philosophy,* Priestley excitedly sketched the prospect of a widening circle of experimenters, continually expanding

99 Priestley (1772a), I, p. xi.   100 *Monthly Review*, 1st ser., 37 (1767), p. 94.
101 Priestley (1775–77), I, pp. xiii–xiv.   102 Priestley (1775–77), II, pp. v, xi.

the range of natural phenomena illuminated by the light of human knowledge and progressively beating back the frontiers of darkness.[103]

Thus the diffusion of knowledge of experimental techniques and the widespread public manifestation of new phenomena were assigned a prime role in the overall process of enlightenment. Like his colleague Erasmus Darwin, Priestley believed that experimental philosophy, "by inducing the world to think and reason, will silently marshall mankind against delusion, and . . . overturn the empire of superstition."[104] The establishment of natural facts in the public realm, and the recruitment of growing numbers of amateur philosophers to replicate them, were means toward the liberation of mankind from unjust secular and religious authorities. It was this context of climactic (indeed millennial) progress that justified Priestley's stepping forth in the character of an experimenter himself.

The rhetoric of Priestley's texts was shaped to insert his own discoveries into the expanding public culture of science. He declared that the analytic and historical style was aimed at "the generality of those persons who take pleasure in reading philosophical writings," and particularly at those who wished to repeat the experiments he described.[105] In order to persuade the general reader of their truth, experiments were related in the temporal order in which they had been performed, with validating particulars of time, place, and the names of witnesses. A neutral descriptive style was used, which Priestley claimed enabled him to communicate facts without any contamination by hypotheses, speculations, or opinions.[106] Exhaustive circumstantial details were given to ease the replication of experiments, which Priestley considered of key importance in gaining public acceptance of his work, and to lead the uninitiated reader gradually into the practice of manipulating apparatus. By these stylistic devices, he strove to satisfy his ambition "not to acquire the character of a fine writer, but of a useful one."[107]

He had discussed some of the features of the style that a useful writer should adopt in the Warrington lectures on oratory and criticism, which were delivered in 1762 and published in 1777. There he had provided a rationale for historical narrative, and an account of its peculiar persuasive power. He suggested that the historical or analytical method was the proper one to use "to communicate truth to others in the very manner in which it was discovered."[108] To recapitulate the temporal order of dis-

103 For example, Priestley (1775–77), III, pp. iii–x; idem. (1779–86), II, p. ix; III, pp. v, xxiii.
104 Richard Roe to Priestley, 3 September 1791, in King-Hele (1981), pp. 215–216, quotation on p. 216. The letter, signed by the Secretary of the Derby Philosophical Society, was probably drafted by Darwin.
105 Priestley (1779–86), II, p. xii.   106 Priestley (1775–77), III, pp. x, xvii, xxx.
107 Priestley (1806), p. 52.
108 Priestley (1777), p. 43 (compare pp. 33–34). For a general discussion of Priestley's rhetorical ideas, see: Howell (1971), pp. 632–647; and Fruchtman (1982).

covery was particularly appropriate for conveying novel or unusual information. It was a potent mode of discourse because it exploited the mental process of the association of ideas in the minds of the readers: Events were naturally better understood and learned if they were linked in the order of their occurrence. Precisely because they took the form of stories, narrative accounts of experiments were thus seen as especially suitable for communicating factual knowledge to a wide readership. As Priestley wrote elsewhere, narrative "is a kind of evidence to which all men are most accustomed, so that it is quite familiar to them; and it is particularly adapted to the great bulk of mankind."[109]

In the *Lectures on Oratory*, Priestley had also indicated the value of circumstantial details in making historical accounts more persuasive. He noted that "stories told with all those circumstances, provided they be not so many as to distract the mind of the hearer, . . . are generally heard with more attention."[110] Of course, Priestley would not have seen himself as simply telling stories in his experimental writings; he was convinced of the truth of what he related. But the Warrington lectures show that he had a fine appreciation of the rhetorical support that even true accounts needed if information were to be communicated effectively to a large public audience. Priestley would have strenuously resisted the description of him as a rhetorician, having exclaimed ". . . sooner would I teach the art of poisoning than that of sophistry"; but this does not mean that his mode of exposition was artless or formed without any deliberation about its appropriateness for his intended audience.[111] On the contrary, the evidence shows that he took the business of writing very seriously indeed, and that he tried hard to make his descriptive style serve the purpose of establishing his experimental claims as public knowledge.

In order to do this, and thus in order to realize the potential of natural knowledge for stimulating moral and material progress, it was essential that experiments be made as widely replicable as possible. This was a further reason for giving considerable attention to detailed descriptions of instruments and procedures. Apparatus was accordingly treated at length with accompanying diagrams, and the descriptions were amplified and revised in successive volumes of the text (Figure 3). Details were given of the construction of a trough suitable for holding water or mercury, the making of glassware, and the management of various sources of heat. The catching, keeping and feeding of mice, which Priestley used to test the goodness of different samples of airs, were discussed at length. Throughout, he emphasized the simplicity of his apparatus, and the relative ease with which it could be made or obtained. He gave the names of suppliers and manufacturers who had provided him with materials

109 Priestley (1795), p. 9.   110 Priestley (1777), p. 85.
111 Priestley (1777), p. 46. On this point, see Fruchtman (1982), pp. 38–39, 47 fn.9.

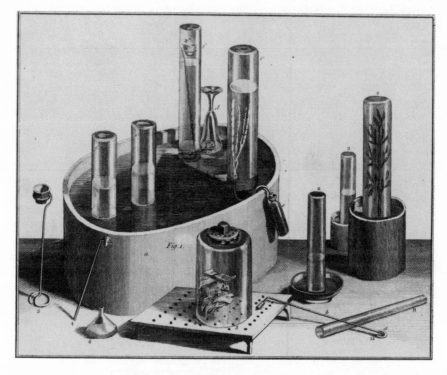

Figure 3. Priestley's experimental apparatus. From Priestley, Experiments and Observations on Different Kinds of Air, *2nd
edn., 3 vols. (London: J. Johnson, 1775–77), vol. I, frontispiece. [Reproduced by permission of the Syndics of the
Cambridge University Library.]*

and equipment. Complaints from readers who could not reproduce any
aspect of the experiments brought further details and clarification in subsequent volumes.[112]

As well as conveying detailed information, Priestley's texts projected
an image of himself as the experimenter. When he described his actions,
he appeared as a character in his own narrative. As an experimenter,
Priestley was shown in a way that furthered the credibility of his descriptions of phenomena and helped advance the model of the scientific community that Priestley as author upheld. The way he portrayed himself
owed something to techniques of authorial representation of the self that
had also been discussed in the lectures on oratory. There Priestley had
noted the importance of "marks of candour" – remarks made by the

112 Priestley (1775–77), I, pp. 6–22; II, pp. xxxiii–xliv; III, pp. 1–4.

author about himself to increase the credibility of what he said.[113] To appear modest and hesitant was, he claimed, a very effective way of getting one's views heard with attention. Accordingly, Priestley was characterized in his texts as honest in his descriptions, modest in the magnitude of the achievement he claimed, and generous in according credit to others. He was portrayed as particularly open to observing the accidental results of experiments and quite liable to be surprised by what he discovered. He was also shown to depend upon simple apparatus and easily accessible resources, which he nevertheless had to husband with careful economy.

Priestley's modesty was shown in a number of ways, making him an exemplar of the moral values proposed for the experimental community as a whole. He explicitly demarcated relations of fact from hypothetical speculation, qualifying the latter with repeated disclaimers of any intention of teaching a systematic theory, and insisting that he was simply and honestly reporting facts.[114] For these he demanded credit, just as he scrupulously accorded credit to all those whose observations he cited, or who had assisted his researches by supplying apparatus or materials. In the economy of merit and credit that Priestley outlined, the communication of facts or artifacts deserved recognition and merited acceptance as part of the progressive endeavor of science; but the construction of theoretical structures only constituted an obstacle to progress.

In line with these principles, Priestley expressed a particular reluctance to take upon himself the responsibility for naming things. Only in the cases of phenomena that were manifestly new, such as nitrous air, would he overcome this reluctance, and then only after consulting with colleagues.[115] He was both advocating and exemplifying a Baconian suspicion of the "idols of the marketplace" – the process whereby the names imposed on things take power over them and shape their interpretation. This he was anxious not to appear to do, because it would undermine the whole effect of the rhetorical apparatus devoted to showing experimental facts just as they were.

It was a further feature of the rhetorical presentation of plain facts that the experimenter was shown as frequently being surprised by the unexpected outcome of his experiments. Like the philosophers whose efforts were recorded in the *History of Electricity*, Priestley modestly described himself as an "instrument in the hands of divine providence."[116] Inci-

---

113 Priestley (1777), pp. 123–124.
114 McEvoy provides a good discussion of the occasions in Priestley's works when he invokes the fact/opinion distinction, though he treats this in terms of its philosophical significance for Priestley himself rather than in terms of its rhetorical functions. See McEvoy (1979), pp. 10–11.
115 Priestley (1775–77), I, pp. 23–24, 109.   116 Priestley (1775–77), II, p. ix.

dents such as the discovery of dephlogisticated air showed how much "more is owing to what we call chance, . . . than to any proper design, or pre-conceived theory in this business." At one point in the narrative, the author interjected, "I wish my reader be not quite tired with the frequent repetition of the word surprize, and others of similar import; but I must go on in that style a little longer. . . ."[117] Going on in this style was one of Priestley's means of demonstrating that his observations of phenomena were not influenced by prior expectations.

One of the most surprising discoveries was the nitrous air test, in which a measured volume of sample air was held in a tube over water, a measured volume of nitrous air added to it, and the two shaken together. When this was done, brown fumes were seen to be produced, which were gradually absorbed by the water. The total volume of air remaining was then measured, and the degree to which this was *less* than the total original volume of the two airs gave a measure of the "goodness" of the sample air. A better quality air – one more suited to support respiration or combustion – would be found to have absorbed a larger quantity of the nitrous air. Of this test, Priestley remarked, "I hardly know any experiment that is more adapted to amaze and surprize than this, which exhibits a quantity of air, which, as it were, devours a quantity of another kind of air."[118]

The nitrous air test was also important to Priestley because it served the requirement of economy. An important part of his portrayed character as an experimenter was his employment of only simple and easily available apparatus, and his careful parsimony in the use of resources. This, along with the provision of exhaustive instructions, was essential for the widest possible replication of his experiments, which Priestley encouraged in order to diffuse knowledge of nature and thereby advance enlightenment. It was vital to the moral structure of public science, as he envisaged it, that experiments should be simple and economical to perform. The nitrous air test was simple enough to be widely replicable, and also dispensed with the time-consuming and troublesome business of keeping mice to test the goodness of airs. This was a distinct advantage in husbandry of experimental resources.[119]

Priestley's moral code, as it was embodied in his experimental texts, gave him the basis for his criticisms of natural philosophers and chemists who failed to live up to his ideal. Lavoisier in particular was attacked for not adhering to the code of economy, simplicity, and modesty. Not only had he repeated Priestley's discovery of the air released from heated mercury calx with an apparatus "more complex and less accurate" than

117 Priestley (1775–77), II, pp. 29, 42.   118 Priestley (1775–77), I, pp. 111, 114–115.
119 Priestley (1775–77), I, pp. 9–11, 115.

Priestley's own, he had also disregarded the injunction not to theorize in advance of a complete collection of experimental facts. Reacting to Lavoisier's theory that pure air should be identified as oxygen (the principle of acidity), Priestley dismissively remarked, "*Speculation* is a cheap commodity. *New and important facts* are most wanted, and therefore of most value."[120] The dispute between Priestley and Lavoisier became a long and relentless one, rooted (as we shall see) in alternative moral commitments and different visions of the public culture of science.

In his conviction that experimental philosophy had an underlying moral imperative, and in his choice of certain discursive methods to enforce the correct moral code, Priestley stands comparison with his seventeenth-century predecessor Robert Boyle. As authors, both men struggled to maintain a distinction between facts and theoretical opinions, and both adopted forms of discourse devoted to exhaustively detailed narration of experimental facts.[121] To both men, a personal reputation for candor, modesty, and ingenuousness was of the utmost importance in securing public credit for their claims. And both, despite significant differences in theological commitments, portrayed the experimental revelation of nature as part of the human duty to worship God in the works of creation.[122]

There are nonetheless significant differences between these two celebrated experimental philosophers. Although they chose similar textual means to build bridges to their audiences, the social contexts in which they operated were very different. Boyle commanded substantial material resources, which he put to use in sustaining his experimental research. He also occupied a central position in the scientific community of his day. Priestley on the other hand made a virtue of his poverty in material resources and learned to operate socially from a position on the edges of the scientific establishment. He had virtually to create a scientific community around himself, though he was able to build upon the thriving intellectual culture of the English provincial Enlightenment to do so.

Priestley was no less successful than Boyle in reaching his audience and establishing public acceptance of his experimental claims. If anything, Priestley's impact was more widespread and more rapid, at least until the 1790s brought a profound change to the prevailing social and political climate. He stimulated a considerable degree of interest in the phenomena of airs, and gained recognition as their most important discoverer. He also succeeded in recruiting apprentice experimenters to work in this

---

120 Priestley (1775–77), III, pp. xxvii–xxxii (quotation on p. xxx).
121 On Boyle's narratives of experiment, see: Shapin (1984); Shapin and Schaffer (1985), esp. ch. 2; Golinski (1987).
122 On Boyle's theology, see: McGuire (1972). On Priestley's: McEvoy and McGuire (1975); Brooke (1984).

new field of research. Humphry Davy spoke from his own experience when he stated that he knew of no book "so likely to lead a student into the path of discovery as Dr. Priestley's six volumes upon air."[123]

Notwithstanding Priestley's well-evidenced success in communicating with his audience, his style of writing was not universally admired. His rhetoric depended for its effect on certain values being shared by writer and readers. The order of discourse was in this sense underwritten by a projected moral order of public science. Readers had to respect the honesty, candor and scrupulousness of the reporter of experimental facts. Without this degree of trust, the rhetoric of factuality had no purchase. William Cobbett, whose hostile appraisal of Priestley's style is cited at the beginning of this section, showed that an unsympathetic reader might take the cautious qualifications and disclaimers for mere verbosity. What Priestley intended as "marks of candour" might appear to others as unnecessary digressions in an already tortuous mode of description.[124]

This point can be amplified by considering the dispute that Priestley entered into with the scientific lecturer Bryan Higgins, and that he documented in his *Philosophical Empiricism* (1775). Higgins attacked Priestley's claimed discoveries and their authenticating rhetoric, thereby exposing aspects of the artificiality of his textual practice and his projected personal character. Though it does not appear to have seriously damaged Priestley's reputation, the argument throws further light on his practices of public discourse.[125]

The dispute arose after Priestley demonstrated some of his discoveries, including the production of dephlogisticated air, to a group that included certain prominent medical practitioners at Lord Shelburne's London residence in May 1775. One member of the audience, Dr. Richard Brocklesby, subsequently put about the story that Priestley had plagiarized experiments that he had seen performed in Higgins's public chemical lectures. Hearing of the rumor, Priestley wrote to Brocklesby and Higgins, and published the resulting correspondence together with his own observations on the affair.

Two aspects of Brocklesby's charge, which was vigorously seconded by Higgins himself, were of concern to Priestley. First was the attack on

123 Davy (1839–40), VII, p. 117.
124 Cobbett's appraisal occurred in the context of a hostile pro-establishment account of Priestley's emigration to America. Compare the sympathetic view of Priestley's style as characterized by "candour, simplicity and energy," in Corry (1804), p. 53.
125 Priestley (1775). The pamphlet is mentioned by Gibbs (1965, pp. 95–97) and McEvoy (1978–79, p. 154), but to my knowledge it has not received detailed discussion to date. For general information on Higgins's work, see: Higgins ([c.1775?], 1795); Gibbs (1972); Averley (1986), esp. pp. 102–107. The manuscript "Notes taken at Dr. Higgins's Course of Lectures on Chemistry" (1780, Huntington Library, San Marino, California), vol. I, ff. 12r–13r, show that Higgins was demonstrating the nitrous air test (without any acknowledgment of Priestley) in 1780.

the status of his discoveries. Higgins's criticism was that Priestley had stolen his experiments from other chemists, claiming as his own property what were actually well-known phenomena. He put the point this way:

... your originality in experiments consists chiefly in the knack of rendering the phenomena, which all practising chemists have observed and understood, perfectly mysterious and surprizing to others.[126]

Higgins was in effect drawing attention to the artifactual nature of Priestley's discoveries – to the fact that experiments had to be presented in a certain way to become discoveries. According to Higgins, one of Priestley's ways of making his experiments into discoveries was to present them as "mysterious and surprizing." Priestley's language of "accident," "chance," and "surprize," which he had deployed to indicate that the experimenter was no more than a tool of divine providence, was here being read as a means of creating a spectacular discovery out of commonly known events. Priestley's authenticating rhetoric was deconstructed by Higgins as a way of appropriating natural phenomena.

Even more serious was the second aspect of Higgins's attack: the critique of Priestley's moral character. Priestley was being accused of lying and of theft. Charges of such seriousness threatened to undermine all his experimental claims, by casting doubt on his truthfulness. Priestley's distress at having his honesty and candor called into question was very evident. "I thought that my character for veracity, at least, was sufficiently established," he complained.[127] By impugning his moral reputation, Higgins was threatening to destroy Priestley's credibility as an experimental philosopher.

Priestley's response was similarly two-pronged. First, he contended that Brocklesby was not sufficiently well acquainted with the techniques of experiments on airs to be able to distinguish his own from Higgins's, but that the distinction existed nonetheless. Brocklesby retreated from asserting that the two men's experiments were identical to the claim that they were "nearly the same," and Priestley was able to pounce upon this with the declaration that "experiments that appear nearly the same with others, may, in reality, be essentially different from them."[128] He went on to support this point with examples drawn from his own experience of the difficulty of replicating experiments precisely. His opponents' charge had nonetheless exposed a potential weakness that arose from Priestley's refusal to claim the prerogative of any special refinements in apparatus. Having declined to designate his equipment as unique in any way, he had left himself open to the accusation that he lacked originality.

Second, Priestley replied with a personal attack on Higgins's character, which made him out to be the antithesis of the public persona he had

126 Priestley (1775), p. 9.   127 Priestley (1775), p. 2.   128 Priestley (1775), p. 11.

constructed for himself. Higgins was shown as socially inferior and sy-
cophantic toward superiors from whom he hoped for patronage. He was
portrayed as jealously guarding his property and permanently anxious
about money.[129] Experiments, which for Priestley were achievements de-
serving of honor, were supposedly seen by Higgins solely as marketable
property. Clearly Higgins's commercial mentality and insecure social position would prevent him gaining the moral credentials that Priestley
deemed essential for entry into the philosophical community.

Priestley might plausibly have been accused of a degree of hypocrisy
here because, as I have shown, he was himself financially insecure, seeking patronage and operating in the commercial market for printed scientific texts. But the moral values he embodied in his textual persona,
and that he saw as uniting the community of experimental philosophers,
included independence from motives of personal material gain. Although
an economic use of resources was praised, mercenary ambitions were
unacceptable among philosophers because they would weaken an individual's independence and credibility. Candor, modesty, piety, simplicity, generosity, and economy were the virtues that were preached and
exemplified in Priestley's writings. They underlay the proposed discourse
of facts, the economy of merit and credit, and the widespread replication
of experiments – the features of the public culture of science that Priestley most strenuously upheld.

The dispute with Higgins, though it apparently had little effect on the
acceptance of Priestley's research, illuminates a tension between the moral
character he chose to present in his writings and the behavior required
by the context in which he worked. The morality he preached as fundamental to the ordering of scientific culture was sometimes undercut by
the demands of the prevailing circumstances. Priestley, like other experimental philosophers in his milieu, had to act with an eye to material
advantage, even while he promoted the ideal of a public science that
should not be distorted by economic motives. He engaged in the cultural
marketplace, believing, no doubt, that such activity was guided by fundamental moral and theological aims; but there were occasions when
commerce refused to be so neatly confined by virtue. Marketplace values
could potentially undermine morality, and at times Priestley experienced
the contradiction quite acutely. As we shall see in the next chapter, the
tension was also to be exposed at a number of points during the assimilation and application of his experimental work.

129 Priestley (1775), pp. 23–38.

4

Airs and their uses

So, in the most favourable cases, even when it is a routine piece of equipment, the black box requires an active customer and needs to be accompanied by other people if it is to be maintained in existence. By itself it has no *inertia.* . . . the black box moves in space and becomes durable in time only through the actions of many people; if there is no one to take it up, it stops and falls apart however many people may have taken it up for however long before.

Bruno Latour, *Science in Action*[1]

As we have seen, Priestley's descriptions of his experiments were written with the purpose of insinuating his factual discoveries into the public culture of science which surrounded him. His work at the writing desk complemented that at the laboratory bench – both parts of the enterprise were essential to enable his claims to become public knowledge. He described experiments in exhaustive detail in order to encourage their widespread replication. To stimulate the efforts of apprentice philosophers, he used a narrative form in which accidental discovery had a prime role and in which the experimenter was portrayed as an unprivileged observer making the most of scarce material resources. In these ways, Priestley's textual practices embodied and perpetuated a set of moral values he saw as essential to a flourishing scientific community.

We shall now work outward from Priestley's texts to consider the context in which they were read and assimilated. The shift in focus is necessary because we have to examine the situations in which his work was applied to understand how it became established knowledge. His rhetorical efforts notwithstanding, there was nothing about Priestley's texts that guaranteed they would be universally and immediately accepted, as the dispute with Bryan Higgins showed. His discoveries gained acceptance only through being put to use in a variety of settings, including scientific lectures, medical therapy, and projects for improving public health. By examining these settings, we can expose some of the work done by other people (public lecturers, instrument makers, and medical practitioners, for example) to extend his work through society at large.

1 Latour (1987), p. 137.

91

We begin by considering the role of the public lecturers, many of them known to Priestley personally, who adopted his work and communicated it to their audiences. They showed his discoveries in experimental demonstrations and encouraged their audiences to view them as tokens of the utility of Enlightenment science. By presenting the demonstrations within a particular rhetorical framework, they made Priestley's science accord with the expectations and aspirations of those who attended. At the same time, fellow experimenters were also closely following Priestley's work. Experimentally minded medical practitioners, natural philosophers, and instrument makers were among those who took up aspects of his researches that they could apply to their own concerns. In appropriating Priestley's work and putting it to use they also played a part in publicly establishing its utility. Priestley's chemistry was made into public knowledge by the extension of his experimental practices outward from his laboratory into the wider society.

It is important to be clear about the nature of this process. Notwithstanding the claims of what Bruno Latour has called "the model of diffusion," the process of extension of scientific techniques cannot be ascribed solely to an originating individual.[2] Priestley alone was not responsible for the widespread acceptance of his discoveries. It was because his aspirations – for the formation of a civic scientific culture, the application of science to practical purposes, and the advancement of enlightenment – were shared by many that other individuals identified their interests with transmitting his experimental work.

The activities involved in extending experimental discoveries into the public sphere were simultaneously practical, rhetorical, and social. Practical labor was required to construct equipment that would reliably reproduce the phenomeno-technics of airs at sites away from Priestley's own laboratory. Training in the tacit skills of experimentation was also found to be necessary, to supplement the instructions conveyed in written texts. Material artifacts and routines of behavior both played a part in constructing the social relationships that sustained and extended Priestley's technical achievements. And the language in which the utility of his chemistry was described was also an important factor in aligning patrons and public audiences in its support. Apparatus, practices, and language must all be considered if we are to understand how experimental science could come to be applied in situations outside the laboratory. Finally, the wider social context must also be brought into the picture. By viewing the broader landscape, we can detect how scientists used prevalent social relations to mobilize experimental techniques, and perhaps discern the goals they were trying to attain by doing so.

2 Latour (1987), esp. pp. 132–144.

We focus particularly on two pieces of apparatus, created as a result of Priestley's experimental labors and subsequently used very widely. The first was a machine for producing "fixed air," and dissolving it under pressure in water. The second was the eudiometer, a device that embodied Priestley's nitrous air test for the goodness of air, and by which the healthiness of atmospheric air could supposedly be assessed. Through the efforts of Priestley and others, these two experimental devices were transformed into instruments. Their form was modified and (to some extent) standardized, they were manufactured and retailed as commodities, and their purchasers were taught a regular procedure for using them. The showing of these instruments in public lectures backed up assertions of the extensive utility of chemistry. And their widespread adoption, particularly in therapeutic applications by medical practitioners, gave further substance to these claims.

As we shall see, the social relations through which they were circulated and the range of practices in which they became implicated determined the fate of these devices. Here, elements of the social context introduced in the last chapter become relevant, particularly the links between experimenters in networks of associates, their continuing attachment to patrons, and the involvement of scientific instruments in the market for consumer goods. The water-impregnation machine assumed a number of forms and was successfully deployed in a variety of commercial and therapeutic situations. The nitrous-air eudiometer, on the other hand, after a period of spectacularly successful application to problems of public health, ceased by the mid-1780s to be considered a reliable instrument. The once-flourishing eudiometrical program staggered to a halt, as confidence in the accuracy and replicability of the reported results collapsed. The contrasting fates of the water-impregnation machine and the eudiometer show how scientific instruments depend upon a particular context to sustain them. The successes and the failures of attempts to extend Priestley's experimental practices beyond the laboratory can thus be understood in terms of the culture in which his science was put to use.

Priestley's chemistry in public education

In the course of the experiments frequent mention being made of Dr. Priestley, Dr. Johnson knit his brows, and in a stern manner inquired, "Why do we hear so much of Dr. Priestley?" He was very properly answered, "Sir, because we are indebted to him for these important discoveries". On this, Dr. Johnson appeared well content; and replied, "Well, well, I believe we are; and let every man have the honour he has merited".

James Boswell, *Life of Johnson*[3]

3 Boswell (1953), pp. 1247–1248.

I hope that the next time that you shall see a man standing by a tub of water, or a bason of quicksilver, with jars and phials, &c. before him, filled partly with air and partly with water, with a lighted candle, and a variety of little implements at hand, and transferring his different kinds of airs, with some degree of dexterity, from one vessel to another, a red colour appearing here, and a white one there, you will not be so ready to affirm that the operator was instituting the very same experiments that you saw at Shelburne-house.

Priestley, *Philosophical Empiricism*[4]

In this section we explore the role of public lecturing in spreading knowledge of Priestley's scientific work. We shall investigate the links he cultivated with a number of scientific lecturers and the ways in which they presented his work to their audiences. By these means, his experiments were interpreted and put to use by an audience that extended beyond the readership of his books to embrace members of the middle classes in London and in many provincial English towns and cities. The work of these lecturers played a vital part in constituting Priestley's discoveries as public knowledge. As the incident quoted from the life of Dr Johnson shows, Priestley's work became recognized through public lectures, even among those who harbored considerable antipathy toward him personally and were unlikely to trust the claims to moral sincerity that underwrote the credibility of his writings. His knowledge-claims, in other words, gained greater credibility as they became detached from him personally by escaping from his texts into the repertoire of an extensive network of lecturers.

Commenting on the growing awareness of Priestley's work in Britain and France in the 1770s and 1780s, his Lunar Society colleague James Keir wrote:

The diffusion of a general knowledge, and of a taste for science, over all classes of men, in every nation of Europe, or of European origin, seems to be the characteristic feature of the present age. . . . [I]n no former age, was ever the light of knowledge so extended, and so generally diffused. . . . [Priestley's discoveries,] being very striking in their appearance, . . . have contributed much to the diffusion of a taste for Chemistry.[5]

In this account, the spreading awareness of Priestley's chemistry was assimilated to a model of the process of enlightenment itself. Knowledge as such was seen as diffusing outward from its origins, like rays of light from a source. Although the diffusion model might be a natural and tempting one to use in connection with knowledge about gases, which themselves diffuse outward from their point of origin, the argument of this chapter will be that it is inadequate for understanding the establishment of Priestley's discoveries in the public realm.

The diffusion model neglects two essential aspects of this process. First,

4 Priestley (1775), p. 78.   5 Keir (1789), pp. iii–v.

it overlooks the work Priestley had to do to make his newly discovered facts travel out from his laboratory. He used personal contact with a number of lecturers to transmit to them the techniques that would allow them to reproduce the new phenomena. In this way, potential problems with replicating his experiments were circumvented by practical training to support the written descriptions of experimental phenomena and apparatus.[6] Second, the diffusion model ignores the role of Priestley's audiences (the lecturers themselves and the subscribers who attended their courses) in interpreting his work. They were not simply passive recipients of Priestley's message; on the contrary, they played an important part in creating his image as a discoverer by actively assimilating his work and appropriating it to their own interests.[7] We can gain some clues as to how this was done by studying the language used in advertisements by the lecturers concerned.

Priestley ascribed considerable importance to scientific lecturers in relation to the Enlightenment project of advancing general education. In his *Miscellaneous Observations Relating to Education* (1778), he advocated the continuing instruction of members of the aristocracy and gentry to fit them for commerce and citizenship. He proposed that a broadly conceived liberal education could provide the foundations for business success or a life of leisure. Such an education could be extended to women as well as men and could include mathematics, natural philosophy, and chemistry:

... it is a pleasing prospect to those who wish well to the flourishing state of commerce, that chemistry has, of late years, been more generally attended to than ever, and that it is daily introduced into more places of liberal education.[8]

Though Priestley distinguished public education from the private kind, such as he had given the Earl of Shelburne's sons, for example, or had helped to provide at the Dissenting Academies at Warrington and Hackney, he declined to draw a sharp line between them. In practice the line between private and public scientific education was crossed quite frequently. For example, lectures on chemistry given at the Warrington Academy by Matthew Turner (from 1762 to 1765) and John Aikin (from about 1766 to 1774) were advertised locally and attended by members of the local population.[9] John Warltire gave public lectures on chemistry

6 On the importance of social relationships and the transfer of "tacit knowledge" in permitting replication of experiments, see: Collins (1985), chh. 2, 3. The problem of making experimental facts travel out from the laboratory is discussed in a rather different idiom by Bruno Latour. See Latour (1983); and idem. (1987), esp. chh. 3, 6. A good criticism of the diffusion model of popularization of science is also given in Whitley (1985).

7 Cooter (1984), esp. chh. 4–8, presents an excellent study of the processes of interpretation of scientific doctrine by audiences.

8 Priestley (1778), pp. 185–229, 262.   9 Gibbs (1960), pp. 112–113.

and natural philosophy in Warrington in 1773, exploiting the audience built up by the extramural activities of the Academy staff.[10]

The significance for Priestley of properly conducted lectures in natural philosophy, whether in public or semipublic settings, was that they brought their audiences into direct contact with experimental demonstrations of natural phenomena. As we saw in the last chapter, he regarded the public manifestation of phenomena produced by instruments like the air pump and the electrical generator as a forceful stimulus to the spread of enlightenment, and hence as a potent means to undermine the illegitimate authority of corrupt religious and political institutions. He first experienced the public impact of these two devices when he found that their use in his teaching greatly expanded the reputation of the small school he was running in Nantwich.[11] As Schaffer and Money have argued, the experimental manifestation of natural powers came to have a direct political and social importance for Priestley.[12] Josiah Wedgwood drew attention to the rhetorical force and consequent political function of Priestley's electrical experiments in this way:

... what daring mortals you are to rob the Thunder of his bolts, – and for what? – no doubt to blast the oppressors of the poor and needy, or to execute some public piece of justice in the most tremendous and conspicuous manner, that shall make the great ones of the earth tremble![13]

The problem Priestley faced was that of realizing the political effect of experimental demonstrations as widely as possible. This was why he cultivated relations with a number of public scientific lecturers. As early as 1772, the following were listed among the subscribers to the *History of Vision:* John Arden (teacher of experimental philosophy at Beverley), William Enfield (tutor at Warrington), Charles Hutton (teacher of mathematics in Newcastle), Mr Nicholson (teacher of mathematics in Wakefield), Caleb Rotheram (Principal of the Dissenting Academy in Kendal), and Adam Walker (lecturer in experimental philosophy). Of these men, Arden, Rotheram, and Walker were to become important disseminators of Priestley's chemical discoveries.

Adam Walker (1730?–1821) appears to have been the first experimental lecturer to have extended the standard natural philosophy syllabus to include the chemical properties of airs.[14] Walker was a schoolmaster in Manchester before beginning his itinerant career in 1766. He mainly con-

10 Gibbs (1960); McKie (1951); Coley (1969).   11 Priestley (1806), pp. 41–42.
12 Schaffer (1983, 1984, 1987); Money (1988–89).
13 Josiah Wedgwood to T. Bentley, 9 October 1776, quoted in Kramnick (1986), p. 12.
14 Accounts of Walker are given in the *D.N.B.* and *European Magazine and London Review,* 21 (1792), 411–413. The contents of his early lectures are outlined in Walker (1766, 1771).

cerned himself with astronomy and experimental physics until 1773, when he announced he would include "Dr. Priestley's late ingenious Discoveries in Fixed, Mephetic [*sic.*], and Inflammable Airs" in a course to be given at York.[15] Five years later he visited Priestley in London and was encouraged to lecture on the new airs at the Haymarket. According to a contemporary account, Walker's lectures on Priestley's work "served to spread the utility of, and excite further attention to, these discoveries."[16] Walker shortly moved to London, where he continued to give annual courses of lectures to great acclaim. He also toured the Midlands and the North in the early 1780s and spoke at some of the major public schools.

Walker's example was quickly followed by others. John Banks, a lecturer in Manchester who also toured to Bolton, Kendal, and Doncaster, was discussing Priestley's work on different kinds of air from 1775 at the latest.[17] Benjamin Donn (1729–1798), who had previously taught mathematical sciences in lectures and private academies in the southwest of England, traveled to Birmingham and Liverpool in 1779. For his courses there he advertised two additional lectures, called for by "special desire" on the "most valuable experiments of Dr. Priestley."[18] John Arden (1702–1791), who moved from Beverley to Bath about 1776, and came to know Priestley while he was living with Lord Shelburne, introduced a new lecture on different kinds of air into his own course at about the same time.[19] John Warltire (1739?–1810), a chemist who assisted Priestley in his experiments at Calne on dephlogisticated air, gave courses specifically devoted to the new discoveries in Bath, Bristol, and Birmingham in the late 1770s. He continued to include the work of Priestley, whom he dubbed "the Newton of the age," in his subsequent courses on chemistry and natural philosophy in Birmingham, Manchester, and Salisbury.[20] An-

---

15 Advertisement in *York Courant*, 16 February 1773.

16 *European Magazine and London Review*, 21 (1792), p. 411.

17 Musson and Robinson (1969c), pp. 107–109; Banks (1775), chh. 2, 5.

18 Donn's advertisements for his Birmingham course, in *Aris's Birmingham Gazette*, 12 April, 19 April, 26 April, and 3 May 1779, made no reference to pneumatic chemistry. But see Inkster (1980), p. 86, fn. 17, where references are given to his advertisements in the Liverpool papers that did promise to cover the subject. By the following year, pneumatic chemistry had made its way into a revised version of his lecture syllabus: Donn (1780), pp. 29–30. On Donn's career, see Robinson (1963).

19 On Arden, see: Musson and Robinson (1969c), p. 104; Gibbs (1960), pp. 113–114; Torrens (1979), pp. 224–225; Turner (1977), pp. 81–95, 150. For the inclusion of Priestley's discoveries in his lectures, see: James Arden (1782), pp. 55–57; and compare with idem. (1774); and with John Arden (1773).

20 On Warltire's career, see: McKie (1951); Coley (1969). For the inclusion of pneumatic chemistry in his courses, see advertisements in *Aris's Birmingham Gazette*, 4 November 1776, and 26 March 1781. Also note an advertisement for lectures in Salisbury, reproduced from the *Salisbury and Winchester Journal*, 13 November 1786, in Torrens (1979), p. 230. For Warltire's view of Priestley, see the letter from Josiah Wedgwood, 1779, quoted in Partington (1961–70), III, p. 327.

other man known to Priestley, the blind lecturer Henry Moyes (1749–1807), probably mentioned his chemical work in lectures delivered in Birmingham in 1782.[21]

It is not possible to say with certainty that Priestley's experiments on pneumatic chemistry were demonstrated by all of these lecturers. The available syllabuses of their courses usually do not reveal which experiments were actually shown, but only what topics were discussed. The evidence, however, suggests that attention was focused on those of Priestley's airs that were seen as having the greatest significance and utility – fixed, inflammable, nitrous, and dephlogisticated airs. Walker, Arden, Banks, and Donn all discussed fixed air, the first three stressing its role in artificial mineral waters and its general medicinal and antiseptic virtues.[22] These four lecturers also mentioned inflammable air and presumably showed its explosive properties. Walker, Arden, and Donn discussed dephlogisticated air and sketched out Priestley's model of the providential economy of nature, with air being phlogisticated by combustion or respiration and restored by vegetation or by contact with water.[23] Arden, Banks, and Donn also mentioned nitrous air, and followed Priestley in emphasizing its importance as a test of the goodness of a given sample of air. Arden showed this test with his own apparatus, and the others probably did the same.[24]

By reproducing his experiments effectively and in some cases quite quickly, the public lecturers had apparently solved the problems that Priestley had anticipated might hinder replication of his work. In the first volume of *Experiments and Observations* (1774), he had voiced these apprehensions:

... notwithstanding the simplicity of this apparatus, and the ease with which all the operations are conducted, I would not have any person, who is altogether without experience, to imagine that he shall be able to select any of the following experiments, and immediately perform it, without difficulty or blundering. It is known to all persons who are conversant in experimental philosophy, that there are many little attentions and precautions necessary to be observed in the conducting of experiments, which cannot well be described in words. ... [L]ike all other arts in which the hands and fingers are made use of, it is only *much practice* that can enable a person to go through complex experiments, of this or any other kind, with ease and readiness.[25]

21 On Moyes, see Harrison (1957). Advertisements occur in *Aris's Birmingham Gazette,* 30 September and 14 october 1782. Note that only very brief mention of pneumatic chemistry (with no specific reference to Priestley) occurs in Moyes (n.d.) and in idem. [1781].
22 Walker [1780?], and subsequent editions of this text; idem. [1796?]; James Arden (1782); Banks (1775, 1789); Donn (1780).
23 Walker [c.1785?], p. 39; James Arden (1782), pp. 55–57; Donn (1780), p. 29.
24 James Arden (1782), p. 56; Banks (1775), p. 36; Donn (1780), p. 30.
25 Priestley (1775–77), I, pp. 6–7.

The problem Priestley was pointing to was that written descriptions would inevitably be insufficiently comprehensive and precise to specify all the necessary conditions for an experiment to be reproduced. Although carefully written to maximize the replicability of the phenomena described, his texts might need to be supplemented by personal contact to inculcate practical skills. By the following year, this lesson had been borne upon him even more forcefully. In the dispute with Higgins the degree to which experiments had been reproduced identically was precisely the point at issue. In his second volume of *Experiments and Observations* Priestley therefore reiterated that skill in the use of apparatus, "cannot be communicated by any verbal instruction, but must be the result of much practice."[26]

Priestley's own experience of experimental work had brought this problem to his attention, and it also suggested the remedy. In 1774 he wrote to Caleb Rotheram about "the great difference between *seeing* and *reading*." He went on, "I have not yet found any person, though ever so good a philosopher, and who has read my papers ever so carefully, but is surprised to see me actually make the experiments."[27] Seeing the experiments could supplement the deficiencies of reading. Thus Priestley saw the importance of cultivating personal contacts and of providing practical experience in his own laboratory in order to get his experiments replicated and thereby make his facts into public knowledge.

Accordingly, he cemented personal links with public lecturers, persuading them to include his discoveries in their courses and providing practical training in the skills they would need to reproduce his phenomeno-technics. Priestley's connections with Walker, Warltire, and Arden are well documented. Walker was certainly known to him personally by 1774 at the latest. On his visit to Priestley in London in 1778, he was given apparatus with which to show the pneumatic discoveries in his lectures.[28] In 1780, Priestley invited him to visit Birmingham and offered to arrange a course of lectures to be given locally. On this occasion, Priestley wrote that his new laboratory was unfinished, so that "I cannot promise to shew you anything, but we shall have many things to talk over. . . ."[29] Warltire had been assisting Priestley at Calne during his work on dephlogisticated air in 1774–75. He was later recommended by his mentor as having "prepared a set of experiments which have given the greatest sat-

---

26 Priestley (1775–77), II, p. xxxiv.
27 Priestley to Caleb Rotheram, 31 May 1774, in Schofield (1966), p. 146.
28 *European Magazine and London Review*, 21 (1792), p. 411.
29 Priestley to Walker, 1780, quoted in Gibbs (1960), p. 115. For advertisements for Walker's 1781 courses of lectures in Birmingham, see *Aris's Birmingham Gazette*, 2 and 30 July, 13 August, 5 and 12 November 1781.

isfaction" to audiences at his lectures.[30] Warltire continued to correspond with Priestley through the late 1770s and gave courses in Birmingham several times after the latter moved there in 1780.[31] Arden was also known to Priestley during his Calne years. The two men showed one another experiments during this period, and both became founding members of the Bath Philosophical Society in 1779.[32] Arden shared Priestley's view of the importance of visual demonstration in the spread of experimental knowledge – in his lectures he wrote that the subject was "better understood by seeing the Experiments performed, than by a long and tedious Application to Books only."[33]

By forging personal connections with these lecturers, giving them practical advice on apparatus and techniques, and providing laboratory training in the relevant skills, Priestley furthered the public recognition of his discoveries. The phenomena he produced in his laboratory could not have become established outside without a lot of additional effort on his behalf. Literary work was part of the labor necessary to transport to other settings the complex of apparatus and techniques that had produced the new phenomena. But texts themselves were not enough without a supporting structure of material equipment and skilled practice. Not infrequently, Priestley complained about failures of other experimenters to replicate his operations precisely: "Whether it be owing to the persons being unaccustomed to these experiments, or some other cause, I find very few who make them with due circumspection."[34] The imparting of that due circumspection by training in the relevant techniques was part of the work Priestley found necessary to make his laboratory experiences into public knowledge.

This process also involved Priestley's audiences in the active appropriation of his work. They took notice of what he was saying because they could interpret it in ways that accorded with their own aspirations and interests. For lecturers, Priestley's work offered them resources that they could deploy to capture and hold the attention of middle-class audiences in provincial towns and cities. In a competitive market, they served their own needs by satisfying and building upon the expectations of their customers. Hence, we can discern elements of the attitudes that audiences

---

30 For Warltire's presence at Calne, see Priestley (1775–77), II, pp. 34–36. Priestley's recommendation occurred in his letter to Matthew Boulton, 28 September 1776, in Schofield (1966), p. 158.
31 See advertisements in *Aris's Birmingham Gazette*, 27 March, 10 April, 23 October 1780; 19 and 26 March, 2 April 1781; 20 May and 24 June 1782; and Money (1988–89).
32 Priestley to Benjamin Vaughan, 26 March 1780, in Schofield (1966), p. 181. On Priestley and Arden in the Bath Philosophical Society, see A.J. Turner (1977), pp. 81–95; Williams and Stoddart (1978), pp. 55–80.
33 John Arden (1773), p. [2].
34 Priestley to Matthew Boulton, [before November 1777], in Schofield (1966), p. 162.

brought to the interpretation of Priestley's discoveries by considering the language in which his work was presented in lectures and in the associated press advertisements.

As the number of lecturers and their geographical spread indicate, opportunities for this kind of cultural entrepreneurship were good in the late eighteenth century. Nevertheless, in the larger population centers, where the best markets lay, lecturers faced a highly competitive situation. They were obliged to advertise themselves assiduously, continually to refine their apparatus and to keep their repertoire up-to-date in order to compete with other lecturers and with alternative attractions. Benjamin Donn, advertising in the Birmingham press in May 1779, pointed out that there had been no fewer than seven courses on natural philosophy in the town since the previous August. The same newspaper gave evidence of such rival attractions as musical and theatrical performances and political debating clubs.[35]

As we have seen, lecturers in natural philosophy had been establishing their presence in provincial cultural life since the early decades of the century, whereas lecturers in chemistry had been fewer in number and (with one or two exceptions) addressed a smaller, more occupationally specialized audience. Priestley's experimental work stimulated new departures in both of these didactic genres. Operators who were already covering natural philosophy (Arden, John Banks, Donn, and Walker) were inspired to expand their range to include the chemistry of gases. Chemical pneumatics followed naturally on from physical pneumatics, demonstrated with the air pump, which had long been part of natural philosophy courses. In addition, those who were already devoting serious attention to chemistry (Warltire and Moyes) found that Priestley's experiments gave them attractive and dramatic demonstrations, which reinforced their claims about the extensive utility of the subject.

Priestley's discoveries were presented as evidence that science was continually advancing and that audiences were being kept in touch with the latest developments. In this way the lecturers exploited and reinforced their audiences' perceptions of involvement in ongoing cultural and social improvement. Banks introduced his 1775 lectures by asserting that "it seems providentially appointed, as some way necessary to the carrying on, or well being of society, that the sciences should be gradually improving."[36] As late as 1789 he was still describing Priestley's experiments on the restoration of vitiated air as recent discoveries, though they were then fifteen years old.[37] Warltire's 1776 lectures on air were said to comprise "the greatest Discoveries that have been made during the pre-

35 Aris's Birmingham Gazette, 3 May 1779.
36 Banks (1775), p. 6.   37 Banks (1789), p. 38.

sent Century." The contents were said to be "of undoubted Importance to Persons of every Rank and Profession."[38] Moyes, presumably antici-pating that the public might doubt that a blind man could get full infor-mation about recent developments, made it known in his advertisements that he was "particularly attentive to all late Discoveries, of which, by Means of an extensive Acquaintance, he never fails to have early Intelli-gence."[39] All the lecturers had to present themselves as closely in touch with the latest experimental work.

By emphasizing the recency and novelty of Priestley's experiments, lec-turers who took them up made a pitch for the attention of a fashion-conscious clientele. They frequently referred to their audiences' "refine-ment," "politeness," and "sensibility." The fact that women formed a significant proportion of these audiences emphasized their civilized and enlightened character, in the view of lecturers such as Walker.[40] The price of the subscription for a course, typically one guinea for men and half a guinea for women, clearly excluded a large proportion of the population, but the publicity held out the prospect of a relaxed and egalitarian am-bience for anyone who could afford to pay. Audiences were flattered by being told that they were participating in cultivated forms of civic life and were reminded how much they already knew about a topic. Walker introduced the subject of natural philosophy by proclaiming, "It is not necessary that a sensible and polite People should be informed how inti-mately these Branches are connected with every ennobling Quality of the Mind."[41]

That experimental demonstrations were to be shown could be pre-sented as a further tribute to the audience's refinement and discernment. Viewing experiments, the public was let in on the reasoning processes of philosophers themselves. Walker promised his audience that he would "not ask your assent to any Proposition before we have exhibited those capital Experiments that have established its Truth."[42] Demonstrations of Priestley's discoveries could be used to redeem this kind of promise.

If experiments were to be used in this way, they could not be presented as spectacular entertainment. Lecturers generally followed Priestley's lead by exhibiting nature in connection with an aesthetic of the sublime. Ex-periments might be striking or awe-inspiring, but they were explicitly not mysterious or magical. Warltire struck the right balance, advertising that one of his courses would be "very entertaining and instructive, being

38 Advertisement in *Aris's Birmingham Gazette*, 4 November 1776.
39 Advertisement in *Aris's Birmingham Gazette*, 30 September 1782.
40 Walker (1807), pp. v–vi.
41 "Exordium" from Walker's introductory lecture, printed in *Manchester Mercury*, 10 December 1771. Compare Warltire's advertisement in *Aris's Birmingham Gazette*, 20 May 1782.
42 Walker in *Manchester Mercury*, 10 December 1771.

upon Subjects continually occurring in common Life, and very full of the most striking Experiments."[43] Chemical lecturers did not want to appear boring or too specialized, but neither did they borrow the rhetoric of showy competitors like "Sieur Herman Boaz," a kind of philosophical conjuror who arrived in Birmingham at the same time as Warltire in April 1780, and promised a display "of so striking and singular a Nature, as to be past all human Conception, and [which] in an Age and Country less enlightened, would have appeared supernatural."[44] Nor was there very much scope for spectacular refinements of apparatus, such as were made much of in the teaching of subjects like mechanics, optics, and astronomy.[45] Priestley had presented his discoveries as modest, unspectacular achievements that required only the simplest of apparatus. Public lecturers confirmed that dimension of his accomplishment in their performances.

More important than spectacle in gaining an audience for chemistry lectures was utility. Lectures on different kinds of air could be used to get over the message that many practical arts depended on chemistry for knowledge of the properties of substances. Priestley's discoveries made this point particularly well for a general audience. Lecturers who introduced chemistry into public courses were not appealing to the interests of technical specialists. The air was not the subject of any particular existing art; nor was it tied to any particular locality, as were most of the chemical trades at this time.[46] The air, after all, was everywhere. What was therefore shown by reference to uses such as the medical virtues of fixed air, or the use of inflammable air in balloons, was the utility of chemical philosophy for humanity as a whole. Philosophers were seen to have captured and tamed a ubiquitous element, and to have made the benefits available to all.

43 Advertisement in *Aris's Birmingham Gazette*, 26 March 1781.
44 Advertisement in *Aris's Birmingham Gazette*, 10 April 1780; quoted in Money (1977), p. 131, which also provides background information on lectures and other cultural events in the Birmingham region at this time. As Money points out, although Warltire appears initially to have feared that he would lose customers to Boaz, and accordingly offered to replace his advertised lectures with a more attractive course, he seems in the end to have secured enough subscribers. There is more information on Boaz in Money (1988–89) (part II), pp. 77–79.
45 For examples of lecturers drawing attention to their apparatus, see advertisements by Walker (2 and 30 July, 5 November 1781), Donn (12 April 1779) and Warltire (10 April and 23 October 1780), in *Aris's Birmingham Gazette*. Equipment for chemical experiments was inevitably more mundane than Walker's "extensive and superb" Eidouranion or "transparent orrery," Warltire's "new Solar Apparatus," or Boaz's "Grand Thaumaturgick Exhibition of Philosophical, Mathematical, Sterganographical, Sympathetical, Sciateroconatical and Magical Operations."
46 Warltire announced a number of courses on practical chemistry which were designed to cover the arts, particularly mining and metalwork trades, practised in the Birmingham area. See *Aris's Birmingham Gazette*, 10 April 1780, 24 June 1782, 27 September 1784.

In this way, those who attended scientific lectures were made to feel involved in the progress of natural knowledge, which was presented as part of continuing material and moral improvement. Lecturers were appealing to their audiences' own experiences and expectations of enlightenment to make Priestley's work meaningful to them. Because the provincial middle-classes saw themselves gaining in gentility and "refinement," they did not wish to be taught technical aspects of the practical arts but to be shown in broad terms how natural philosophy and chemistry were stimulating economic and social progress. For similar reasons, though they wished to be shown experiments, they wanted to be assured that such entertainment was "rational" and instructive and that their reasoning powers were also being engaged.

The language that lecturers used in their advertisements exploited the cultural values of prospective customers to excite an interest in chemical experiments on air. It thus indicates at least some of the expectations that the lay public brought to the interpretation of Priestley's work. Despite the difficulty of recovering it from the historical record, the interpretive activity of his audiences deserves to be recognized as part of the process of the constitution of Priestley's discoveries. His status as a discoverer was conferred upon him by a public that, thanks in large part to entrepreneurial lecturers, took a significant interest in his work. Recognition of his accomplishment derived not only from Priestley's own efforts, but also from those of his audience.

Priestley's part of this process comprised literary and social activity to communicate descriptions of his experiments and the practical skills that would enable them to be repeated. In his view, philosophers were merely fortunate observers, the tools of divine providence, whose discoveries could be reproduced by anyone who chose to do so. It was vital that experimental phenomena should be made manifest to public audiences as part of the overall program of enlightenment. Communication of experiments within the philosophical community and to a wider public was nothing less than a moral imperative. As Money has put it:

... since the moral perfection of Man was to be achieved through an ever increasing knowledge of nature, ... all men should know all things equally. For Priestley, all knowledge was one; ... and the way to it lay, not through theory, but through the candid presentation of experiential facts to the opened minds of all. ... In such a Baconian instauration, public performance was not a mere adjunct of natural philosophy, useful if properly controlled, but its very essence.[47]

But Priestley's science did not become accepted simply because he wanted it to be. It did so because it was produced in a receptive context, in which patrons, allies, and audiences could be readily recruited. Contrary to the

47 Money (1988–89), pp. 69–70.

suggestions of the diffusion model, Priestley's audiences also played a part in constituting his discoveries. The work of the experimental lecturers and the understanding of their courses by those who attended them helped to make his discoveries into established public knowledge.

The birth of pneumatic medicine

The most significant realm for application of Priestley's chemical discoveries was medicine. His work encouraged the development of programs of research and therapy that extended the emphasis of Enlightenment medical discourse on the environmental causes of disease. For many decades, variations in the qualities of mineral waters in different locations had been surveyed in an attempt to match their chemical composition with their supposed therapeutic benefits. When "airs" or "spirits" were discovered in certain waters, this field of enquiry promised to connect with meteorological investigations of the effect of air on health and disease. As the program of pneumatic medicine developed, further scrutiny was devoted to possible aerial factors in the origins of ill-health, such as miasmata and effluvia from marshes, cesspits, and graveyards. Improvements in the water supply and ventilation in towns, hospitals, prisons, ships, and other locations were called for as part of broader schemes for social reform.[48]

Priestley's work stimulated the further development of this field in several respects. The discovery of new gases suggested novel therapeutic methods. The new airs could be administered by respiration, by application to the skin, or by dissolving them in water. Preparations of solutions of gases connected pneumatic medicine with previous work on the medical benefits of mineral waters.

In this section I shall discuss how the medical uses of the new airs were developed. In contrast with previous work on mineral waters, pneumatic medicine was mostly articulated by an identifiable group of medical practitioners. In each area of application, we find Priestley cultivating connections with those who were making use of his work. Many of his medical followers shared his ambitions for social and political change; all sought to reform medical practice along the lines suggested by experimental science. Priestley's pneumatic chemistry provided them with important tools to advance their campaign.

Numerous links of friendship and association in local scientific societies connected the pneumatic practitioners with one another. Many of

48 Relations between pneumatic medicine and public health reform in the eighteenth century have been discussed in: Jordanova (1987); Riley (1987); and Corbin (1986). My treatment also owes much to Schaffer (1990).

them were patronized by the influential Sir John Pringle, a lifelong enthusiast for public health reform. Under Pringle's benign guidance, pneumatic medicine was a favored realm for experimental research. It enjoyed
a degree of approval and encouragement from the metropolitan scientific
establishment that it was subsequently to forfeit.[49]

Divisions, however, also occurred among those who applied Priestley's
work in this field. As we shall see, a number of medical practitioners and
instrument makers competed with one another for the credit of having
designed an efficient apparatus to impregnate water with fixed air. Commercial rivalry and claims to intellectual property entered into these disputes, as they did in the field of eudiometry. Although Priestley himself
opposed the influence of monetary rewards on experimental science, it
proved impossible to detach improvements in therapy and instrumentation from involvement in the commercial markets for medical treatment
and manufactured goods. In pneumatic medicine, as in many other areas
of social life, traditional ideals of virtue and the moral economy were
being undercut by the increasing prevalence of commerce.[50]

Priestley's discoveries of new types of air, when first announced in the
early 1770s, fed directly into a branch of medical research that was already flourishing under Pringle's patronage. Pringle had worked on the
problem of putrefaction as a cause of disease since 1750, when he published his *Observations on the Nature and Cure of Hospital and Jayl
Fevers*. In 1752 he produced a book on diseases of the army, which identified the dangers of field hospitals and camps pitched next to putrid
marshes. In both works he pointed to the connections between local meteorological conditions and the outbreak of epidemics. His interests also
embraced techniques to combat scurvy, then a scourge of the British navy
in its voyages of exploration and colonial expansion. Pringle viewed scurvy
as a putrid disease, and throughout his career upheld the importance of
good ventilation in ships and buildings to prevent the buildup of putrid
effluvia. He directed attention at certain "antiseptics," particularly fermenting vegetable substances, which he viewed as antidotes to putrefaction.[51]

David Macbride (1726–1778), a former naval surgeon and Secretary
of the Dublin Medico-Philosophical Society, took up Pringle's theme of
fermentation as a counteractive to putrefaction in his *Experimental Essays* (1764). Macbride identified a cementing principle, responsible for
the cohesion of bodily tissues, the loss of which led to putrefaction. Putrid diseases could be countered by antiseptics, which restored the ce-

49 Singer (1948–50).
50 Pocock (1985). Schaffer (1990) discusses analogies in this connection between notions
of aerial and moral virtue.
51 Singer (1948–50), esp. pp. 229–247. Carpenter (1986), chh. 3, 4.

menting principle. Among these, Macbride identified fixed air, a "subtle gas" that was capable of extinguishing fire and suffocating animals, but could also preserve the body from putrefaction. Fermenting materials that released fixed air were therefore likely to alleviate or cure putrid diseases. Macbride thus endorsed the practice of issuing fresh vegetables and malt to sailors as antiscorbutics. His methods, along with others, were judged to have been applied with success by Captain James Cook in his first voyage to the South Pacific in the late 1760s.[52]

Macbride also connected his perspective with the work of Joseph Black on fixed air. Black had shown that calcareous earths, such as magnesia alba and lime, were strongly bound to fixed air. When they were deprived of it they became caustic and soluble in water. Upon restoring their fixed air they became mild and insoluble. To Macbride, this confirmed that fixed air was the cementing principle and was acidic in nature. His *Experimental Essays* made Black's work widely known for the first time, by connecting it with the prevalent medical concern with the problem of the environmental causes of putrid diseases.[53]

Macbride's work gained widespread recognition in the 1760s and 1770s, not least from Pringle, who used his period as President of the Royal Society to encourage the adoption and development of his theory of putrefaction. Pringle also endorsed Priestley's work as a contribution to this field of research, bringing it to the attention of many experimentally inclined medical men. He engineered the award to Priestley of the Society's Copley Medal in 1773, in acknowledgment of the work reported in the paper "Observations on different kinds of air," in the *Philosophical Transactions* of 1772.[54]

In his presidential discourse on the occasion of the award, Pringle placed Priestley's discoveries in a lengthy historical tradition of work on the physical, chemical, and medical qualities of air. After briefly invoking studies of its physical properties by the great seventeenth-century natural philosophers Galileo and Boyle, he focused on more recent investigations of its chemical nature. Pringle's aim was to describe a series of steps leading progressively toward the recognition of the identity of fixed air and its isolation from certain compounds and mineral waters. Thus Stephen Hales, who had shown in his *Vegetable Staticks* (1727) how "air" could be released from various solids and liquids, was taken to have anticipated the discovery of fixed air, though Pringle admitted that he had not recognized it as chemically distinct from normal atmospheric air. In 1765 the Cumberland physician William Brownrigg (1711–1800) had more accurately identified the "aerial spirit" present in certain mineral

52 Macbride (1764). On the development of Macbride's work, see Scott (1970); and, for background, Carpenter (1986), pp. 65, 76–77.
53 Macbride (1764), ch. 5.   54 Singer (1948–50), pp. 248–250; McKie (1961).

waters with the mephitic exhalations known to miners as "choak-damp." Black and Henry Cavendish had further explored the release and capture of fixed air by solid earths and by water and had recorded its acidic and mephitic properties.[55]

Pringle's Copley Medal address was a *pièce d'occasion*, designed to construct a pattern out of more than five decades of experimentation that would point teleologically toward his current concerns. As a historical account it might be criticized for ignoring the importance of apparatus in the development of pneumatic medicine. It can be argued that the significant innovation was not so much the idea of a mineral spirit in certain waters, as the development of the instrumentation that allowed this spirit to be extracted, collected, and distinguished from common atmospheric air. It was the invention and refinement of the pneumatic trough, an apparatus to hold glass jars inverted over water, that provided the technical means to collect and manipulate gases. Without this apparently simple piece of technology, eighteenth-century pneumatic chemistry would not have been possible.[56]

The refinement of the pneumatic trough, from Hales through Brownrigg to Cavendish and Priestley, went along with the development of chemical tests to differentiate the gases collected. The possession of a technology for testing airs distinguished Brownrigg's achievement in 1765 from those of Hales, or indeed from his own earlier work. In a paper read in 1741, Brownrigg had anticipated being able to collect and differentiate the exhalations of various mineral waters, but had not possessed the technical means to do so that he later acquired.[57] Similarly, Cavendish's application of these techniques to the analysis of water from the well in Rathbone Place in 1767 contrasts with the description of the same water in 1756 by Charles Lucas. Lucas had realized that the water lost its acidic spirit on standing exposed to the air, but could do nothing to capture and identify the volatile substance.[58]

Pringle's review of the progress of pneumatic medicine also ignored the skepticism that surrounded attempts to identify aerial components of waters. Hales referred to "this much neglected volatile Hermes, who has so often escaped thro' [chemists'] burst receivers, in the disguise of a subtle spirit."[59] Despite improvements in the techniques for capturing and displaying spirits, they remained evasive and insubstantial entities. Shaw admitted that his mineral spirit was an incoercible substance, ". . . so that catching it in a Bladder, bringing it over a Retort, or supplying an

55 Pringle's address is reprinted in McKie (1961), pp. 2–12.
56 Parascandola and Ihde (1969); Badash (1964).
57 Brownrigg (1765), pp. 236–237, 240. Cf. Eklund (1976), pp. 535–536; Coley (1982), pp. 131–133. On Brownrigg, see Wood (1948–50, 1948–51).
58 Cavendish (1767); Lucas (1756), pp. 142–145.   59 Hales (1727), p. 180.

Assembly Room with it, seem all Chimerical."[60] The continuing incoerc-
ibility of these spirits through the 1740s and 1750s reinforced doubts as
to the utility of chemical analysis for revealing the essential virtues of
mineral waters, doubts that were strengthened by disputes over other
supposed constituents of waters such as sulfur.[61] Even as late as 1789,
after the work of Brownrigg, Cavendish, and Priestley had exhibited the
aerial component of mineral waters to most observers' satisfaction, the
eccentric physician James Graham maintained that:

The soul, spirit, or specific qualities of most natural productions mock, scorn,
and elude, the detection of even the most ingenious and most accurate chemist,
physician, or philosopher, that ever did, or ever will exist.[62]

Pringle's Copley address disregarded all such doubts, presenting instead a story of the progressive revelation of factual truth. Priestley's
discoveries were displayed as the natural (almost inevitable) culmination
of a succession of preliminary experimental work. Priestley had found a
way of artificially impregnating water with fixed air, so as to give it the
virtues already found in certain natural mineral waters. Artificial mineral
waters containing this antiseptic quality could thus be produced at sea,
and Pringle hoped they would prove useful in the treatment of scurvy.

Priestley's other discoveries were also related by Pringle to their potential medical applications. Nitrous air, for example, had proved to be a
powerful antiseptic – in fact, a more powerful one than fixed air. Its use
in tests for the goodness of atmospheric air was also very valuable. Pringle quoted Francis Bacon in this connection: "These are noble experiments that . . . teach men to choose their dwelling for their better health."
Finally, Pringle drew attention to Priestley's discovery of the role of vegetation in restoring air corrupted by combustion or respiration. He endorsed Priestley's view of this phenomenon as a crucial element in the
providential economy of nature:

From these discoveries we are assured, that no vegetable grows in vain, but that
from the oak of the forest to the grass of the field, every individual plant is ser-
viceable to mankind; if not always distinguished by some private virtue, yet making a part of the whole which cleanses and purifies our atmosphere.[63]

By linking Priestley's work with that of Macbride on the antiseptic
qualities of fixed air, Pringle pointed the way to its assimilation by medical practitioners. Some were already using fixed air therapeutically. In
appendices to the three volumes of *Experiments and Observations on
Different Kinds of Air*, published between 1774 and 1777, Priestley assembled testimonials as to the medical efficacy of the gas. A number of

60 Shaw (1735), pp. 10–11.   61 Coley (1982), pp. 133–137; Rousseau (1967).
62 Graham (1789), p. 19. Compare also Randolph (1745), sigs. a3r–a4r.
63 McKie (1961), pp. 7, 9, 10–11.

the pioneers of pneumatic medicine contributed to these appendices, describing how they had used fixed air in the expectation, inspired by Macbride, that it would constitute an antidote to putrefaction. Thomas Percival, a physician at the Manchester Infirmary, made patients with pulmonary diseases respire the air, having heard of the success of this method from the Birmingham physician William Withering (1741–1799).[64] Adam Walker applied fixed air externally to ease an inflammation of his wife's breasts.[65] More frequently it was taken internally, dissolved in water or some other drink, or was injected directly into the gut via a clyster inserted into the anus. The former method was used by Percival to treat a case of the stone and by William Hey, a Leeds surgeon, to treat putrid diseases. Both of them used Priestley's apparatus for dissolving the air in water, wine or beer.[66] Clysters were used by Percival, Hey, Matthew Dobson of Liverpool, and John Warren of Taunton to treat various putrid diseases.[67]

Most, though not all, of the treatments were said to have been successful. Percival described one case where clysters of fixed air had produced an alleviation of the putrefaction, but the patient died from having lost too much blood.[68] A more typical case was that described by Warren: The patient, a gentleman aged twenty-three "of great temperance, and of a good constitution," had had a nervous fever for ten weeks when he began to show signs of putridity. Peruvian bark was given and his drinks were ordered to be impregnated with fixed air. Nevertheless, the fetor of his breath and body became very objectionable and that of his stools "absolutely intolerable," so that nurses could only be persuaded with difficulty to attend him. Clysters of fixed air were ordered to be injected every three to four hours, and within five days the putridity and fever had completely vanished. Warren concluded, "He is now perfectly recovered, and a living miracle of what fixed air, under Divine Providence, is capable of effecting on the human oeconomy, in cases of the worst and most putrefactive nature."[69]

By printing accounts such as this, Priestley was cementing his connections with a group of self-consciously reformist medical practitioners, who were applying the latest scientific advances to therapeutic practice. William Hey (1736–1819) had worked his way up in the course of his career from surgeon's apprentice to senior surgeon at the Leeds Infirmary and was subsequently to become mayor of the city for two terms of office. He was President of the first Literary and Philosophical Society in

64 Priestley (1775–77), I, pp. 300–314.   65 Priestley (1779–86), I, pp. 464–466.
66 Priestley (1775–77), I, pp. 288–291, 304; II, pp. 360–368.
67 Priestley (1775–77), I, pp. 292–299, 305–311; II, pp. 369–379.
68 Priestley (1775–77), I, pp. 305–310.
69 Priestley (1775–77), II, pp. 375–379 (quotation on p. 379).

Leeds from 1783 to 1786. Hey had known Priestley since 1769 and was proposed by him for election as a Fellow of the Royal Society in 1775. Throughout his career, Hey promoted experimentation with new therapeutic methods; he lobbied for the foundation of the Infirmary in the 1760s and was still pursuing an experimental approach with electrical therapy in the 1790s.[70]

The career of Thomas Percival (1740–1804) paralleled in a number of respects that of Hey, with whom he was acquainted. A graduate of the Dissenting Academy at Warrington, Percival was a moving force in the foundation of the Manchester Infirmary and of the city's Literary and Philosophical Society, of which he was President almost continuously from its foundation in 1781 until his death. He had already espoused the therapeutic application of fixed air and held out the promise of substantial medical improvements arising from Priestley's researches in the second volume of his *Essays Medical and Experimental* (1773).[71]

Matthew Dobson was an Edinburgh graduate who became a physician in Liverpool, an FRS, and a member of the Manchester Literary and Philosophical Society. He published his own *Medical Commentary on Fixed Air* (1779), which advocated its use against putrid diseases and cited cases from his own experience and from those of Hey, Percival, and Macbride.[72] Other medical men who explored the promise of pneumatic therapy and who showed an interest in Priestley's work included John Warren, Taunton physician and Edinburgh graduate; John Haygarth (1740–1827), then physician at the Chester Infirmary; Nathaniel Hulme (1732–1807), a London physician and former naval surgeon who had published on scurvy; William Falconer (1744–1824), a former colleague of Haygarth's at Chester, who moved to practice in Bath in 1770 and subsequently published a revised edition of Dobson's work; and the Manchester surgeon-apothecary Thomas Henry (1734–1816), who was to follow Percival as President of the Literary and Philosophical Society.[73]

In most cases, these men were of relatively humble social origins, but they shared substantial ambitions and carved out successful careers in the medical world. Hey, Henry, and Hulme began their education as apprentices to apothecaries or surgeons. Of those who were graduates, only Haygarth had an Oxbridge degree, and he had also studied under William Cullen at Edinburgh; most had pursued their studies in Scotland

70 D.N.B.; William Hey, "Medical and Surgical Cases" (12 vols., MS, Leeds University Library, Special Collections), esp. vol. 10.
71 D.N.B.; Percival (1772–73), II, pp. 76–79, 81–85.
72 For Dobson, see: [Simmons] (1783); Dobson (1779).
73 Haygarth, Hulme, Falconer, and Henry are listed in the D.N.B.; Warren is in [Simmons] (1783).

or on the Continent. Dobson, Falconer, Hulme, and Warren had obtained degrees at Edinburgh. Percival, who studied for a while in the Scottish capital following his years at Warrington, took his MD at Leyden. Dobson, Percival, and Haygarth remained loyal correspondents of Cullen, to whose teaching they paid fulsome tributes.[74] The group was united by numerous links of friendship and by collaboration in the pursuit of shared intellectual interests. In 1770, Percival reported to Cullen that he, Dobson, Haygarth, and John Bostock were meeting regularly in Warrington, "for our mutual improvement." Percival and Haygarth worked together collecting mortality statistics and public health data for Manchester and Chester.[75] The group subsequently played a central role in the Manchester Literary and Philosophical Society, which elected Priestley, Falconer, and Haygarth as Honorary Members.[76]

Pneumatic medicine was explored by these practitioners as part of a cluster of interests centered on the role of the environment in the maintenance of good health. Experimental therapeutic methods connected with subjects such as sanitary reform and urban planning, via notions of good and bad air and the dangers of putrid effluvia. The pneumatic practitioners saw themselves as prophets of a forthcoming scientific enlightenment in medicine, when the causation of disease by material elements in the natural environment would be recognized and controlled. They believed, as Adam Walker explained in his *Philosophical Estimate of the Causes ... of Unwholesome Air* (1777), that it was proper for an age of "philosophy and enlarged sentiment" to locate in the air the causes of diseases that a previous age of "religious tyranny" had ascribed to divine retribution.[77]

As well as promoting pneumatic therapy by publishing descriptions of it, Priestley contributed an important therapeutic technique – namely, his method of impregnating water with fixed air. He developed the apparatus for dissolving fixed air in water in 1772, at the urging of his patrons the Duke of Northumberland and Sir George Savile. At their suggestion, the method was presented to the Lords of the Admiralty, who were sufficiently impressed to command that trials should take place at sea on the effectiveness of impregnated water against scurvy. On the orders of the Admiralty, the apparatus and instructions for its use were taken on Captain Cook's second voyage to the South Pacific.[78]

74 Thomson (1832–59), I, pp. 460–462, 627–628, 635–636, 638–642.
75 Percival to Cullen, 11 November 1770, in Thomson (1832–59), I, p. 635; Percival (1774–76); Haygarth (1778). Haygarth's paper in particular makes extensive use of Priestley's concepts of vitiation and restoration of air.
76 Nicholson (1923–24), p. 105. According to correspondence between Percival and Haygarth in 1785 (in Percival (1807), I, pp. ci–ciii), members of the Manchester Society donated 50 pounds to support Priestley's experimental work in that year.
77 Walker (1777), p. 29.   78 Gibbs (1965), pp. 80–82; Coley (1984), pp. 36–37.

The same year, 1772, Priestley published a description of his method and apparatus in a short pamphlet.[79] There he gave a detailed description of a device that was simple to construct from easily-available materials. A diagram showed the arrangement: Chalk and acid were mixed in a bottle, and the resultant fixed air was led out by a tube via a bladder and bubbled through water in another bottle inverted over a trough. The bladder allowed the operator to control the flow of fixed air and to exert extra pressure to facilitate its dissolution. It also gave flexibility to the connection between the vessels so that the bottle containing the chalk and acid could be shaken. The pamphlet was published, Priestley wrote, in the public interest, without hope of reward, and accordingly was made comprehensible "even to those who have no previous knowledge of the subject."[80]

Medical men were clearly part of the audience to which Priestley thought the invention would appeal. In his pamphlet he reviewed the research of Brownrigg, Pringle, and Macbride, which had suggested that certain mineral waters gained their curative powers from the fixed air they contained. Priestley's device could produce artificial waters that also possessed these medicinal virtues. To emphasize the medical potential of the innovation, he cited therapeutic applications of it already made by Percival and Hey.[81]

By making his invention public, Priestley had secured credit for it, but he had also surrendered the right to control its subsequent development. As might have been expected, improvements were rapidly announced by other experimenters, and variations of Priestley's apparatus were put to use in contexts that he had not envisaged. In a paper read at the Royal Society in December 1774 and published the following year, the Bristol apothecary John Mervin Nooth suggested a refined version of the apparatus. This comprised three glass vessels mounted one on top of another. The bottom one contained chalk and acid, producing fixed air that passed up through a valve into water in the middle vessel. The top vessel was an overflow reservoir that maintained pressure on the water and fixed air in the middle one. Nooth claimed that his version, though admittedly more complex in construction, required less skill from the operator than had Priestley's method: "[A]lthough in the hands of the doctor, the apparatus was sufficiently convenient, it must be confessed, that the conduct of the process required more address than generally falls to the share of those that are unaccustomed to such experiments."[82]

In a forty-page section of the second volume of his *Experiments and Observations on Different Kinds of Air* (1775), Priestley made a detailed response to Nooth's claims, which he clearly felt constituted a challenge

79 Priestley (1772b).   80 Priestley (1772b), esp. pp. 3, 5–6.
81 Priestley (1772b), pp. 2, 17–21.   82 Nooth (1775) (quotation on p. 59).

to his credit for the invention. He ridiculed Nooth's assertion that the bladder in his apparatus could taint the water unpleasantly, responding that even those with the most delicate sense of taste had never made this complaint to him. He also insisted that Nooth was mistaken in believing that methods of dissolving fixed air in water had been known before his 1772 pamphlet. Although retrospectively such methods might seem an obvious step from the researches of Brownrigg, that step had not in fact been taken. Priestley was insistent that the credit for the discovery was his. He had always made clear that he was prepared to do without monetary reward; what he wanted was what "is strictly my due, *the sole merit of the discovery.*"[83]

Priestley also rejected Nooth's statement that particular skill was required to use his version of the device, remarking that "many persons, altogether unused to experiments, have, to my knowledge, succeeded in it very well, . . . without any assistance besides what they got from the printed directions."[84] This was a vital point for him. Although he was prepared to allow that Nooth's device might be more useful in certain contexts – for example, in a private household – he resented the accusation that his own method required superior expertise or resources. This accusation was a challenge in the terms of Priestley's own ethics of public science, because it suggested he was failing to make his discoveries as widely replicable as possible. On the contrary, he insisted that it was his own method that was simpler, less expensive, and hence more widely reproducible. His device, he said, used "no other vessels but such as are in constant family use." He explained that:

My apparatus costs little or nothing, because no vessels are made for the purpose; and both the chalk and acids are made to go as far as possible. . . . Whereas Dr. Nooth's method requires a peculiar and expensive apparatus, and more waste is unavoidable in the use of it.[85]

Here Priestley was reasserting the moral imperative underlying his vision of the public culture of science. For knowledge to be established and rendered useful it had to be widely replicated. Lack of resources should not be allowed to hinder those who wished to confirm discoveries or put them to use. Perhaps unfortunately, it was not possible to enforce this ideal in the circumstances in which his method was acquiring its utility. Having yielded his discovery up to the public, Priestley found that the ideal of accessibility and simplicity was being undermined. Instead, the

---

83 Priestley (1775–77), II, pp. 263–303 (quotation on p. 276). Compare p. 298: ". . . after the discovery of the first method of accomplishing this end, all subsequent methods may be called little things; and they may be endlessly diversified, without any great claim of merit."

84 Priestley (1775–77), II, p. 297.   85 Priestley (1775–77), II, pp. 273, 297.

device was adapted and refined to serve numerous particular interests in a variety of different contexts.

Thus Thomas Henry and John Haygarth developed variants of the apparatus for use in ships at sea. In a tract dedicated to the First Lord of the Admiralty, Henry described how his machine would produce large quantities of pure sparkling water, at little cost and without skilled operators. He promised that the innovation would be a significant contribution to solving the problem of scurvy.[86] Aiming for a different kind of customer, Adam Walker devised a mechanism by which water and fixed air were agitated together in a wooden barrel. By this means, large quantities of impregnated water could quickly be produced for audiences at public lectures.[87]

Other versions of Priestley's apparatus and artificial mineral waters produced by his process were marketed as consumer products. Nooth's machine was further modified by J.H. Magellan, and the device was manufactured by William Parker of Fleet Street, a glassware and instrument maker. The Nooth-Magellan-Parker machine became the most widely used means of water impregnation; it was said in 1777 that more than 1,000 such instruments had already been sold to customers as far afield as the East Indies.[88] Under the name "gasogene," this apparatus became a familiar feature of many middle-class homes until well into the nineteenth century. As an alternative, it was possible from 1793 to purchase artificial mineral waters manufactured on a large scale by a high-pressure process by J.J. Schweppe in his factory off Cavendish Square.[89]

Medical men also developed versions of the device for various therapeutic applications. Thomas Percival and William Falconer used impregnated water in experiments to dissolve urinary calculi. Thus Priestley's discovery yielded a successor to many other supposed "lithontriptics," or solvents for "the stone," which had been investigated in the course of the century.[90] Some doctors asserted rights of invention over variations or products of Priestley's process. Nathaniel Hulme, for example, preached the virtues of a solution of fixed air that he prepared by mixing acid and alkali salts in water in a sealed container. According to the title of his tract, the solution constituted, *A Safe and Easy Remedy . . . for the Relief of the Stone and Gravel, the Scurvy, Gout, &c.*[91] Priestley's friend William Bewley popularized an impregnated solution of fixed alkali (sodium carbonate), which became known as aqua mephitica alkalina or Bewley's

---

86 Henry (1781). See also Henry and Haygarth (1785).
87 Walker's device is shown in Gibbs (1965), p. 70.　88 Magellan (1777), esp. p. 3.
89 Coley (1984), p. 37; Gibbs (1972), p. 204.
90 Percival (1772–73), I, pp. 71–80; Falconer (1776). For previous work on lithontriptics, see: Coley (1984), pp. 37–38; Viseltear (1968).
91 Hulme (1778).

julep and was subsequently recommended by Falconer, Dobson, and Beddoes.[92] As with analyses of the contents of mineral waters, authors in this field were not slow to claim individual credit for the application of scientific knowledge to medicine.

Opportunities for doctors to assert their expertise in this area were, however, limited by the ease with which patients could use artificial mineral waters to treat themselves. Priestley appears to have favored such self-medication, to judge from his decision to print a letter by Withering in the second volume of his *Experiments and Observations on Natural Philosophy* (1781). Withering gave extensive details of his own version of the water impregnation machine, which he claimed was cheaper and easier to make than those currently available. Readers who wished to prepare their own mineral waters for domestic medical use were advised to construct their own apparatus, rather than purchasing one of Parker's elaborate glassware models.[93]

The key to the utility of Priestley's device for impregnating water with fixed air was the variety of modifications that could be made to it and the range of settings in which it could consequently be deployed. Variations were used by public lecturers and medical practitioners; they could be made by commercial instrument makers or by unskilled craftsmen. The solutions of fixed air that the apparatus produced were used by doctors or for self-medication by unskilled patients and sailors. Thus Priestley's invention entered into the commercial market for consumer goods and medical therapies. It was deployed to entertain audiences at lectures and to attract the patronage of naval authorities searching for ways of preventing scurvy at sea.

The success of the apparatus resulted from the wide range of individuals who adopted it to serve their own interests – medical practitioners, entrepreneurial instrument makers, public lecturers, and patients themselves. The device proved robust enough to be used in many places: Dr William Ewart was reported to have taken it on a voyage to China, and it was also noted as a household accessory of a typical female hypochondriac in the 1790s.[94] This robustness was clearly related to the instrument's orientation toward a relatively unproblematic end. Although its therapeutic benefits might be contested, it was widely agreed to work insofar as it produced a solution of fixed air. This fairly straightforward measure of the device's utility made it adaptable for use in a large variety of circumstances.

92 Dobson (1787); Coley (1984), p. 37. Matthew Boulton was reported to be taking the "aerated alkali" in 1783 mixed with a little warm milk (James Watt to Joseph Black, 5 May 1789, in Robinson and McKie (1970), pp. 176–177).
93 Priestley (1779–86), II, pp. 389–394.
94 Beddoes [1794a]; *The Medical Spectator*, 1 (1792), 19.

In this context, Priestley's conception of a public scientific culture in which property rights and privilege would not restrict access to knowledge and its applications seemed an irrelevant ideal. His invention was successful precisely because it served many interests and could be made many people's property. Looking back upon his career he continued to regard this as a happy discovery, one that had proven an unqualified benefit to mankind as a whole. But it is clear that it did not acquire its extensive utility in the way he had originally envisaged. Priestley gave his invention to the public with a show of benevolent liberality, but its actual utility was constructed through the efforts of many others as they turned his device (and its descendants) to their own purposes and used it to advance their own interests.

The analysis of air

The development of research in eudiometry provides another example of experimenters' extending Priestley's discoveries in ways he could neither anticipate nor control. Techniques for measuring the relative purity of samples of air, derived from his original nitrous air test, were applied in many different contexts. But, in contrast with the water-impregnation machine, the outcome in this case was not a successful invention, but a failed instrument. Eudiometrical apparatus and procedures diversified and became more complex. Individual practitioners staked their claims for the superiority of their own methods, and as a result no consensus emerged. In the absence of agreed-upon techniques, replicable results were not achieved. Instead, disputes about the propriety of various methodological refinements multiplied. The nitrous air eudiometer failed to live up to the original expectations vested in it as an instrument.

This failure can be read as another sign of the limitations of Priestley's moral order for the public culture of science. His ideal of simplicity and economy in the design of experiments was frequently referred to, but did not enable eudiometrical practitioners to decide between competing methods. Priestley himself was unable or unwilling to create agreement by imposing new standards of rigor or precision. And when Henry Cavendish tried to do so, his attempts were without success. Some experimenters, apparently relying on Priestleian values of simplicity and individual autonomy, resisted being disciplined in more rigorous practices and refused to accept refinements in instrumentation.

Priestley had immediately noted the potential of the nitrous air test when he described it in the first volume of his *Experiments and Observations* (1774). There he noted that the contraction in volume of a mixture of sample air and nitrous air suggested the possibility of measuring the goodness of the atmosphere in various locations. Samples of air from each place could be situated on a scale between pure dephlogisticated air and completely phlogisticated (respired) air.[95] For Priestley, the test was a direct measure of the degree of phlogistication of a sample. "In fact, it is phlogiston that is the test," he wrote.[96] As such, it was taken to confirm and embody the phlogiston theory, from which Priestley had derived explanations of a number of phenomena observed in the course of his research on gases.

The test was not, however, tied inseparably to phlogiston. As we shall see, it proved capable of appropriation and reinterpretation by antiphlogistic chemists, and continued to be applied well after the overthrow of the phlogiston theory. Its appeal was based on the promise of an instrumental means of assessing the purity of air. It appeared to be an objective substitute for Priestley's previous techniques for assessing aerial virtue – for example, by using the sense of smell or by timing how long mice could survive in a measured volume. As with all instruments, however, its utility was to depend on subsequent application and interpretation and on the contexts of practice in which it was taken up.[97]

Priestley's test was very soon replicated in Italy, where research on the purity of the atmosphere took off rapidly. There Felice Fontana (1730–1805) and Marsilio Landriani (c. 1746–c. 1816) developed different forms of apparatus for performing the test on samples of air. Landriani's "eudiometer" rather than Fontana's "evaerometer" became the standard name for this kind of instrument, but the rivalry between the two men continued long after the matter of nomenclature had been settled. Each claimed priority in the invention of a practical eudiometer, and accusations of plagiarism, dissimulation, and quackery flew between them.[98]

Both Fontana and Landriani were trying to use their instruments to enlist the support of patrons, respectively the Grand Duke of Tuscany and the Austrian ruling house in Milan. They promoted eudiometry as part of ambitious programs of environmental and social control, of the kind that appealed to the rulers of many states in *ancien régime* Europe. Fontana discussed in detail the possibility of linking the quality of air in certain places with the productivity of agriculture, or the incidence of famine and epidemic diseases. If such relations could be established, surveillance of the atmosphere in a certain territory would yield the power

95 Priestley (1775–77), I, pp. 111–116.   96 Priestley (1775–77), I, p. 208.
97 The general claim that scientific instruments require particular contexts of practice to sustain their use has been persuasively argued by sociologists of modern science. See: Collins (1975); idem. (1985), ch. 2; Latour (1987), ch. 3; Pinch (1986), esp. pp. 212–214.
98 The Italian context is discussed in Schaffer (1990); and in Riley (1987).

to control many aspects of human life. Eudiometry would play its part in government-directed programs of improvement.

Fontana published numerous designs of eudiometers to attract the attention of his patrons. Elegance of craftsmanship and convenience of display were the main qualities stressed in these designs, many of which were never put into production.[99] Landriani similarly projected a number of versions of the instrument. One of the few that was made was presented to Count Firmian, a minister of the Austrian government of Lombardy, in 1775. The device had every appearance of an accurate and reliable scientific instrument, with an elaborate cut glass and metalwork construction and a precise scale. Proponents of eudiometry prophesied that devices like this would soon become as common as the thermometers and barometers they were clearly designed to resemble.[100]

Both Fontana and Landriani reported to English pneumatic chemists on their "eudiometrical tours" of the Italian countryside and both sent samples of their instruments to Priestley. Landriani contributed a letter to the third volume of Priestley's *Experiments and Observations*, describing his travels around Italy to assess the relative salubrity of the air in different locations. In general, he confirmed the opinions of local people as to where the best air was to be found – for example, at the tops of mountains. Fontana wrote to Priestley about his surveys in Tuscany, measuring the purity of the air near marshes, on hilltops, and at the coast.[101]

In England also, eudiometrical field trips caught on in the late 1770s. Unsurprisingly, given his interest in the environmental causes of disease and the therapeutic value of certain types of air, Sir John Pringle was an early enthusiast for eudiometry, which he had in effect foreseen in his 1773 Copley Medal address.[102] Priestley himself did not attempt to gain support for this type of research and did not construct any elaborate eudiometrical instruments, but he did persuade acquaintances like Boulton and Percival to supply him with samples of air from various manufacturing areas for comparison with air from the countryside.[103]

Most eudiometrists, on the other hand, went out-and-about for themselves. William White, a York physician with an interest in problems of public health, published a survey of his city and its environs in the *Philosophical Transactions* in 1778. He measured the purity of the air in his

99 On Fontana, see Knoefel (1979, 1984).   100 Landriani (1775), Gerardin (1778).

101 Landriani's letter is in: Priestley (1775–77), III, pp. 380–381. Fontana's was published as Fontana (1779). Knoefel (1984), p. 141, records that the Italians sent copies of their instruments to Priestley; their presence in his laboratory is noted in McKie (1956–57).

102 Singer (1948–50); McKie (1961), p. 9.

103 Priestley to Matthew Boulton [not dated], in Schofield (1966), pp. 161–162; Priestley (1778–86), I, pp. 269–273.

home, in the streets, on bridges over the River Ouse, and in marshy areas. He confirmed the claims of Pringle and others as to the dangerous mias-mata that emanated from marshes and made them very undesirable sites for habitation or military camps.[104] On a trip to London in 1779, Fontana assessed the quality of the air in city streets, in rural Islington, and on the top of the dome of St Paul's Cathedral. He failed to find anything to choose between them, though Adam Walker had claimed two years previously that the air was actually better at the level of the street than at the rooftops.[105] In 1780 the Dutch physician Jan Ingenhousz (1730–1799) reported to Pringle on eudiometrical measurements made in the course of a journey down the Thames from London, across the North Sea, and around the Low Countries. His unsurprising conclusion was that the air was generally of better quality at sea or at the coast and that inland air was liable to become particularly corrupted in hot and sultry weather.[106]

These investigators differed markedly in the degree to which they were concerned about standardizing eudiometrical measurements. To White, the eudiometer was an unproblematic instrument, capable of yielding a direct measure of the contamination of the air by phlogiston, which he viewed as the principle of putrefaction. It was therefore more reliable as an aid to public health research than the sense of smell.[107] William Falconer wrote to Cullen in 1777 that he had constructed a eudiometer on Priestley's principles and found it to "answer exceedingly well, and re-ducible to a standard."[108] Ingenhousz and Fontana, on the other hand, realized that procedures would have to be standardized if the instrument were to yield meaningful measurements.[109] In 1779, Fontana wrote:

The least variation of circumstances causes very great variations in the results of the experiments, . . . so that the purest common air would appear to be noxious and phlogisticated air; and the dephlogisticated air would appear less good, and even noxious. . . . I have not the least hesitation in asserting, that the experiments made to ascertain the salubrity of the atmospherical air in various places, in different countries and situations, mentioned by several authors, are not to be depended upon; because the method they used was far from being exact, the elements or ingredients for the experiment were unknown and uncertain, and the results very different from one another.[110]

Fontana's own method was designed to supply a standard apparatus and a regular procedure for using it, which could rectify this confused situation. The "Fontanist" method, as it was described for example in Ingenhousz's *Experiments upon Vegetables* (1779), gained several adher-

---

104 White (1778, 1782).   105 Fontana (1779), pp. 447–450; Walker (1777), p. 32.
106 Ingenhousz (1780).   107 White (1778), pp. 197, 200–201, 204–205.
108 Falconer to Cullen, 8 June 1777, in Thomson (1832–59), I, p. 648.
109 Ingenhousz (1780), p. 362; cf. idem. (1776), esp. p. 258.
110 Fontana (1779), pp. 445–446.

ents, in Britain and on the Continent.[111] Ingenhousz himself, and the experimental philosopher Tiberius Cavallo (1749–1809), developed eudiometrical techniques that were proclaimed to be refinements of Fontana's method. But this failed to produce general agreement among practitioners of eudiometry. Other experimenters upheld alternative procedures, and results continued to show dramatic disparities. As further sources of possible errors in measurement were identified, confidence in the accuracy of any of the reported results declined.

According to Ingenhousz, Fontana's method was capable of circumventing the problems that had caused other investigators to despair of the possibility of accurate eudiometrical measurements. Provided the procedure was carried out "constantly in the same way," he asserted, "we may with as much precision judge of the degree of purity of the common air, as we are now able to judge of its degree of heat and cold by a good thermometer."[112] Fontana's apparatus comprised a large graduated tube, suspended in a tall narrow water bath, and a smaller tube with a sliding top to provide a standard measure of air. The large tube was filled with two measures of atmospheric air initially. Then three measures of nitrous air were added, one by one. After each measure of nitrous air was introduced, the tube was shaken for thirty seconds, then allowed to settle for sixty seconds, and the total volume in the tube was then measured.

Ingenhousz went on to list no fewer than twenty possible sources of error that the experimenter must guard against while performing the operation, including accidentally communicating heat to either of the tubes from his hands, failing to standardize the pressure when volumes of gases were being measured, allowing water or air-bubbles to adhere to the sides of the tubes, or failing to account for variations in temperature and pressure during the experiment. Even alerted to all these possible errors, the novice experimenter would need considerable practice before he could expect to achieve acceptable results. Ingenhousz assured his readers that Fontana himself had spent "some years assiduous labour before he reduced this method to that degree of accuracy which it has now acquired in his hands."[113]

Despite all these precautions, no consensus was formed among experimental philosophers as to the validity of the Fontanist method. Priestley himself, in a letter written in March 1780, expressed surprise that, notwithstanding his enumeration of twenty possible sources of error, Ingenhousz had not included the variability caused by differences in the supply

---

111 Ingenhousz (1779), esp. pp. xxi–xxii, 149–184, 276–287. Other Continental Fontanists are mentioned in Knoefel (1979), p. 32 and fn.
112 Ingenhousz (1779), p. xxii.
113 Ingenhousz (1779), pp. 160–180 (quotation on p. 160).

of nitrous air. The need to ensure consistency in the quality of this air was commonly stressed in Landriani's eudiometrical procedures. Priestley wrote that: "... it is altogether without reason that the Abbé Fontana (in Dr Ingenhousz) pretends that the measure of good and bad nitrous air comes to the same thing in his method of applying the test. I am astonished, and provoked, at the little care with which some persons make experiments, and the confidence with which they report them."[114]

Charges of lack of care and overconfident claims haunted every eudiometrist, Fontanist or otherwise. In 1777, Magellan had denounced all the instruments of Fontana and Landriani as "more or less liable to considerable objections." He favored a simple graduated tube, like that originally used by Priestley, into which the atmospheric air and the nitrous air were introduced after having been mixed in another vessel. This method, he thought, "the readiest of all, whenever no great nicety is required in observations of this kind."[115] In his 1781 *Treatise on the Nature and Properties of Air*, Tiberius Cavallo attacked Landriani and Magellan on behalf of the followers of Fontana. Magellan's "mistakes," as specified by Cavallo, included failing to allow a standard length of time for the gases to mix and settle, not compensating for differences of pressure on the gases, and not admitting more than one volume of nitrous air. Cavallo went on to describe a version of Fontana's method, claiming that its accuracy in skilled hands "can hardly be believed by persons who have not observed it."[116]

Magellan defended himself against Cavallo's strictures in the third edition of his *Description of a Glass-Apparatus* (1783). He maintained that the method he had adopted from Priestley had the advantage of allowing the airs to mix thoroughly, before their combined volume was measured, whereas Cavallo and other Fontanists "throw one air after the other" into the large graduated tube, without allowing them to mix. Admittedly, the Fontanists produced fairly consistent results, whereas Magellan confessed that he usually had to average several measurements that varied quite widely; but the apparent uniformity was actually an illusion. Magellan compared the experiments of the Fontanists to the conjuring tricks of contemporary showmen and illusionists: "Is not this similar to the tricks of Jones, Comus, Breslaw, and Katterfelto, who make things appear what in reality they are not?"[117]

Magellan was protesting against what he saw as an unnecessary elab-

---

114 Priestley to Benjamin Vaughan, 26 March 1780, in Schofield (1966), p. 181. Landriani himself stipulated that the quality of the nitrous air used was a significant variable (see Landriani (1775)).

115 Magellan (1777), pp. 15–47 (quoted from the third edition, Magellan (1783), pp. 31, 33).

116 Cavallo (1781), pp. 315–327, 328.

117 Magellan (1783), pp. 54–69 (quotations on pp. 67–68).

oration of methods that removed eudiometrical procedures from the range of competence of an unskilled practitioner. Experiments would then appear as a kind of prestidigitation, producing results by methods that would be obscure to the ordinary observer and not repeatable by another experimenter. Against this, Magellan invoked Priestley's authority and stressed the importance of simplicity: "Simplicity in philosophical experiments, and cheapness of the instruments required for their processes, are two of the most desirable circumstances, *caeteris paribus*, in the investigation of natural phaenomena."[118] The Lunar Society chemist James Keir had already stipulated in 1779 that the best eudiometers were the simplest ones.[119]

Paradoxically, Cavallo also endorsed the Priestleian ethic of simplicity. He agreed that "Complex instruments are not only expensive, and subject to be easily put out of order, but they occasion very frequent mistakes."[120] The Fontanist program, however, was based upon the necessity for relatively complex apparatus and procedures and the recognition of multiple sources of errors. It sought to avoid those errors by establishing acceptance of a particular device and imposing quite rigid discipline upon the experimenters who were to use it. Hence the emphasis by writers such as Cavallo on the need to adopt a precisely uniform procedure, on the importance of training and acquiring skills, and on the need to defer to those with greater expertise. Cavallo enjoined the novice experimenter, if he felt inclined to challenge the validity of the method, to "defer his judgment till after several repeated trials, and even after consulting those persons who, being acquainted with the subject, may either point out the cause of his mistake, or suggest to him some new way of trying the questioned fact."[121]

Fontana's apparatus provides an example of a scientific instrument that failed to find the support needed to sustain its use. To produce acceptable results with it, operators would need to discipline their behavior in line with a rigorous procedure and to acquire a substantial expertise in recognizing possible sources of error. Members of the experimental community in Britain in the late 1770s and early 1780s were not prepared to accept these conditions. On the contrary, Magellan and other disciples of Priestley continued to point out errors in the Fontanist method itself, such as those stemming from variability in the quality of nitrous air. Magellan's reaction to Cavallo suggested that, for upholders of Priestley's values of simplicity and economy, the complexity and expense of the Fontanist apparatus and the high level of skills required to operate it inclined them not to accept its validity.

118 Magellan (1783), pp. 54–55 fn.   119 Keir (1779), p. xxii.
120 Cavallo (1781), p. 354.   121 Cavallo (1781), pp. 333, 350–351, 365.

The dispute between Magellan and Cavallo also reflected a deeper rift opening up in the social structure of English science. Pringle was succeeded as President of the Royal Society in 1778 by Sir Joseph Banks, who rapidly consolidated his position to become the most powerful patron of science in Britain. By 1783, Banks was involved in a personal feud with Magellan, whose role as an independent channel of communication with Continental savants challenged his centralizing ambitions and whose attempts to thwart his institutional reforms at the Royal Society he resented.[122] Cavallo, on the other hand, had become one of Banks's clients, being nominated by him to receive a research grant funded by a bequest to the Society in 1781.[123] It appears likely that Banks supported Cavallo in the dispute with Magellan. The recruitment and disciplining of observers, which Cavallo's Fontanist method of eudiometry would require, was in line with the techniques Banks himself used to enlarge and consolidate his "learned empire." Certainly, Banks used his position to block the publication of eudiometrical research by Arthur Young, which was transmitted to the Society by Magellan in 1783.[124]

Banks's capture of the commanding heights of English science provided little encouragement for continuing the program of eudiometrical research. Though Cavallo continued to advocate nitrous air eudiometry through the 1790s, Banks had no other clients willing to work in this field.[125] Pringle did continue to patronize eudiometry, until his death in 1782, though his influence in the scientific community was now on the wane. Ingenhousz's career had owed a lot to Pringle; and it was perhaps partly a sense of loyalty to his former patron that led him stubbornly to continue to defend the Fontanist method. In various Continental publications continuing into the late 1780s, Ingenhousz ignored the rising tide of skepticism and reported further improvements in the design of nitrous air eudiometers and in the accuracy of their results.[126]

In Britain, however, faith in the program of nitrous air eudiometry had virtually collapsed by this time. Priestley himself had not been willing to recruit and discipline eudiometrical observers. Nor, apparently, was Henry Cavendish. Although Cavendish's paper, "Account of a New Eudiometer," published in the *Philosophical Transactions* in 1783, did suggest new and more rigorous standards of apparatus and method, the suggestions were not taken up by others in the experimental community. Cav-

122 Carter (1988), pp. 199–200; Richard Kirwan to Banks, 8 April 1788, in Dawson (1958), p. 493. On Magellan's friendship with Priestley, see: Benjamin Franklin to Priestley, 4 May 1772, and Priestley to Alessandro Volta, 5 August 1779, in Schofield (1966), pp. 101–102, 174–175; and Guerlac (1961), pp. 37–39, 51, 56–57.
123 *Philosophical Transactions*, 71 (1781), 509.
124 Betham-Edwards, pp. 150–151 (I owe this reference to Schaffer (1990)).
125 Cavallo (1798), pp. 3–5.
126 Ingenhousz (1779) was dedicated to Pringle. See also: Singer (1948–50), pp. 143–144; Ingenhousz (1785); idem. (1787), esp. I, pp. 195–197, 214–215, 226.

endish proposed that the accuracy of Fontana's method could be improved upon by adopting a new arrangement. A measured quantity of the atmospheric air sample was to be added by a steady stream of droplets to a graduated tube containing nitrous air over water. The graduated tube was to be shaken constantly while the sample air was being added. Cavendish described a vessel for measuring out standard volumes of the gases and an inverted funnel for producing the stream of droplets. He also identified a further, hitherto unnoticed, source of error in the experiment. Different kinds of water varied greatly in their ability to absorb the nitric acid produced when the gases mixed. This factor alone, he claimed, gave rise to much greater variation in the result than any differences in atmospheric composition that had yet been recorded. Hence, pure distilled water should always be used in the apparatus.[127]

Cavendish's paper represented a further attempt to standardize the practices of eudiometrists and hence to facilitate replications of their measurements. His recommendations would require a considerably enhanced level of expertise. He called for craftsmanship in the production of apparatus, skill in its operation, precision in measuring a range of variables, and some mathematical ability in compensating for changes in temperature and pressure. Even if all experimenters were to acquire these competences, however, it was not apparent that the results would be worthwhile. Cavendish explicitly abandoned the original aims of the eudiometrical project, recording that he had not found any differences between the purity of air in different locations that were larger than the expected margins of experimental error. In many cases, the sense of smell remained a more precise indicator of impurities in the air than any eudiometrical instrument.[128]

That Cavendish should have ended up endorsing the sense of smell, which eudiometry was originally supposed to replace, indicates how little was recognized as having been achieved by nearly ten years' work in the field. Those most forcefully arguing for greater sophistication in methods, such as Cavendish himself, tended to endorse a skeptical conclusion: The differences in the quality of the atmosphere that had been reported were artifacts resulting from one or another methodological failing in the experiments. This was the view that Priestley himself had been drawn toward as early as 1779.[129] In the same year, Fontana had sounded a distinctly skeptical note about the degree of atmospheric diversity that might be uncovered by eudiometrical research:

When all the errors are corrected it will be found, that the difference between the air of one country and that of another, at different times, is much less than what is commonly believed, and that the great differences found by various observers are owing to the fallacious effects of uncertain methods.[130]

127 Cavendish (1783), esp. pp. 115–117.   128 Cavendish (1783), pp. 134–135.
129 Priestley (1779–86), I, pp. 270–271.   130 Fontana (1779), p. 447.

In a rather defensive conclusion, Fontana declared it possible that some differences in atmospheric quality might be established and could prove significant for human life. But the original ambition of uncovering substantial variations in atmospheric composition and correlating these with medical effects had clearly been abandoned. Disciples such as Ingenhousz might continue to pursue the eudiometrical grail, but the quest had been deserted by many of its leading figures. In 1783, another of these, Jean Senebier of Geneva, augmented the now considerable skepticism. Experiments reported in his *Recherches sur l'Influence de la Lumière Solaire* showed that mixtures of gases over water could continue to diminish in volume for weeks. His conclusion was that these observations demonstrated "the inexactitude and perhaps the uselessness of eudiometers which require the use of nitrous air."[131]

As Senebier was well aware, eudiometrists were already beginning to explore alternative methods. In 1777, Alessandro Volta had pointed out the possible eudiometrical applications of the reaction in which a sample air was mixed with "inflammable air" and ignited by an electric spark. Again, the diminution in volume could be taken as a measure of the purity of the sample air. Volta refined his "sparking eudiometer" in the following decade, and the device was also taken up by Cavendish, Antoine Lavoisier, and others. It became central to arguments surrounding Lavoisier's antiphlogistic theory, when it was discovered that water was formed after ignition of the gases. In Lavoisier's terms, the reaction was interpreted as the formation of water by combination of hydrogen (the inflammable air) with oxygen in the sample. Versions of Volta's apparatus continued to be used by Lavoisier and other chemists to assess the oxygen content of particular gaseous mixtures.[132]

Other methods were also developed for this purpose. In the early nineteenth century, Humphry Davy, T.C. Hope, and W.H. Pepys all developed new types of eudiometers, exploiting different chemical reactions to give measures of oxygen content. Apparatus that exploited the nitrous air reaction continued to be used in chemical education. F.C. Accum advertised Fontana's eudiometers for sale in connection with his course of chemical lectures in the 1800s, and the device was also mentioned in James Parkinson's popular *Chemical Pocket Book* (1799). Parkinson noted, however, that the efforts of Priestley, Ingenhousz, and Fontana had produced only inconclusive results, because of their failure to account properly for the various sources of error.[133] Notwithstanding the occasional appearance of nitrous air eudiometers in didactic settings, it is clear that

131 Senebier (1783), pp. 297–322.
132 Volta to Priestley, 10 December 1776, in Priestley (1775–77), III, pp. 381–383; Partington (1961–70), III, pp. 326–328.
133 Parkinson (1807), pp. 41–42; Accum (1804b), p. viii.

the instrument was no longer in active use for research. The connections with medical reform programs, which had originally sustained the aim of determining differences in atmospheric quality in different locations, had been severed. The project had failed to achieve anything like its originators' aims and had unraveled in disillusionment and disappointment.

The failure of nitrous air eudiometry can be ascribed in part to the problems of building a network of disciplined practitioners to use the instrument. The program had originated in Priestley's discovery of the nitrous air test, but it developed in a way that he made no effort to control. Like all his discoveries, the test was presented as a providential accident – a surprise occurrence. He turned it over to the experimental community as a potentially useful tool, but one for which he sought little credit and accepted no responsibility. It was characteristic of Priestley that he never sought to impose his authority upon the eudiometric debates by proposing more rigorous procedures, greater accuracy in measurement, or more refined apparatus. Perhaps he was wary of the compromises with political or commercial power that an attempt to build a network of eudiometrical informants would require.

On the other hand, Fontana and his followers, including Ingenhousz and Cavallo, did attempt to regulate eudiometrical research in this way. They realized that improved apparatus was needed and that it should be sustained by more rigid disciplining of observers so that experimental practices could be made uniform. The Fontanist program failed, however, a failure that stemmed in part from the opposition of experimenters who declined to be drilled in Fontanist procedures. Magellan and Keir reacted as if they felt that the Fontanist method contradicted Priestley's ethic of scientific culture. Elaborate apparatus could imply expense and needless complexity; disciplining observers suggested the cultivation of elite expertise. Priestley, in contrast, had always stressed simplicity, cheapness, and the equality of all observers before the providential self-revelation of nature. As we have seen, these were crucial to his understanding of the purpose of natural philosophy, which was capable of serving the aim of public enlightenment only if its discoveries were replicated, demonstrated, and experienced as widely as possible.

Whereas the key to the success of Priestley's invention for impregnating water with fixed air lay in the variety of modifications and contexts in which it could be deployed, much the same factors were implicated in the collapse of research in nitrous air eudiometry. Eudiometrical apparatus was widely circulated: Models were sold by commercial instrument makers such as William Parker,[134] and presented by experimental philos-

---

134 Magellan (1777), p. ii. Ingenhousz (1785), p. 342 fn., mentions some Continental suppliers.

ophers such as Fontana and Landriani to their patrons. But the accompanying procedures were not standardized. Some experimenters resisted the rather weak attempts that were made to discipline practices of measurement, and hence the instrument could not be made to work in a uniform and sustained way.

The fates of the water-impregnation apparatus and the eudiometer both signify limitations of Priestley's model of the public culture of science. He had considerable success in communicating his factual discoveries to various audiences, but did not succeed in enrolling practitioners in disciplined practice of the kind that would be necessary to sustain more sophisticated instrumentation and procedures. Appealing to the moral ideals of economy, modesty, and liberality, Priestley ignored the roles of self-interest and social ambition in propelling scientific accomplishment. Although his ethical vision of scientific culture appealed to some who shared his social aspirations, it could not encompass or harness individual competition in a milieu that was strongly structured by relations of patronage and by the prevalence of the market. We shall see that pneumatic chemistry and medicine faced further troubles in the late 1780s and 1790s, when Priestley's followers were confronted with the dual challenge of Lavoisier's alternative mode of chemical practice and a turning of the political tide against their Enlightenment ideals.

5

The coming of the
Chemical Revolution

Hence we see, in religious, moral and metaphysical controversies, as well
as in those that relate to physical objects, converts are seldom produced by
the direct force of right reasoning; but in an indirect method, from the
repetition of their adversaries' arguments, with a view to confute them. It
seems necessary, when the mind is thus misled, that an equal force of habit
should be practically generated in favour of the opposite side of a controversy, before impartiality can be produced.

William Nicholson, *Dictionary of Chemistry* (1795)[1]

We have seen how Joseph Priestley and his followers established a distinctive form of public experimental science in Enlightenment England.
We now move on to consider the fate of that science in the last decades
of the eighteenth century, when it faced the challenge of revolutionary
innovations in chemical doctrine advanced by Antoine Laurent Lavoisier
(1743–1794). For many years, Priestley and such allies as Richard Kirwan, James Keir, and William Nicholson resisted Lavoisier's radical claims.
Underlying the lengthy controversy about chemical theory can be de-
tected concerns about the way chemistry was to be constituted as a public
activity.

Lavoisier's novel claims were made in the form of assertions as to the
results of certain experiments; they were perceived, however, as theoretical interpretations rather than as straightforward facts. His British opponents denied that Lavoisier's claims were facts at all, by casting doubt
on his form of chemical practice. His work was seen as exemplifying
novelties of instrumentation and experimental method, and implying a
new model of the relationship between experimenter and audience. In
the course of the controversy, therefore, issues were raised concerning
the propriety of using certain apparatus, the legitimacy of drawing conclusions from a small number of experiments, and the acceptability of a
particular structure of relations within the experimental community. Lavoisier's substantive claims became enmeshed in arguments over the ways

---

1 Nicholson (1795), p. vi.

in which meaning could be ascribed to chemical experiments and the context in which they should be performed.

Priestley, who strenuously resisted the proposed conceptual changes, also resented what he saw as moral dangers inherent in the method and social profile of Lavoisier's chemistry. For him and for other British chemists, debates about Lavoisier's chemistry were thus connected with currently contested questions of the proper form of public science. Issues of chemical language, for example, could not be separated from wider moral and political questions.

Lavoisier's theory and its reception in Britain

To understand the British resistance to Lavoisier, one must grasp that his self-proclaimed revolution in chemistry was an event of many dimensions. To agree with Lavoisier's claims was to adopt radically new views of many chemical entities and operations. Elements and compounds, combustion and calcination, the nature of gases and acids, were all to be conceived of in fundamentally new terms. Acceptance of the new system also entailed acceptance of new procedures of experimentation, new instruments, and a new nomenclature for chemical compounds.

Such a fundamental change in perspective could not be established overnight, and in Britain the debate was particularly lengthy and wide-ranging. The leading English chemists were well aware of Lavoisier's work by the mid 1770s, but many continued to resist his theory until the end of the following decade. Priestley remained defiantly opposed to the French system until well after his emigration to the United States in 1794. His persistence and resourcefulness in finding new experimental results with which to challenge the French claims dragged the debate out considerably longer in Britain than in Continental countries. In Germany, by contrast, controversy surrounding Lavoisier's theories was relatively brief. It began in the 1790s and was then concluded fairly quickly, as German chemists concentrated their attention on a designated crucial experiment that resolved the debate.[2]

Being lengthier in Britain, the dispute ramified into consideration of a larger number of issues. Priestley's sustained opposition to Lavoisier raised a series of wide-ranging questions concerning the methods of science, its discursive representation, and its social function. As John McEvoy has shown in a series of illuminating discussions of the Priestley-Lavoisier dispute, questions of method and style and issues concerning the social context in which chemistry was practised were not far beneath the sur-

2 Hufbauer (1982), chh. 7, 8.

face.[3] Issues of fact could not be disentangled from questions of experimental procedure, instrumentation, language, and so forth. Concentrating on these matters enables us to connect the British response to Lavoisier's chemistry with the themes already developed in this book. We can discern how the debate was coextensive with discussion of the proper forms of communication in science and its legitimate role in civic culture.

In the discussion that follows I shall therefore be picking a selective path through the thicket of the Chemical Revolution. I shall be concentrating on the ways in which Lavoisier sought to make his experiments persuasive and on British responses to them. My discussion will rely on a substantial body of research on Lavoisier's achievement, which has recently expanded our knowledge of its range, its temporal development, and its reception in different national contexts. In recent historiography, the focus of attention has widened out from 1772 (dubbed by Henry Guerlac "Lavoisier's crucial year") to periods both earlier and later in his career. At the same time, historians have become more aware of his work on a wide range of theoretical and practical problems. Whereas the Revolution used to be identified solely with the overthrow of the phlogiston theory of combustion and calcination, it is now seen as embracing important innovations in theories of acidity, respiration, and chemical composition.

For example, it has been argued that Lavoisier's concept of the gaseous state crucially shaped his work on heat and gases. René Fric and J.B. Gough have traced the origins of this work to notes made in 1766 on an essay by the German chemist, J.T. Eller.[4] In these notes, Lavoisier speculated on the nature of air and proposed that it might be a compound of water with fire. Subsequently he articulated the view that each gas was a distinct chemical entity, owing its gaseous form to a (temporary) combination with a certain amount of the substance of heat – caloric. Such a view, as Bernard Langer has pointed out, differentiated Lavoisier from Priestley, who believed all airs shared basic chemical properties and were distinguished only by being combined with more or less phlogiston.[5] In this respect, as Evan Melhado has noted, Lavoisier's views were also different from those of most other French chemists.[6] The majority of them adhered to the doctrines of G.E. Stahl, according to which heat or fire (identified with phlogiston) entered into chemical union with other bodies, thereby altering their properties. Lavoisier, a chemist with a lifelong interest in the other physical sciences such as astronomy, physics, meteorology, and geology,[7] aligned himself with the more physical theory, according to which gases were regarded as normal matter that had been

3 McEvoy (1975), esp. ch. 9; idem. (1988a, 1988b).
4 Fric (1959); Gough (1968, 1971).   5 Langer (1971).   6 Melhado (1983, 1985).
7 Guerlac (1975); McKie (1952); Rappaport (1967).

disaggregated by the addition of heat without concurrent chemical change. Gough, together with Robert Siegfried and R.J. Morris, has traced the importance of this perspective through to the 1770s and 1780s, when Lavoisier undertook his mature work on pneumatic chemistry and calorimetry.[8] Following a similar track, Robert Kohler has suggested connections between his ongoing interest in the nature of elastic fluids and the work on combustion and calcination that has long been regarded as his major achievement of those decades.[9]

From this perspective, Lavoisier's renowned criticisms of the phlogiston theory might appear as almost incidental products of an overall drive to understand the nature of gases. Such an interpretation would, however, run the risk of giving an unbalanced assessment of his overall goals.[10] Notwithstanding the orientation of this recent research, the period c. 1772–1777 still appears crucial in Lavoisier's career. At the beginning of this period, drawing on his studies of calcination of metals and the combustion of sulfur and phosphorus, he established the fixation of air in these processes. He went on to follow up Priestley's experiments on dephlogisticated air, repeating the operation for producing it by heating red mercury calx. In 1776–1777 he determined that this "purest part of the air" was what combined with solid substances in the course of their calcination or combustion. He also disclosed its role as the portion of the atmosphere consumed in respiration.[11] He further connected this pure air with the theory of acidity, a long-standing concern of his chemical research, by ascribing to it the power to make substances acidic.[12] It was for this reason that pure air or eminently respirable air was subsequently named oxygen – the acid-generator. By the end of 1777, Lavoisier was in a position to launch his first comprehensive attack on the phlogiston theory from the standpoint of an alternative theory of combustion, respiration, and acidity.

Persuading others was, however, a different matter. Lavoisier still had no converts among qualified specialists by the early 1780s. Leading French chemists, such as Claude-Louis Berthollet, Antoine Baumé, and Louis Bernard Guyton de Morveau, accepted some of his findings, particularly the fixation of pure air in combustion and calcination, but continued to see a role for the release of phlogiston in those processes.[13] Crucial for converting Berthollet and Guyton, as well as for recruiting Antoine François de Fourcroy to his campaign, was Lavoisier's experimental work in the early to mid 1780s. In 1782–1783 he worked in collaboration with

8 Morris (1969, 1972); Siegfried (1972).   9 Kohler (1972, 1975).
10 This point has been well made by Perrin (1986, 1989).
11 This aspect of Lavoisier's work has been considerably illuminated by Holmes (1985), who has shown how close the interactions were between his different areas of interest.
12 Crosland (1973); LeGrand (1972).   13 LeGrand (1975).

the physicist Pierre Simon de Laplace on heat exchanges in chemical reactions.[14] They employed a new instrument, the ice calorimeter, to measure the heat released from combustion, calcination and respiration, all
of which they saw as releasing caloric from atmospheric oxygen and fix-
ing the basis of oxygen in chemical combination. In 1785, Lavoisier staged
his most impressive experimental demonstration, the public analysis of
water into its constituent gases (inflammable air and pure air – hydrogen
and oxygen) and their synthetic combination to form water again.[15]

After the 1785 demonstration, Lavoisier's persuasive task was rendered easier among French chemists, as a number of the most prominent
members of the community swung over to his support. With their backing he launched a more overt and confident attack on the phlogiston
theory – the essay 'Réflexions sur le phlogistique," published in 1786.
This was followed by a collaborative work, *Méthode de Nomenclature
Chimique* (1787), in which he was joined by Guyton, Fourcroy, and Berthollet as coauthors. In this essay, Lavoisier's chemistry assumed its most
comprehensive form to date; it was presented as a system of nomenclature by which all compounds could be named so as to indicate their composition.[16] In 1789, Lavoisier put forth a further systematization of his
perspective in a textbook, *Traité Élémentaire de Chimie*, designed to play
a part in courses of chemical education. Armed with this weapon, his
disciples continued to fight to establish his system of chemistry in French
educational institutions, even after his death in 1794.

The process by which Lavoisier's chemistry became accepted in countries outside France has recently been illuminated by a number of studies.[17] Hufbauer's work gives a particularly detailed picture of the reaction to Lavoisier's theory in Germany.[18] Unfortunately, the reception of
the new chemistry in Britain has not so far received this kind of comprehensive study, though Carl Perrin, Arthur Donovan, and John Christie
have laid the groundwork for an understanding of the situation in Scotland.[19] They have all drawn attention to the specific circumstances of the
communication of French chemistry to the Scots – the personal links that
were important channels of intelligence and the institutional settings in
which the new perspective was received. They also note that the Scots
had developed their own distinctive outlook on chemical theory, having
articulated concepts of heat and gases that had already been applied to
many phenomena in natural philosophy.

Perrin's study emphasizes institutional and personal factors linking
Scottish and French chemists – for example, the career of James Hall,

14 Guerlac (1976).   15 Daumas and Duveen (1959); Duveen and Klickstein (1954).
16 Siegfried and Dobbs (1968); Crosland (1962), chh. 5, 6.
17 Snelders (1988); Lundgren (1988); Gago (1988).   18 Hufbauer (1982).
19 Perrin (1982); Donovan (1979); Christie (1979).

who spent time in Paris in the early 1780s and subsequently promoted the French theory in Edinburgh. The role of Joseph Black is also shown to have been significant. His conversion to Lavoisier's theory, announced in 1789, though possibly based upon an incomplete understanding, was a signal to all Scots that the French chemistry could no longer be dismissed. Christie's account brings into the picture aspects of the wider intellectual context that shaped the Scottish reaction to Lavoisier. He describes a phlogistic renaissance in Scotland in the 1770s, when the concept of phlogiston (frequently identified with the Newtonian ether) was applied to explain numerous natural phenomena. Adair Crawford and P.D. Leslie were using phlogiston in explanations of animal heat at this time, while James Hutton was applying it to geological processes. Although for some Scots, familiarity with the phenomena of heat and gases inclined them in favor of accepting Lavoisier's theories, for others (such as Hutton) phlogiston was too theoretically vital and ubiquitous to be given up.

This commitment to the articulation of the theory of phlogiston in a wide range of contexts formed an important dimension of the climate in which antiphlogistic chemistry was received in Britain. In England as well as in Scotland one can discern a phlogistic renaissance in the 1770s and 1780s, with the concept being developed and extended in many areas of natural philosophy. English chemists tended to see phlogiston as a ponderable material entity, albeit one that was extremely subtle and expansive.[20] Priestley, Richard Kirwan, and Henry Cavendish all envisaged phlogiston as a chemical constituent of many substances that was bound by the forces of affinity and released in combustion or calcination. This view was shared by such lesser-known English writers as Charles Hopson and Bryan Higgins.[21]

Phlogiston understood in this way was a central feature of English pneumatic chemistry in the 1770s. In the last chapter we saw how it was implicated in the program of nitrous air eudiometry, with its connections with pneumatic medicine. In the early 1780s, phlogiston acquired a new lease of life when it was identified by Kirwan with the basis of "light inflammable air" (the gas Lavoisier was to call hydrogen). This identification was soon accepted by Priestley and others. Throughout the 1780s, Kirwan was to insist that phlogiston was "no longer to be regarded as a mere hypothetical substance"; on the contrary, it was a demonstrated empirical phenomenon.[22]

20 Ziemacki (1974); Langer (1971).
21 Hopson (1781), esp. pp. 5–7. Bryan Higgins's views are mentioned in William Higgins (1791), pp. xi–xii.
22 Kirwan (1789), p. 9. See also: Priestley (1783), pp. 400–402; and compare Priestley's letter to Lavoisier, 10 July 1782, mentioning experiments with inflammable air "that seem to prove that it is the same thing that has been called *phlogiston*" (quoted in Guerlac (1976), p. 222, fn. 61).

For this reason, English chemists could not accept that Lavoisier's studies of combustion had the implications he claimed. Instead, Kirwan proposed an influential alternative view – namely, that phlogiston released by a burning body combined with dephlogisticated air to form fixed air, which then united chemically with the residue of the solid to form a calx or acid. This account had the advantage of accommodating the weight gain that was agreed to occur in instances of combustion and calcination, while maintaining the existence of phlogiston. Kirwan defended this interpretation in his work, *An Essay on Phlogiston and the Composition of Acids* (1787).

A crucial additional factor in the controversy as it unfolded during the 1780s was the question of the composition of water. In early 1784, Cavendish announced that a mixture of inflammable and dephlogisticated airs would produce water when ignited by an electric spark. The experiment, previously performed by Priestley and Warltire, emerged from the tradition of eudiometry, which had been given a new direction by Alessandro Volta's invention of the electric-spark eudiometer. Cavendish was investigating the diminution of airs by phlogistication; sparking with inflammable air was the currently favored way of achieving this. The production of water in the reaction was an unexpected result, but one that rapidly became an unavoidable fact in the continuing debate.[23]

Lavoisier repeated Cavendish's experiment before his account was even published, having been informed about it by Charles Blagden in 1783. Two years later he staged the public demonstration designed to show that water could be decomposed to (and recomposed from) its constituent elements, hydrogen and oxygen. From Lavoisier's standpoint, Cavendish's result offered a means of explaining two classes of phenomena that had previously constituted troublesome anomalies. The inflammable air generated by metals when they dissolved in acids could now be explained as a product of the decomposition of water, and the reduction of lead calx and other calces by inflammable air could be understood in terms of the combination of the gas with oxygen from the calx to synthesize water.

Many British chemists remained unconvinced by this explanation. Kirwan proposed that inflammable and dephlogisticated airs yielded water only in conditions of extreme heat. At lower temperatures they would combine to produce fixed air, as happened in the normal process of combustion.[24] Kirwan sustained this view through the 1780s, only finally relinquishing the phlogiston theory in 1791, after having failed to demonstrate his claim that the two gases could combine to form fixed air.

---

23 Cavendish (1784).
24 Kirwan (1784a) gives his view that "phlogistic processes" involve the generation of fixed air. For general accounts of Kirwan's work, see: *Dictionary of Scientific Biography*; Donovan (1850).

Priestley's view was different, but he too refused to agree with Lavoisier's interpretation. Having replicated Cavendish's experiment, he concluded that the water produced by the reaction of the two gases was present in their composition in the gaseous state – the reaction seemed to show that all airs contained some proportion of water. There was, however, another product – nitrous acid – a compound of the ponderable bases of the gases. Performing the experiment with well-dried samples of inflammable and dephlogisticated airs (as he still preferred to call them), Priestley produced both water and nitrous acid, the latter in significant quantities. Predictably, Lavoisier and his allies tried to cast doubt on Priestley's experiments by asserting that the acid had been produced as a result of contamination of his sample of oxygen by nitrogen from the atmosphere. But Priestley consistently denied such contamination, and his reputation as an experimenter was such that Keir, Nicholson, and other English chemists accepted his word.[25]

In 1791, Priestley brought forward a further series of experiments to show that inflammable and dephlogisticated airs could form either nitrous acid or water, depending on the relative quantities involved. He wrote excitedly to Josiah Wedgwood that this result showed that he and Lavoisier had both been correct in their description of experimental facts but that the phlogiston theory would survive. He promised that his report of the experiments would "decide this long contest." He was still insisting as late as 1800 that when proper proportions of inflammable and dephlogisticated airs were ignited by an electric spark, "a highly phlogisticated nitrous acid is instantly produced."[26]

Although most chemists came round to the view that the gases Priestley used in this experiment were indeed contaminated,[27] he continued to sustain the debate through the 1790s by referring to another experimental result – the production of inflammable air from finery cinder (a type of iron calx) heated with charcoal. The problem for the antiphlogistic theory was to explain how inflammable air could be produced without any source of water to be decomposed. William Cruickshank resolved the matter to most chemists' satisfaction in 1801, by showing that the inflammable air in this case was not hydrogen but an oxide of carbon (now called carbon monoxide). Even then, Priestley criticized Cruickshank's findings and continued to defend the phlogiston theory.[28] By this

25 See Keir (1789), pp. vi–vii; and [Nicholson], "Preface by the Translator," in Fourcroy (1790), I, pp. v–xxi. See also the comments supporting Priestley by J.H. Magellan in his translation of Cronstedt (1788), I, p. lii.
26 Priestley to Wedgwood, 26 February 1791, in Schofield (1966), p. 255. The published paper was Priestley (1791a). For the later remarks, see Priestley (1800), p. 45.
27 Discussions of this include: [Pearson] (1794), pp. 47–56; and William Higgins (1791), pp. 4–5.
28 Cruickshank (1801a, 1801b). For Priestley's defiant response, see Priestley (1803).

stage, however, he was virtually without allies among recognized chemists, and the overthrow of the phlogiston theory had been all but universally accepted.

The instruments of persuasion

Though his position had evidently declined in credibility by 1800, Priestley in the 1780s might have been taken for the leader of the opposition to Lavoisier. Many British chemists sided with him in disputing the French claims, and this persistent antagonism made serious demands on Lavoisier's persuasive resources. At issue in the controversy were not just the facts of the matter but how and by whom experiments were to be conducted and the results interpreted. Questions of fact led rapidly to other questions, including those of experimental method and the proper forms of chemical discourse. The participants came to appreciate how debates about the facts are essentially debates as to how scientific practice is to be carried on.[29]

Lavoisier's experimental work in calorimetry and his public demonstrations of the analysis and synthesis of water were reported to English chemists as part of his strategy to win them over to his point of view. The translation with critical commentary of Kirwan's 1787 *Essay*, prepared by Lavoisier's wife and a number of his new-found supporters, was a direct attempt to shift opinion on the other side of the channel. Subsequently, the nomenclature and the *Traité*, both quickly translated into English and much commented upon thereafter, played a significant part in the English campaign.

In the course of this controversy, Lavoisier's instruments themselves became subjects of discussion. As historians have noted, much of his work made use of novel, sometimes elaborate and expensive, apparatus.[30] Three instruments in particular – the balance, the calorimeter, and the gasometer – were used to yield measurements of quantities of weight, heat, and volume, which were accorded crucial significance in support of factual claims. Precisely because those claims remained controversial, however, Lavoisier's instruments continued to be subject to critical scrutiny. Their interpretation, even the decision as to whether they had worked properly, remained a contested matter.

Lavoisier's readiness to deploy new instrumentation and to borrow it from disciplines other than chemistry had been characteristic of his work

---

29 This theme has been explored in such historical and sociological studies of scientific controversies as: Shapin and Schaffer (1985); Pinch (1986); and the studies collected in Collins (1985).
30 For a helpful recent discussion, see: Levere (1990).

since his early researches in mineralogy and geology. Already in the 1760s he had been using thermometric and barometric measurements in geological surveys and developing hydrometric methods of mineral-water analysis. As Holmes, Perrin, and Donovan have pointed out, his command of instrumentation derived from the other physical sciences was an important reason for his success in chemistry.[31] Lavoisier was always more interested in expanding disciplinary boundaries in search of applicable theoretical and instrumental resources than in defending the traditionally recognized borders of chemistry.

Such a strategy required significant financial and material resources. Lavoisier was fortunate in this respect in being able to command substantial private wealth. With his own assets and with some assistance from the Parisian Académie des Sciences, he was able to have expensive apparatus made for him. For example, the instrument-maker Pierre Mégnié was paid 1814 livres for several commissions between December 1783 and September 1785. Of this the Académie paid no more than 300–400 livres, the remainder coming from Lavoisier's own pocket. During the same period Lavoisier and Laplace were also paying the tinplate worker Naudin about 600 livres for the two ice calorimeters they used in their experiments on the nature of heat. From 1788 another instrument maker, Nicholas Fortin, was employed by Lavoisier. He was paid further substantial sums, including 600 livres for a highly accurate beam-balance.[32]

To Priestley and his followers, expenditure of this scale on scientific apparatus was not only undesirable but reprehensible, because it foreclosed the possibility of Lavoisier's experiments being replicated by others who lacked his wealth. In Priestley's view, this implied a failure to submit discoveries to public validation by the relevant community. He regularly charged Lavoisier with using such costly apparatus that his experiments were impossible to reproduce. The 1785 experiment seemed to Priestley to require "so difficult and expensive an apparatus, and so many precautions in the use of it, that the frequent repetition of the experiment cannot be expected; and in these circumstances the practised experimenter cannot help suspecting the . . . certainty of the conclusion." Summarizing his position in 1800, Priestley still maintained, ". . . till the French chemists can make their experiments in a manner less operose and expensive . . . I shall continue to think my results more to be depended upon than theirs."[33]

Lavoisier's balances were prime examples of refined and costly apparatus, introduced with the justification that more sophisticated standards of experimental measurement were required in chemistry. His use of these

31 Perrin (1989), p. 63; Donovan (1988b); Holmes (1985), pp. 281–282.
32 Daumas (1955), pp. 135, 141, 149.   33 Priestley (1800), pp. 77, 50.

devices has been held up by modern historians as a symbol of his willingness to recruit for chemistry the methods of the more exact physical sciences. As Daumas, Gough, and Lundgren have pointed out, chemists before Lavoisier used balances in a relatively imprecise manner, largely for practical purposes such as weighing drugs rather than in the service of fundamental theoretical claims.[34] Lavoisier, on the other hand, had a number of precision balances constructed, which he applied to rigorous chemical research. This was a feature of his career from his use of an apparatus built by Chemin in 1770 to disprove claims for the transmutation of water into earth to his commissioning of Fortin to build the large beam balance in 1788. Possession of Fortin's instrument gave Lavoisier almost unrivalled mastery of weight measurement. Only two other devices in the world, one built by Ramsden for the Royal Society and the other by John Harrison for Henry Cavendish, were regarded as comparable in precision.[35]

In the 1789 *Traité* the balance occupied a prominent position as the first of the chemical instruments described. There Lavoisier stressed the importance of weighing in chemical operations and mentioned the difficulties of standardization of weights in different locations. He mingled descriptions of his own apparatus with injunctions to other chemists to procure for themselves similarly precise instruments. In this way he gestured toward the problems of replication that Priestley and the English chemists had raised. But his proposed solution was simply to look forward to the day when other chemists could come to share his resources of advanced apparatus.[36]

Such a proposal was hardly likely to be acceptable to Lavoisier's English opponents. They were, however, obliged to concede the superiority of Lavoisier's balances over any others. Dispute therefore focused, not on the accuracy of the apparatus, but on its relevance to resolving the question of the existence of phlogiston. Priestley's comments indicated he was fully aware of the balance's rhetorical importance in displacing the imponderable phlogiston from chemistry. He remarked on "the great use that the French chemists [make] of scales and weights," but pointed out that they had not been able to weigh either caloric or light. He continued ironically, "Why may not *phlogiston* escape their researchers when they employ the same instruments in their investigation?" Nicholson echoed Priestley's challenge, pointing out that phlogiston was ". . . not the only material substance of which the essence is too subtle for our observation." By bringing into question the ability of the balance to detect all

---

34 Gough (1988), pp. 20–21; Daumas (1955), pp. 113, 136; Lundgren (1990).
35 Daumas (1955), pp. 132–139. Cf. Kirwan (1789), p. ix fn. (a note by the editor of Kirwan's text, William Nicholson).
36 Lavoisier (1789), II, pp. 327–341; idem. (1790), pp. 295–303.

substances, Priestley and his allies attempted to undermine Lavoisier's appeal to it in his campaign against phlogiston.[37]

In the case of the calorimeter, the name given to the apparatus designed by Lavoisier and Laplace in 1781, dispute focused more immediately on its reliability as a measuring device. The instrument was employed with what was claimed to be a high degree of accuracy, and its results were accorded a critical importance in Lavoisier's writings. There were, however, severe problems of replication, and consequently many British experimenters refused to accept that the instrument worked in the way its inventors had claimed.

Lavoisier and Laplace described their instrument and recounted the experiments performed with it in their *Mémoire sur la Chaleur* of 1783.[38] As Lissa Roberts has pointed out, the apparatus was initially presented as an unnamed machine for measuring heat. The authors introduced the instrument to the reader in connection with an account of the progressive improvement in thermometers over the previous decades. The implication was that instruments for measuring heat could be expected to achieve the same degree of accuracy as those for measuring temperature. The construction of the apparatus was described by introducing an idealized fiction: a hollow sphere of ice within which chemical operations could be conducted. The amount of heat generated by chemical change could be assessed by measuring the amount of ice that melted from the inner surface of the sphere, the central region being insulated from any external heat by the outer layers of ice.

Of course this was a fictional ideal; Lavoisier and Laplace were obliged to admit that their device was only an approximation to this, and one with certain evident problems. The central reaction chamber was surrounded by a chamber packed with ice, which was surrounded in turn by an insulating compartment also packed with ice. The water produced by melting ice in the middle chamber was collected at a tap at the base of the apparatus. Some loss of water was inevitable at this stage. Furthermore, the interior of the apparatus had to be accessible from the outside, so that heating or cooling from external sources could not be completely avoided. The authors insisted that these problems could be overcome by skillful use of the device. For example, the apparatus should be used only when the ambient temperature was a few degrees above freezing. Higher temperatures would put at risk the effectiveness of the insulation, whereas if the ambient temperature were below freezing, heat from the reaction would be used to raise the temperature of the surrounding ice without melting it.

These acknowledged difficulties reflected the many problems experi-

37 Priestley quoted by McEvoy (1988b), p. 203. Nicholson's remarks are in his "Preface" to Fourcroy (1790), I, p. x.
38 Lavoisier and Laplace (1982). For discussion, see: Roberts (1991a); and Guerlac (1976).

enced in the course of Lavoisier and Laplace's experiments with the device. But such problems were given only cursory acknowledgment by the time the apparatus came to be presented in Lavoisier's *Traité*. There the machine was given the name "calorimeter" and was portrayed as an unproblematic instrument, an "Apparatus for measuring Caloric" similar to the other measuring devices described. It was put forward as an instrumental confirmation of the caloric theory of gases and hence as warrant for the antiphlogistic account of the production of heat and light in combustion.[39]

This description did little justice to the severe limitations that the history of the machine had revealed. There appear in fact to have been no generally agreed-upon replications of Lavoisier and Laplace's experiments in the years after they were first performed. On the contrary, problems with managing the machine were widely discussed for at least a couple of decades. In 1784, Wedgwood tried to design his own version, inspired by knowledge of the possible inaccuracies of the French machine. He was particularly concerned with the efficiency with which all of the melted ice could be drained from the middle compartment, because he believed that a significant quantity of water would adhere to the surface of the ice that remained unmelted. Kirwan and Black appear to have been persuaded by Wedgwood's objections, and withheld their assent to Lavoisier's claims. Beddoes also regarded Wedgwood's reservations as highly damaging, and Nicholson reiterated the grounds for skepticism as late as 1808.[40]

Others who tried to make the ice calorimeter work reported insuperable problems. An anonymous writer of 1797 asserted that "little reliance ... can be placed on the accuracy of this much-boasted process of the French chemists," notwithstanding that the results were presented "with all the precision of the new school." In 1800, F.C. Accum concluded that "this instrument ... is by no means entitled to the pompous praises of complete accuracy which have been so liberally bestowed upon it." Even J.H. Hassenfratz, who had assisted Lavoisier and Laplace in their experiments, subsequently failed to manage the device efficiently. Although the apparatus passed into courses of laboratory instruction in chemistry, it was little used by subsequent researchers. Thomas Thomson appears to have been correct when he recorded in 1830 that the ice calorimeter "has failed in the hands of every one who has attempted it since the publication of [the original] ... experiments."[41]

39 Lavoisier (1789), II, pp. 387–402; idem. (1790), pp. 343–356.
40 Lodwig and Smeaton (1974); Beddoes's comments are in Bergman (1785), pp. 350–351.
41 Lodwig and Smeaton (1974), pp. 4–6, 8–9, 13, 15; [Anon.] (1797), pp. 20–21. Compare the remarks in the journal *The Chemist*, 1 (1824), p. 249, where the ice calorimeter is said to be "neither a very correct instrument, nor of great utility."

The debate over the calorimeter exposed how Lavoisier's instruments, supposedly capable of measuring with an unprecedented degree of precision, were at issue as much as his substantive empirical claims. The strategy of the authors of the *Mémoire sur la Chaleur* had been to gain acceptance of the utility of their instrument as a measuring device without at first suggesting any of its theoretical implications.[42] Some credit had indeed been gained: Thomas Thomson, notwithstanding his later skepticism about the calorimeter, had originally been disposed to accept Lavoisier and Laplace's results, "from the well-known accuracy of these philosophers."[43] But that accuracy was soon called into question as attempts to reproduce the results failed. English experimenters, placing a high value upon widespread reproduction of experiments, remained unpersuaded by an instrument that could not be reliably replicated.

At issue then, as McEvoy has pointed out with specific reference to Priestley, was not the desirability of precise quantification, but whether precision could be achieved and what it would mean.[44] The issues were intimately connected because the acceptability of Lavoisier's conclusions was integrally bound up with judgments of the reliability of his instruments and procedures. Hence Priestley charged Lavoisier with using "too much of correction, allowance, and computation" in calculating the results of experiments adduced in support of the antiphlogistic theory.[45] When such claims were contested, the instruments and measurements deployed to support their plausibility were bound to be brought into question.[46]

The arguments following the 1785 demonstration showed this clearly. This public display in Paris on 27 and 28 February comprised two simultaneous experiments. In one, water was synthesized by mixing measured quantities of hydrogen and oxygen and igniting them electrically in a reaction vessel. In the other, steam was decomposed by passing it through a red-hot iron gun barrel. Hydrogen gas was collected from the other end and the increased weight of the barrel due to the formation of iron oxide

42 Roberts (1991a).   43 Lodwig and Smeaton (1974), p. 6.
44 McEvoy (1988b), esp. pp. 204–205.   45 Priestley (1800), p. 44.
46 Compare the discussion of the precision of Lavoisier's experiments, in Holmes (1985), pp. 183–198. Holmes uses the evidence of manuscript remains to explore the theoretical assumptions made by Lavoisier and Laplace and the reasons why they adjusted some of the results of their experiments. He makes a good case for the necessity of exercising judgment in the interpretation of experimental results in any science. My purpose here is rather different: I am not concerned to judge the propriety of Lavoisier's interpretive maneuvers, but to record their degree of persuasive success and failure and the objections made to them at the time. That experimental work always involves interpretive assumptions and that such assumptions can always be criticized seems to me well established. The relevant questions here concern how effective the presentation of the results was. Who was convinced and who resisted the author's claims? It is by answering this kind of question that I hope to relate debates over the outcome of experiments to the problem of the constitution of science as a public culture.

was measured. The setting of the two experiments was designed by La-voisier to recruit allies to his cause from among the leading French chem-ists. To this end, more than thirty witnesses were selected, including many of the leading chemists and physicists of the Académie des Sciences. A whole battery of authentication measures was adopted, including signing the papers on which the results were recorded and sealing the apparatus when it was left overnight.[47]

The emphasis on precision and quantification in the experiments was meant to enhance their rhetorical effectiveness. Precautions included careful calibration of the specially constructed gasometers in which oxygen and hydrogen were to be held, so that the volumes of the gases could be calculated at the prevailing temperature and pressure. A result was achieved for the proportions of the two gases in the composition of water, which Lavoisier asserted was correct to nearly one part in two hundred. In sub-sequent references to the experiment he emphasized that its conclusive-ness hinged upon the precision with which all the operations had been carried out:

One of the points of the modern doctrine, which would appear to be most solidly established, is the formation, the decomposition and the recomposition of water; and how would it be possible to doubt it, when one sees that in burning together 15 grains of inflammable gas and 85 of vital air, one obtains exactly 100 grains of water, [and] that one can, by way of decomposition, recover exactly these two principles and in the same proportions?[48]

Such supposedly accurate measures were, however, insufficient to per-suade those British chemists who continued to resist Lavoisier's notion of water as a compound. In his 1788 translation of Fourcroy's *Elements of Natural History and Chemistry*, Nicholson commented:

If it were allowable to place much dependence on the quantities of products in experiments of this nature, wherein different philosophers disagree, the results of M. Lavoisier, which answer to the proportions of the two principles determined in the composition of water, would add much force to his conclusions. But the advocates for phlogiston contend, that the water unites totally with inflammable or calcinable substances, and sets their phlogiston at liberty.[49]

Priestley's own view of Lavoisier's demonstration was that its com-plexity had increased the chances of error. Comparing the equipment used in Paris with his own, he wrote that the former was "extremely complex, as a view of their plates will shew, and mine was perfectly simple, so that nothing can be imagined to be less liable to error." Simple

47 Daumas and Duveen (1959); Duveen and Klickstein (1954); Daumas (1955), pp. 142–150; Holmes (1985), pp. 211–214, 237–238.
48 Lavoisier, "Rapport sur les nouveaux caractères chimiques" (1787), quoted in Langer (1971), p. 223.
49 Fourcroy (1788), I, p. xvi.

experiments, which did not demand elaborate calculations of quantities, were to be preferred for their reliability. Furthermore, Priestley had performed the experiment numerous times, always producing nitrous acid as well as water from a mixture of dephlogisticated and inflammable airs. He summed up:

> ... it appears to me not a little extraordinary, that a theory so new, and of such importance, overturning every thing that was thought to be the best established in chemistry, should rest on so very narrow and precarious a foundation; the experiments adduced in support of it being not only ambiguous, or explicable on either hypothesis, but exceedingly few.[50]

In his 1789 introduction to the second English edition of Kirwan's *Essay on Phlogiston*, Nicholson again took up the question of the precision of Lavoisier's measurements in the 1785 demonstration. He acutely noted the rhetorical role played by assertions of accuracy in Lavoisier's accounts of his experiments: "... a reference to weights in the experiments of Mr. Lavoisier is made to constitute a great part of the arguments adduced to prove the composition of water, and its decomposition." Some of the measurements given, however, particularly for the weights of gases, laid claim to what seemed to Nicholson a quite implausible degree of exactness. He insisted that difficulties with weighing the gasometers would prevent gases being weighed with anything like the precision Lavoisier claimed; he charged the French chemist with "an unwarrantable pretension to accuracy" in his gravimetric experiments. Lavoisier had recorded weights to six, seven, or even eight significant figures, whereas Nicholson estimated that only three figures could be asserted reliably. He concluded that Lavoisier's assertions that a scrupulous exactitude had been maintained, and hence that a demonstrative order of proof had been achieved, were highly questionable:

> If it be denied that these results are pretended to be true in the last figures, I must beg leave to observe, that these long rows of figures, which in some instances extend to a thousand times the nicety of experiment, serve only to exhibit a parade which true science has no need of: and, more than this, that when the real degree of accuracy in experiments is thus hidden from our contemplation, we are somewhat disposed to doubt whether the *exactitude scrupuleuse* of the experiments be indeed such as to render the proofs *de l'ordre demonstratif*.[51]

---

50 Priestley (1800), pp. 48, 76–77. Compare Nicholson's remarks, that Lavoisier's experiments were "extremely nice; such that the utmost accuracy of observation was necessary to distinguish the result. In such cases mistakes are easily made; nay, it is scarce possible to avoid them" (Nicholson, "Preface" to Fourcroy (1790), I, p. xiii). Also note the comments by Bryan Higgins in Higgins, ed. (1795), p. 175. Higgins charged Lavoisier with using an experimental apparatus which was "complex and expensive," and in which the weights of gases could not be accurately measured.
51 Kirwan (1789), pp. viii–xi.

Demonstration, authority, and community

As this passage indicates, the arguments between Lavoisier and his British opponents spilled over from questions of fact and method to questions of the rhetorical form of science. In the course of advancing his factual claims, the French chemist was also taken to be proposing a new model of scientific communication and persuasion. In describing the 1785 experiment, he insisted that exact measurements *had* been achieved and therefore that the experiment constituted a demonstration of the composition of water, "if in any case the word Demonstration may be employed in natural philosophy and chemistry." He went on:

The proofs which we have given ... being of the demonstrative order, it is by experiments of the same order, that is to say by demonstrative experiments, which [*sic.*] they ought to be attacked.[52]

In advancing this view of the demonstrative potential of chemical experiments, Lavoisier was borrowing a model of proof from geometry and those physical sciences that had already attained a mathematical form. His claim was that experiments conducted skillfully with reliable instruments and precise measurements could provide demonstrative evidence of certain claims. Upon these the geometrical structure of a newly rationalized chemistry could be constructed. As he explained in the Preface to the *Traité*, a science could be built up step-by-step on the basis of facts that were each "a consequence, following immediately from an experiment or an observation."[53]

Priestley and other British chemists, on the other hand, resorted to a more traditional view of the standard of knowledge that could be achieved in chemistry. Nicholson spoke for Kirwan, Priestley and others when he accused the French chemists of the "specious pretence" of "representing the antiphlogistic theory as being not a theory, but merely a plain statement of facts."[54] On the British model, theoretical knowledge (which could only ever be of a provisional nature) could be gained only from ordering and interpreting a large body of observational and experimental data. On these grounds, Priestley justified his preference for his own more numerous experiments over Lavoisier's "exceedingly few" ones. The priority should be to accumulate experimental facts, which were solid and permanent whereas theoretical hypotheses could be expected to come and go. Even those British chemists who came to accept Lavoisier's theory had a strong awareness of what William Higgins described as ". . . the impossibility of persuading us by experiments alone, . . . so prone are

52 Lavoisier, quoted in Kirwan (1789), pp. 59–61.   53 Lavoisier (1789), I, p. viii.
54 Nicholson, "Preface" to Fourcroy (1790), I, p. xiv.

we to reconcile every phenomenon we see to our manner of thinking."[55] In the light of this, it made sense to weigh the quantity of experimental evidence on either side of a disputed question.

Kirwan gave expression to this kind of view when he gave the following response to Lavoisier's charge that he was refusing to accept the demonstrated consequences of experimental facts:

> It is very true that I do not content myself with the immediate inferences from *single* facts. In my opinion, the book of nature should be interpreted like other books, the sense of which must be collected not from single detached passages, but from an attentive consideration of the whole, so that what is defective in one text may be supplied from another.[56]

A similar conception of the mode of demonstration appropriate to chemical discourse was articulated in Keir's preface to *The First Part of a Dictionary of Chemistry* (1789). Keir, a longstanding colleague of Priestley's in the Lunar Society of Birmingham who shared his aspirations for political reform, also attacked the pretensions implicit in the French claim that chemical discourse could be remodeled along the lines of geometry. Chemical knowledge was not solidly enough established, Keir insisted, to permit deductive reasoning from principles. The French system could therefore offer only a pretense of command of the subject, one that would inevitably be limited by the tendency to interpret facts from a particular theoretical point of view.[57]

This was the rationale for Keir's use of the dictionary format, one that (as he noted) had been favored by the French chemist Macquer in the generation before Lavoisier.[58] The arrangement of chemical facts in this form enabled the reader to perceive their multiple connections and to order the phenomena in a way that served his purposes, without submitting his judgment to domination by a particular hypothesis. The dictionary thus played its part in an envisaged community in which attested experimental facts would be communicated between credited individuals. Keir saw himself as writing for a public audience eager to learn facts, whose independent interpretation of those facts was not to be intruded upon. He wrote with enthusiasm of the spread of enlightenment and of the diffusion of scientific knowledge. "[I]n no former age, was ever the light of knowledge so extended, and so generally diffused," he declared. Participation in the process of cultural enlightenment imposed a respon-

---

55 William Higgins (1791), p. viii. This kind of view was sometimes used as a post-hoc rationalization of the switch from phlogistic to antiphlogistic interpretations – it was suggested that both could be reconciled equally well with the experimental evidence. This is not, of course, the sentiment of Lavoisier, but it is expressed by Nicholson in the passage cited at the beginning of this chapter, and seems to be suggested by Kirwan in his (1807), II, pp. 542–543.

56 Kirwan (1789), p. 304.   57 Keir (1789), pp. ix–x.

58 On Macquer, see especially Wilda Anderson (1984), ch. 3.

sibility upon authors to respect the autonomy of their readers' judgments. Keir concluded his preface:

To enable these judges, the public of all nations and of all times, to decide with a full knowledge of the question, every view in which the subject can be considered, every argument for and against ought to be presented to them.[59]

We can now see how questions of the public function of science were involved in the controversy between Lavoisier and the British chemists. Lavoisier's model of an order of discourse based on demonstrative experiments would require them to be performed in a certain way. Instruments would have to be expensive and made by skilled craftsmen. Witnesses would be assembled for demonstration experiments, and in view of the difficulties of using precision instruments, the possibilities for replication would be limited. The audience would be expected to accept the proposed implications of a declared experimental result. As Lissa Roberts has put it:

Lavoisier did not rhetorically open the door to his laboratory, as it were, so that others might autonomously exploit his machinery. He sought to induce active participants to enter an experimental space prefabricated as an entwined continuum of material technology and systematically theoretical interpretation.[60]

The resistance of British chemists to Lavoisier's new instruments and procedures was an indication of how different his experimental and rhetorical practices were from traditional expectations. Historians such as W.R. Albury and Wilda Anderson, who have discussed Lavoisier's discursive methods extensively, have identified the roots of his distinctive approach. They have focused upon his debts to geometry, to the mathematized physical sciences, and to the epistemology of Condillac.[61] They have also pointed out how Lavoisier mobilized material objects, discursive resources and social relations toward the single end of securing acceptance of his novel theoretical claims. As part of the same rhetorical process, audiences of various kinds were assembled at experimental demonstrations and in the classrooms where the *Traité* and other Lavoisian textbooks were used. As Anderson has stressed, Lavoisier's discursive innovations were useful only in relation to a system of discipline, in which the texts were supplemented by social and institutional measures. By these means, the new system made its distinctive contribution to chemical didactics, aided by the reforms of French educational institutions that followed the outbreak of the Revolution.[62]

Priestley's outlook placed him in principled opposition to this mode of scientific activity. He rejected the use of complex and expensive instru-

59 Keir (1789), pp. iii–iv, xx.   60 Roberts (1991a), p. 217.
61 Albury (1972); Wilda Anderson (1984). See also Roberts (1991b).
62 Wilda Anderson (1984), pp. 127–128. Cf. also Bensaude-Vincent (1990).

ments as an assertion of private rights over what should be the public apparatus of experimental philosophy. The cultivation of skills of measurement and calculation appeared to him a restriction on the accessibility of science to the lay population. And for the French chemists to claim in addition a geometrical degree of certainty in the exposition of their doctrine was a further insult to the autonomous judgment of other observers. Thus Priestley perceived Lavoisier's revolution in chemistry primarily as an assertion of power over the philosophical community. He rejected the imposition of discipline upon practitioners, the impairment of their independent powers of reasoning, and the curtailment of the possibilities of replication that the proposed transformation of chemistry implied. Far from identifying Lavoisier's innovations with the revolutionary events in French politics after 1789 (of which he largely approved), Priestley saw him as reverting to dogmatism and despotism. Addressing Lavoisier's surviving disciples in 1796, two years after their mentor had been guillotined during the Reign of Terror, Priestley wrote:

... you will agree with me, that no man ought to surrender his own judgment to any mere *authority*, however respectable. . . . As you would not, I am persuaded, have your reign to resemble that of *Robespierre*, few as we are that remain disaffected, we hope you had rather gain us by persuasion, than silence us by power.[63]

These terms – judgment versus authority, persuasion versus power – encoded Priestley's fundamental concerns about Lavoisier's chemistry. What were presented as improvements in the demonstrative capacity of chemical discourse were perceived by the English experimenter as attacks on the autonomy of his reason. Lavoisier seemed to be treating his audience as incapable of independent judgment rather than trusting them as observers who could draw their own conclusions from experiments. Priestley feared that his ideal of science in the service of public enlightenment would be undermined; he perceived direct connections between issues of propriety in scientific discourse and those concerning the role of science in the public realm.

Many of the same concerns surfaced in the extended controversy that followed the publication of the new chemical nomenclature in 1787. For Lavoisier the new nomenclature was another instrument, analogous to material instruments in the laboratory, to be used in the construction of a demonstrative discourse.[64] He proposed that the introduction of new terminology for elements and compounds would constrain teachers and students of chemistry to adopt his view of composition. The nomenclature constituted a further component of the system of discipline by which

63 Priestley (1796), p. 17; cf. Priestley (1800), p. xi.
64 Albury (1972), pp. 110–185, esp. quotations from Lavoisier on pp. 115–116, 157. See also Wilda Anderson (1984), pp. 124–137.

the new theory would be transmitted. As Maurice Crosland has shown, the terminology proposed in the *Méthode de Nomenclature Chimique* was forcefully advocated by disciples of Lavoisier in Britain, such as James St. John, George Pearson, and Robert Kerr. The dissemination of the new chemical language appears to have aided the acceptance of Lavoisier's doctrine especially among students of the subject. By 1800, the use of the new terms was all but universal in Britain among those recognized as qualified chemists.[65]

There was, however, a strong current of resistance, particularly from older members of the chemical community, and much of this was motivated by concerns similar to those we have seen articulated by Priestley, Keir, and Kirwan. In the late 1780s, Cavendish and Keir recorded their resentment of the new nomenclature, in which the French chemists had insinuated their own particular hypothesis under the cover of a language designed for general use. Cavendish denounced the "very mischievous" new terminology, a symptom of "the present rage of name-making." If a name was in use and well understood, Cavendish complained, only confusion could follow from attempts to change it. Keir similarly saw the new nomenclature as an attempted appropriation of chemical language to serve the interests of a sect that was trying to advance its own peculiar opinions. Such a hypothetical language could not be considered "as the general language of chemistry."[66]

Behind these objections lay the view, articulated prominently by Priestley, that language should be the common currency of a community devoted to purely factual discourse. A language framed to serve a particular theory, such as that of the French chemists, would be inappropriate for use in such a setting. The British chemists aimed their sights at a language that would serve their vision of the experimental community, one in which individual observers contributed factual narratives on a basis of equality of credit. As Nicholson put it, a nomenclature should thus be "founded on the most incontrovertible facts only." To Keir, what seemed illegitimate about the new nomenclature was precisely what Lavoisier and his colleagues had considered its strength – namely, that "we cannot speak the language of the new Nomenclature, without thinking as its authors do."[67]

Resistance to the new terminology continued through the 1790s. In 1794, Bryan Higgins found that several members of his Society for Philosophical Experiments and Conversations objected to the use of Lavoisier's terms. The following year, Nicholson declined to preempt the judg-

65 Crosland (1962), pp. 193–206.
66 Henry Cavendish, draft letter to Charles Blagden, 16 September 1787, in Cavendish (1921), pp. 324–326; Keir (1789), pp. ii, xiv–xx.
67 Nicholson (1790), p. vii; Keir (1789), p. xv.

ment of a still-undecided public by employing the new nomenclature in his *Dictionary*. Some of the expressions of resistance to the new system used explicitly political language, comparing the chemical nomenclature with such authoritarian acts of the French revolutionary regime as changing the names of institutions and measurements. Thus Stephen Dickson in 1796 criticized the French chemists for having done irreparable injury to the science, though he went on to say that it was "as little probable, that they will establish all their new chemical names, as their new political or chronological ones, in other countries than their own."[68]

Even those British chemists who perceived the need for reform of chemical nomenclature were obliged to make allowance for national sensitivities on the issue. In 1802, Kirwan wrote that he could largely accept the Lavoisian terminology, though he criticized aspects of its application to mineralogy. He also restated some of the familiar arguments against novelty in this field in terms that echoed the conservative rhetoric of his fellow Irishman, Edmund Burke. Kirwan attacked the French chemists for inflexible application of supposedly rational principles, in much the same way as Burke had attacked the revolutionary regime. The sacrifice of convenience and long habit to rules and principles was, Kirwan said, "a truly harsh, intolerant and despotic maxim." Nor did the supposed didactic utility of the new language persuade him: "The science is not to be charged with a cumbersome train of words merely to gratify the indolence of beginners," he declared.[69]

In the course of making this criticism, Kirwan resorted again to the model of the discursive order of chemistry that had informed Keir's *Dictionary*. Taking Lavoisier up on his assertion that "it is the series of facts that constitutes science," he retorted: "I should rather say it was a knowledge of the relation that subsists between the facts that occur."[70] Chemistry was still seen by Kirwan as a very uncertain science, knowledge of which would inevitably consist in knowledge of a wide variety of particular facts and possible relations between them. The same conception was prominent in Nicholson's resuscitation of Keir's abandoned dictionary project in 1795.[71] British chemists continued through the 1790s to rely upon this model of the way chemistry should be organized in voicing reservations about Lavoisier's project. They also continued to prize the values of observational autonomy and independence of judgment. Edward Peart, for example, in his *Anti-Phlogistic Doctrine Examined* (1795), announced that "nothing but a prejudice preventing con-

---

68 Bryan Higgins, ed. (1795), p. 13; Nicholson (1795), p. vii; Dickson (1796), p. 27. See also [Chenevix] (1820), p. 405.
69 Kirwan (1802), esp. pp. 53–54, 70–71, 73.   70 Kirwan (1802), pp. 72–73.
71 Nicholson (1795), esp. pp. iii–vi.

viction, or mental imbecility, can be urged as an excuse for any one who will retain the anti-phlogistic doctrine after duly considering it."[72]

Connected with this was a continuing admiration for Priestley, even among some who found his stubborn adherence to the phlogiston theory incomprehensible. George Pearson praised Priestley's ability to convey information with respect to *facts*, notwithstanding their differences on matters of interpretation.[73] Other writers referred to the moral virtues that Priestley had instantiated and that had underwritten the credibility of the facts his discourse was designed to communicate. T.L. Rupp concluded a paper that severely criticized some of Priestley's experiments by expressing the hope that he had said nothing "disrespectful of this great man, whom I sincerely esteem and admire both as a man and as a philosopher."[74] Peart declared that he felt able to disagree with Priestley's opinions precisely because he had expressed them as such and had clearly demarcated them from narratives of facts. Priestley remained, in Peart's view, a philosopher,

... whose general knowledge, whose moral character, and whose various and unremitted labours in the extensive fields of science deserve the warmest approbation and best wishes of every man, who is actuated by the true love of wisdom, of virtue, or of mankind.[75]

The continuing celebration of the Priestleian virtues of economical experimentation, resistance to the authority of theory, and confidence in personal experience of facts, found its most idiosyncratic expression in the writings of Robert Harrington. Harrington, a surgeon practising from the early 1780s in Carlisle, wrote a series of highly eccentric and badly organized works on chemistry and natural philosophy between 1781 and 1819.[76] He doggedly upheld an exotic version of the phlogiston theory, which he accused Priestley, Cavendish, and Kirwan of having stolen from him and then abandoned. Lavoisier and other eminent chemists were mercilessly attacked in rambling and sarcastic tirades. When he failed to gain a hearing for his outpourings under his own name, Harrington published *A Treatise on Air* (1791) under the pseudonym Richard Bewley,

72 Peart (1795), p. 41.   73 [Pearson] (1794), p. 56.
74 Rupp (1798), p. 162.   75 Peart (1795), p. 41.
76 Among his ten books see Harrington (1781, 1796, 1819). An impression of the poor reputation that he had among contemporaries can be gained from reviews in the *Monthly Review*, 1st ser., 66 (January to June 1782), 98–102; 74 (January to June 1786), 449–451; 2nd ser., 6 (September to December 1791), 435–439; 22 (January to April 1797), 107. On the other hand, it is remarkable that much of the first volume of the *Medical Spectator* (2 vols., London, 1792, 1793) is devoted to a positive account of Harrington's theory. The editor, believed to have been Dr John Sherwen, a physician at Enfield, claimed that Harrington was unknown to him personally but that he had been led to take an interest in his theories by a correspondent. (See *Medical Spectator*, 1 (1792), iii–iv, 115–132; 2 (1793), 3–34 [first pagination].)

shamelessly heaping praise on his own theories and complaining about their lack of public appreciation.[77]

Priestley was almost the only contemporary chemist who earned any praise from Harrington, though he did not entirely escape belligerent censure. He was hailed as the first experimenter and his defense of phlogistic chemistry against the "despicable combination" of French chemists was held up for admiration. The new chemical nomenclature and the claims of Lavoisier and his allies to an unprecedented accuracy in experimentation were exposed to ridicule:

We must all be sensible, that in those nice chemical processes of airs, how very uncertain and deceitful they are; and, more particularly, how the *great theorists* are anxious to bring their experiments to tally with their theory. . . . The difficulty of ascertaining the exact quantities and weights of these airs consumed, – the vanity of establishing their wonderful theory, – the far too great a scale it was upon, – their anxiety for the revolution it would produce; – on all these accounts I dispute their experiments *in toto*.[78]

As well as relying in some respects upon the democratic epistemology that Priestley and his followers had articulated, Harrington's colorful rhetoric captured elements of the conservative invective that was increasingly being turned against Enlightenment pneumatic chemistry in the 1790s. Characterizing the leading French chemists as a conspiracy – a "philosophical combination" – was one aspect of this. Another was the identification of pneumatic chemistry as a species of hysteria – a French mania or kind of enthusiasm. As we shall see in the next chapter, this kind of attack was focused especially on the research on breathing nitrous oxide conducted by Beddoes and Davy in the last years of the century. Harrington picks up this theme in his own inimitable style:

[Dephlogisticated nitrous air] is not respirable, but only capable of making our wise theorists feet, dance, and caper like *Merry Andrews* to a French cotillion, while it makes their heads roam at large into their *aerial* regions, their sensations being most ecstatic, (and I should say, *lunatic;* or, still more apropos, *Laputastic.*) – See Dr. Beddoes and Davy's Reports. No stage quack ever exhibited such Wonders! WONDERS!! WONDERS!!![79]

In this respect, Harrington points us toward the topic that we shall consider next: how the reactionary political climate of the 1790s proved increasingly hostile for Priestley's pneumatic chemistry and for the program of enlightened public science that his disciples had developed in connection with it.

77 [Harrington] (1791).
78 Harrington (1804), pp. 65, 101 fn., 217 (quotation on p. 171).
79 Harrington (1804), pp. 60, 70, 101 fn., 208 fn., 275 (quotation from p. 96).

6

"Dr. Beddoes's breath": Nitrous oxide and the culmination of Enlightenment medical chemistry

[Chemistry] led to the heart of Matter, and Matter was our ally precisely because Spirit, dear to Fascism, was our enemy.

Primo Levi, *The Periodic Table.*[1]

This is supposed to be the age of *aerial* philosophy; I wish it were the age of common sense, for at present it has taken an aerial flight; and, unfortunately, candour and justice have flown away with it. *O tempora! O philosophia!*.

Robert Harrington, *The Death-Warrant of the French Theory of Chemistry.*[2]

In a work written in 1792, the physician and chemist Thomas Beddoes (1760–1808) commented on the much-debated experiment in which Lavoisier claimed to have synthesized water from oxygen and hydrogen gases:

What for instance is it, that prevents me from being as certain, that water consists of hydrogene and oxygene airs, as of any proposition in Euclid? – nothing surely but the incompetency of my senses. In the first place, I cannot perceive whether these airs do not previously contain a large quantity of water; secondly, the heat that appears, and of which I have no adequate perception, perplexes me; and thirdly, the occasional appearance of an acid in the water. Now if I could perceive the small quantity of azotic air present separately uniting with a certain portion of the oxygene air to form acid, while the hydrogene air unites with the rest to form water; if I could see that the airs previously contain only a little or no water

1 Levi (1985), p. 52.   2 Harrington (1804), p. 47 fn.

beforehand, and if there was no heat and light, I should have demonstrative evidence.[3]

Beddoes's remarks touched accurately on the disputed questions surrounding the interpretation of this experiment. The production of heat was thought by some chemists to be inadequately explained by Lavoisier's caloric theory, whereas the prior presence of water in the gases and the appearance of nitrous acid among the products constituted the grounds for Priestley's objections to Lavoisier's claim. Such a grasp of the relevant issues was to be expected of Beddoes, a former Edinburgh medical student who had studied chemistry under Black and then become a lecturer on the subject at Oxford.[4]

Also to be expected perhaps was Beddoes's qualified acceptance of Lavoisier's theory. At Oxford he distinguished himself by his extensive knowledge of Continental works on natural philosophy and medicine and by strenuous efforts to bring them to the attention of what he clearly felt was a moribund academic institution.[5] As he showed subsequently when he became a regular contributor to the *Monthly Review*, Beddoes was very well informed about current chemical work in France and Germany. It comes as no surprise to find, from occasional remarks in these reviews, that he was willing by the early 1790s to give a large degree of credence to Lavoisier's theory.[6] Black's declared readiness to do the same in 1789 might have been a factor in Beddoes's decision.[7]

The way in which Beddoes assimilated Lavoisier's theoretical perspective shows some of the distortions inherent in any process of translation between different cultural contexts. Particularly striking are the different understandings of the concept of demonstrative evidence. Although Beddoes was prepared to recognize Lavoisier's experiments as demonstrative (or nearly so), he understood the criterion in a thoroughly empirical way. To Beddoes, it was immediate sensory experience that was demonstrative, and not the sequence of steps that might connect it with theoretical conclusions. If Lavoisier could satisfy doubts as to what was actually perceived in his experiment, then he would have rendered it conclusive, in Beddoes's view. To Lavoisier, on the other hand, experiments were to be made demonstrative by being embodied in structures of discursive reasoning. This was why geometry provided him with his model of scientific rationality. It was, he wrote, "... when chemical proofs lead us

3 Beddoes (1793a), pp. 108–109. The composition of this work can be dated by the Dedication to Davies Giddy, signed by Beddoes at Oxford, September 1792.
4 Beddoes's involvement in debates on the phlogiston theory in the Chemical Society at Edinburgh is reported in Kendall (1949–52), pp. 349–350, 391.
5 Stansfield (1984), ch. 4.
6 *Monthly Review*, 2nd ser., *19* (January to April 1796), 194–198; 23 (May to August 1797), 139–142. See also: Stansfield (1984), ch. 7 and Appendix 1 (pp. 254–260).
7 Beddoes [1794a] was dedicated to Black.

to certainty – when starting from an evident principle we arrive at an evident consequence – . . . [that] we, like the geometers, make use of the word 'demonstration'."[8]

Beddoes's remarks occurred in the context of a work, *Observations on the Nature of Demonstrative Evidence,* dedicated to the deconstruction of what he took to be mystificatory views of the nature of geometry. Beddoes argued that geometry was based solely on empirical knowledge, and that it possessed no certainty greater than that of the experience on which it was based. Traditional teachers of mathematics were accused of attempting to elevate geometrical reasoning above experience, and hence of making mysterious a subject that was purely empirical and could be taught as such. This empirical knowledge was in turn identified with intuition. Beddoes wrote: "Intuitive and experimental knowledge are one and the same thing."[9]

In comparison with Lavoisier, then, Beddoes showed little concern for the discursive context in which experience was to be rendered in order to make experiments persuasive. As we have seen in the last chapter, Lavoisier manifested a considerable capacity for organizing the experimental space, for wielding instruments, situating his audience, and constructing a language of representation in order to make his experiments carry his intended meaning. Beddoes on the other hand emphasized the autonomy of individuals' intuitive experience, which he regarded as possessing a unique persuasive value. As he wrote in the dedication to his book, "the full evidence of intuition . . . must always carry conviction to the mind of the individual."[10] In this respect, Beddoes followed Priestley's epistemological principles, even though he accepted the substance of Lavoisier's doctrine.

This conviction of the primacy of individual sensory experience was indicative of Beddoes's attachment to the values of the English provincial Enlightenment. His text contained a substantial chapter praising the linguistic theory of John Horne Tooke, whose historicist perspective on the development of languages was earning him the respectful attention of many of Beddoes's reformist and radical contemporaries.[11] Beddoes himself was an openly avowed radical and consequently faced increasing opposition in the conservative political climate that came to prevail in Britain after the outbreak of the French Revolution. As Trevor Levere and Dorothy Stansfield have shown, he resigned his position in Oxford under pressure from the government and the academic establishment. Hopes of an endowed chair in chemistry at the university were dashed when the Home Office was informed that Beddoes, the prime candidate, was "a

8 Lavoisier, manuscript notes c.1780–81, quoted in Albury (1972), p. 133.
9 Beddoes (1793a), pp. iii, 25, 87.   10 Beddoes (1793a), p. iii.
11 On Horne Tooke and his reception, see Olivia Smith (1984), ch. 4.

most violent *Democrate* and . . . takes great pains to seduce Young Men to the same political principles with himself." Refusing to compromise or to remain silent about his convictions, he subsequently appeared in Priestley's company on a Home Office list of "Disaffected & seditious persons."[12] Beddoes's commitments to political liberty, to freedom of speech and association, and to the destruction of hierarchy and social privilege, continued to shape his career in medicine and chemistry after he left Oxford in 1793.[13]

In this respect, Beddoes will be seen to have continued Priestley's project of developing chemistry as an Enlightenment public science. He shared many of the progressive social and political aspirations that motivated much of Priestley's scientific work, and also invested them in the field of pneumatic medicine that his predecessor had founded. Beddoes, however, had the misfortune to make his career in the 1790s, a time of profound crisis in British politics and culture. Ideas about the proper forms of civic activity (including science), which had become common during the decades of the Enlightenment, were subjected to conservative challenge. Reactionary thinkers disputed the desirability of widespread public education and the plausibility of the expectation that science would solve problems of health and welfare. Priestley, Beddoes, and others were subjected to criticism and ridicule for their espousal of chemistry and other sciences as means of advancing the material and moral progress of humanity.

These disputes embraced many aspects of the social functions of chemistry and the other sciences, but they are conveniently focused by consideration of a single episode. This was a series of experiments on the respiration of nitrous oxide (laughing gas), reported by Beddoes and his young protege Humphry Davy in 1799. Like certain other phenomena of the period that Robert Darnton has dubbed "the end of the Enlightenment," nitrous oxide was regarded by its proponents as a symbol of the potential utility of Enlightenment natural philosophy.[14] To the conservative opponents of this philosophy, the effects of the gas were symptoms of the delusion and hysteria with which the Enlightenment had culminated. In relation to these experiments, Beddoes's emphasis on intuition could be read as endorsement of a dangerous kind of enthusiasm. The case of nitrous oxide thus provides an emblematic episode in which we can discern some of the political and cultural connotations of debates concerning the status of science at this time. It also comprised a formative episode in the early career of Davy, whose subsequent attempts to cultivate a public audience for chemistry will be traced in the next chapter.

---

12 Levere (1981a) (quotation on p. 65); cf. Stansfield (1984), ch. 5.
13 For another informative account of this, see Levere (1984).   14 Darnton (1970).

The Pneumatic Institution

After his departure from Oxford, Beddoes settled in Bristol, where he devoted himself to the development of pneumatic medical treatment. He took up residence in the Hotwells area of the city, a traditional health resort, which offered the experimental physician a large number of prospective patients. There he attempted to raise subscriptions toward the foundation of a "Pneumatic Institution," where a comprehensive range of therapeutic gases would be offered to those suffering from a variety of ailments. To this end, he brought out a number of publications in the mid 1790s. The *Observations on the Nature and Cure of Calculus, Sea Scurvy, Consumption, Catarrh, and Fever* (1793) surveyed some of the complaints for which pneumatic therapies held out hope of alleviation or cure. Among the remedies Beddoes recommended was the use of carbonated vegetable alkali ("Bewley's julep") as a dissolvent for the stone. He also conjectured, on the basis of experiments by his friend Edmund Goodwyn and the surgeon Thomas Trotter, that scurvy was due to a gradual deprivation of oxygen from the body. Consumption, on the other hand, appeared to be due to an excess of oxygen in the system, and patients were known to have derived benefit from breathing a mixture of air and carbonic acid.[15] The following year (1794), Beddoes's *Letter to Erasmus Darwin* reiterated the benefits of pneumatic therapy in cases of consumption. A representative instance was that of the son of George Crump, a surgeon-apothecary in Albrighton, Shropshire. The boy was prescribed a mixture of airs (of lower oxygen content than normal atmospheric air) and a special diet; he apparently derived some relief from the therapy, his father reporting that "the poor boy used frequently to ask me for some of Dr. Beddoes's breath."[16]

In the following year, Beddoes began what was to be a longstanding collaboration with the inventor and industrialist James Watt. Watt's daughter Jessie was afflicted by a hysterical fever and, although Beddoes made efforts to save her by pneumatic therapy, died in June 1794. Watt was nonetheless impressed by Beddoes's confidence in the possibilities of pneumatic medicine, and pledged his support for the project. He helped to raise subscriptions for the Pneumatic Institution from among his Midlands acquaintances, collaborated with Beddoes on the *Considerations on the Medicinal Use of Factitious Airs* (initially published in two parts in 1794, and expanded to five parts in 1795–6),[17] and manufactured apparatus for administering the gases. With Watt's support, Beddoes's

---

15 Beddoes (1793b), esp. pp. 22 fn., 45–54, 114–128. On Goodwyn and Trotter, see Carpenter (1986), pp. 88–91.

16 Beddoes [1794b], esp. pp. 33–38. 17 Beddoes and Watt (1794–96).

enterprise slowly gathered momentum, and in March 1797 he was able to advertise the opening of the Institution, where free pneumatic treatment was offered for consumption, asthma, palsy, scrofula, venereal diseases ". . . and other Diseases, which ordinary means have failed to remove."[18]

Detailed accounts of the circumstances in which the Pneumatic Institution was founded have already been given by Levere and Stansfield.[19] What warrants emphasis here is the degree to which Beddoes's project was rooted in an Enlightenment model of public science, a model, as we have seen, that was articulated by Priestley in connection with his dissemination of his own chemical discoveries. Beddoes, we should note, also developed his program in conjunction with overt calls for social reform and for reform of the medical profession. He attempted to place himself at the head of a network of progressive medical practitioners who were continuing the project of pneumatic medicine initiated by Priestley. As one might expect, Beddoes found most of his support coming from the same circles of enlightened provincial intellectuals who had supported his predecessor, centered around the remaining members of the Lunar Society of Birmingham. As Priestley had already discovered, such an alignment could shelter the experimenter from isolation from the metropolitan scientific establishment. In the course of the 1790s, Beddoes found himself ostracized by the Royal Society and by its President, Banks. In this situation, he deployed well-tried techniques for communicating his discoveries. He published a series of factual narratives of pneumatic treatment, including detailed descriptions of the preparation and application of the gases, to permit widespread replication of his experiments. With Watt's assistance, he also distributed apparatus for administering the gases to sympathetic medical practitioners and to patients themselves.

The connection between the development of pneumatic medicine and hopes for political and social reform was made quite explicitly by Beddoes – for example, in the Preface to the *Observations on Calculus*:

. . . now, when the human mind seems, in so many countries, about to be roused from that torpor, by which it has been so long benumbed, we may reasonably indulge the expectation of a rapid progress in this, the most beneficial of all the sciences. An infinitely small portion of genius has hitherto been exerted in attempts to diminish the sum of our painful sensations; and the force of society has been exclusively at the disposal of Despots and Juntos, the great artificers of human evil. Should an entire change in these two respects, any where take place, every member of society might soon expect to experience, in his own person, the consequence of so happy an innovation.[20]

18 Advertisement quoted in Levere (1977), p. 46.
19 See particularly: Levere (1977); Stansfield (1984), ch. 9; and Stansfield and Stansfield (1986). Cartwright (1967) is also useful.
20 Beddoes (1793b), p. iv.

In other works, calls for reform were more precisely targeted at the medical establishment. Beddoes took the viewpoint of a new breed of practitioners, distrustful alike of unqualified "empirics" and of those who rested on the laurels of university degrees. He articulated his claims in an Enlightenment idiom, attacking "the parade of mystery with which [medicine] is usually enveloped" and denouncing those physicians who "have contrived to retain a privilege which the priesthood has lost." The rational alternative, in Beddoes's view, was a physic based on secure scientific knowledge, especially of chemistry and physiology.[21]

One component of the scientific medical theory that Beddoes endorsed, and through which he sometimes rationalized the effects of the gases used in pneumatic medicine, was Brunonianism, the doctrine of the heterodox Edinburgh physician John Brown (1736–1788). Beddoes published an edition of Brown's works in 1795, quoting on the title page the judgment of his friend Erasmus Darwin that the works in question were of "great genius." The basis of Brown's theory was the claim that diseases could be divided into two classes, "sthenic" and "asthenic," according to whether they were caused by excessive or insufficient excitement of the nervous system. Too much or too little excitement could be counteracted by the appropriate therapy, respectively either depressant or stimulant. The gases used in pneumatic medicine were described in these terms by Beddoes: Carbon dioxide was a depressant, for example, whereas oxygen was a stimulant. Beddoes was not a consistent devotee of Brunonianism, nor did he manage to convince many of his colleagues of its correctness, but his remarks indicate that he did sometimes see it as a partial answer to his dreams of a scientific theory of the body and its disorders.[22]

For Beddoes, medical reform was perceived as part of the social and political reform for which he never ceased to argue. Through the 1790s he produced a series of cheap pamphlets on topics such as the desirability of peace with France and the affront to freedom of speech represented by the Pitt government's "gagging bills."[23] In his *Essay on the Public Merits of Mr. Pitt* (1796), he defended the involvement of a physician in political affairs: Doctors were professionally interested in the causes of pleasure and pain, he argued, and this concern necessarily required attention to the social context of human life, as well as to individuals.[24]

21 Brown (1795), p. cvi; Beddoes [1749b], p. 29 fn.; idem. (1799), pp. 34, 38; idem. (1808), pp. 17–18; and Beddoes's views paraphrased in letter from Robert Southey to John May, August 1799, quoted in Cartwright (1952), p. 73. Porter (forthcoming) provides the most detailed study of Beddoes's ambitions to reform medical practice.
22 Brown (1795). On Beddoes's Brunonianism, see Stansfield and Stansfield (1986), pp. 280, 295. For comments on his views by Erasmus Darwin, see: Darwin to Beddoes [summer 1794?], in King-Hele (1981), p. 254.
23 Beddoes [1794c] (priced at 2 pence), [1795] (priced at 3 pence).
24 Beddoes (1796), pp. 11–13, 17–18.

Pneumatic medicine was seen as an integral part of the process of medical enlightenment that Beddoes sought to bring about. Pneumatic therapy was to be made available not just to medical practitioners but to patients themselves. The expressed aim was that the apparatus for breathing gases "would soon come to be ranked among the ordinary articles of household furniture."[25] By diffusing apparatus and techniques, the public would be enabled to take control of their own medical treatment. This would reduce their subordination to physicians, whose social authority was largely undeserved in Beddoes's view. A more equal distribution of medical knowledge would bring progress in the development of new techniques. Beddoes noted that "The more widely any species of knowledge is disseminated, the more rapidly we may expect that it will make advances."[26] Hence he saw the popularity of the first edition of *Considerations on Factitious Airs* as a good omen, both for the progress of medical knowledge and for the introduction of more egalitarian relations between doctors and patients. The sales of the work indicated, he wrote, "a rising disposition in mankind to take what belongs to their welfare into their own consideration; and to emancipate themselves still further from the danger and servility of implicit confidence."[27]

Given his belief in the political significance of pneumatic medicine, it is unsurprising that Beddoes should have sought support for his project in ways similar to those Priestley had used to find audiences for his work. Withering and Darwin, who had been members of Priestley's Midlands circle in the 1780s, were recruited to endorse Beddoes's campaign, as was Percival.[28] He found further support among a group of younger doctors who were keenly experimenting with gases. As in previous decades, the doctors who were most sympathetic proved to be largely based in the provinces and educated outside Oxbridge; many were pursuing careers in the new provincial hospitals.

Robert Thornton (1768?–1837) was one physician whose work was prominently represented in Beddoes's writings. A Cambridge graduate, Thornton had also studied chemistry at Guy's Hospital medical school. He went on to develop a chemical theory of respiration and applied it to therapeutics in his widely-read book, *The Philosophy of Medicine* (1796), on the basis of which he appears to have built a substantial London practice. He was an admirer of Darwin's work, and unashamedly liberal in his politics.[29]

Richard Pearson (1765–1836) was one of a number of Edinburgh graduates whose pneumatic therapy was described in Beddoes's works.

25 Beddoes [1794b], p. 55.   26 Beddoes (1797), p. 23.
27 Beddoes and Watt (1796), preface to the 2nd edn., p. 9*.
28 Beddoes (1793b), p. 128; idem. [1794a], [1794b].
29 Beddoes [1794a], pp. 22–24, 35–38. See also the *D.N.B.* for biographical details of Thornton.

From 1792 to 1801 he was physician at the General Hospital, Birmingham, where he wrote his *Short Account of the Nature and Properties of Different Kinds of Airs* (1795). Another Edinburgh MD and hospital physician, whose work was cited by Beddoes in support of his own, was John Ferriar (1761–1815), a member of the Manchester Literary and Philosophical Society who in 1789 was appointed physician to the Manchester Infirmary. Thomas Garnett (1766–1802), who reported to Beddoes from Harrogate on pneumatic techniques, was also an Edinburgh graduate. In the course of his career he published chemical analyses of mineral waters and was appointed Professor of Natural Philosophy and Chemistry at the Royal Institution in 1799. Other hospital-based physicians who experimented with pneumatic techniques at that time included John Alderson (1757–1829), physician at the Hull Infirmary, and William Saunders who apparently invited Beddoes to try his techniques on his patients at Guy's Hospital.[30]

The physicians who were receptive to Beddoes's work, like their predecessors who had welcomed Priestley's pneumatic chemistry in the 1770s, comprised an identifiable subset of the medical profession as a whole. They were those Davy was to describe as "the liberal and enlightened physicians of England," who shared aspirations for reform of medical practice and an interest in the techniques of experimental science.[31] Toward the end of the eighteenth century they were able to take advantage of the growing number of hospitals being established in London and the provinces. These institutions provided many practitioners with security of employment and an appropriate setting for experimental work.

It would be wrong, however, to imply that all hospital physicians were sympathetic to what Beddoes was doing. Bartholomew Parr (1750–1810), an Edinburgh graduate and physician at the Devon and Exeter Hospital, took a much more negative view. In a volume of *Essays by a Society of Gentlemen at Exeter* [1796], Parr wrote that pneumatic techniques might indeed prove beneficial in certain cases of cancer, scurvy, consumption, and skin complaints, but that Beddoes's own experiments and writings had "contributed to retard the progress of this kind of investigation, by carrying it farther than observation will support him."[32]

Notwithstanding Parr's criticism, Beddoes could rely on a large degree

30 Cases by these practitioners are reported in Beddoes and Watt (1794–96), part I (Thornton, Pearson, Ferriar, Garnett), part III (Thornton, Alderson, Pearson), part IV (Thornton, Alderson). Biographical information can be found in the *D.N.B.* On Saunders, see also Stansfield (1984), p. 158.

31 Davy, Draft Prospectus for the Pneumatic Institution [April 1799?], in Royal Institution, Davy MSS [hereafter cited as Davy MSS (RI)], box 20b, pp. 11–16 (quotation on p. 13).

32 [Anon.] (1796), p. 282. I have used the University of Wisconsin copy of this work, which is annotated with the names of several of the contributors. Some of these attributions are based on the brief account of the society in [Anon.] [1798], pp. 220–221. On Parr, see also the *D.N.B.*

of support from the Edinburgh medical and scientific community. When part III of the *Considerations on Factitious Airs* appeared in 1795, the subscribers to the Pneumatic Institution listed there included Edinburgh professors Joseph Black (Chemistry), Andrew Duncan (Institutes of Medicine), and Alexander Monro II (Anatomy). Other leading members of the scientific community in the city, such as Sir James Hall, the surgeons Benjamin and John Bell, and the surgeon and lecturer Andrew Fyfe, also backed him. The Royal Medical Society, of which Beddoes had been president while he was studying in Edinburgh, also subscribed. In addition, former Edinburgh students comprised a large proportion of the medical practitioners who contributed to get the Pneumatic Institution off the ground. Of the forty-four subscribers listed as MDs, at least thirteen and possibly as many as twenty-one were Edinburgh graduates; others may have attended the university for a while in the course of their education, as Beddoes himself had done.[33]

In the same list, Beddoes demonstrated the depth of backing for his enterprise among the Midlands enlightened gentry that had previously encouraged Priestley. Watt, of course, was one of the Lunar Society subscribers, as were his business partner Boulton, and Keir, whom Beddoes had come to know in 1791. Richard Lovell Edgeworth, a friend of Darwin and other members of the Lunar circle, became an acquaintance of Beddoes when the latter moved to Bristol. The connection was strengthened when Beddoes married Edgeworth's daughter Anna in 1794, and his new father-in-law became a subscriber. Darwin helped out by gathering subscriptions among his acquaintances in Derby, including leading physicians and manufacturers in the area. Darwin, Watt, Boulton, and Thomas Wedgwood encouraged other members of their families to subscribe, and no less than seven Wedgwoods paid up. Thus support for pneumatic medicine was extended over two generations of the Midlands elite.[34]

Like Priestley, then, Beddoes was finding his strongest supporters among the leading figures of the English provincial Enlightenment, though he had the additional asset of the Edinburgh connection to draw upon. As with Priestley also, such support provided compensation for a degree of isolation from the Royal Society. Relations between Beddoes and Banks

---

33 The list appears in Beddoes and Watt (1794–96), part III, pp. 111–112. 189 subscribers were listed. In addition to the 44 MDs, there were 19 surgeons, but only one apothecary (Thomas Henry, the Manchester chemist). Edinburgh graduates have been traced through [University of Edinburgh] (1867). Unfortunately there remain a large number of uncertain identifications, given the sparse information on individuals in both of these sources.

34 On Beddoes's relations with members of the Lunar Society, see Stansfield (1984), chh. 5, 6. For Darwin's efforts, see his letters to Watt, 17 November and 30 November 1794, in King-Hele (1981), pp. 266–269.

had apparently been reasonably cordial through 1791, when Banks transmitted two papers by Beddoes to the Society. The following year, however, Beddoes began to perceive that he was being ostracized by the President.[35] This was undoubtedly an accurate perception, in view of what appears to have been a campaign by Banks to pack the Society's Council with members of the political establishment and to exclude troublesome radicals.[36] Although Beddoes tried to lobby Banks for a contribution to the Pneumatic Institution, he was not surprised when this was not forthcoming. The speculation among Beddoes's friends was that Banks had "seen Beddoes's cloven *Jacobin* hoof and it is the order of the day to suppress all *Jacobin innovations* such as his is already called."[37]

In the face of the indifference of the national scientific institution, Beddoes naturally turned to his audience in the provincial enlightened community. His task was to convey his pneumatic therapeutics to them in a way that would encourage them in turn to sustain his initiative. The mode of communication he adopted relied considerably on the tactics Priestley had employed in his appeal to a very similar audience in previous decades. Beddoes had to convince readers of his works that his enterprise was a thoroughly public one – that it was not designed to serve a private or party interest and all the knowledge produced would be communicated without any attempt at secrecy. It was essential, as he put it, to show that the Pneumatic Institution "can scarcely be suspected as a private or party job."[38] In this connection, it was not a matter for shame, indeed it was something of a boast, that no money had been donated by "the ministry."[39] The Institution's independence from the organs of government helped to guarantee its public status.

Beddoes's style of writing served the purpose of communication within a public scientific culture conceived in this way. It was essential that faithful narrative accounts of symptoms and treatment should be given, with authenticating details of time and place and the names of patients, doctors, and witnesses. A tight rein was to be kept on expressions of speculative theory or idealistic hopes. Readers could then be invited to judge of the authors, "how far their reasonings are distinguished by philosophical scepticism, and their reports by the austerity of truth."[40] As Beddoes realized, the credibility of the knowledge produced by the Institution would

35 Stansfield (1984), p. 155.
36 Miller (1989), pp. 162–163. See also the correspondence between Priestley and Banks in 1790, over the rejection of Thomas Cooper's candidacy for FRS, in Dawson (1958), pp. 687–688.
37 James Watt, Jr. to John Ferriar, 19 December 1794, quoted in Levere (1984), p. 196.
38 Beddoes and Watt (1796), part I, p. 19*.   39 Beddoes (1799), p. 46.
40 Beddoes and Watt (1796), part I, p. 13*. Compare Beddoes's remarks in letters to Watt, about the problems of finding a "plausible form" of writing to engage public attention and support, cited in Stansfield and Stansfield (1986), pp. 282–283.

depend upon the credit ascribed by the public to its procedures and personnel. Accordingly, he assured readers, "all imaginable precautions would be taken to authenticate facts and give them publicity."[41] This was particularly important in view of the widespread distrust of medical claims thought to be fraudulent or exaggerated. For this problem, the solution lay in the accreditation of factual claims in as public a manner as possible. Beddoes explained:

The fidelity of medical narrations is of immense importance. But the publications of the fraudulent and the undiscerning have almost destroyed all confidence in reports of successful treatment. No means, therefore, of securing authenticity, should be neglected. The whole business should be conducted in the most open manner possible, secrecy of any sort being manifestly incompatible with a design, calculated for the universal benefit of mankind.[42]

As well as a particular type of written narrative, Beddoes's project relied on the circulation of breathing apparatus to communicate its results to a lay audience. An instrument designed by Watt was described in detailed specifications, which were revised in each successive edition of the *Considerations on Factitious Airs*. The device comprised a furnace, in which chemical reactions would be performed, a cooling chamber through which the gases were to pass, and a gasometer to keep them under constant pressure. The aim of the description was to enable the apparatus to be used even by those who, as Watt put it, "have not been accustomed to chemical experiments." Hence, he went on, "clearness has been aimed at, even at the hazard of prolixity."[43] In addition, detailed plates, over which Watt and Beddoes took some pains, would allow for copies of the instrument to be made by other craftsmen. The renunciation of property rights over the design constituted a significant act of public benevolence by Watt and Boulton, who were normally very active in protecting their patents on designs of steam engines. In light of this, Beddoes complimented Watt on his generosity and asserted, "you have succeeded so far as to enable any one, who chooses, to procure elastic fluids with perfect ease, and in the utmost abundance."[44]

Notwithstanding the benevolent gesture of the publication of the designs, it is clear that Watt and Boulton engaged in the manufacture of the breathing-machines as a commercial operation. They were making and advertising them from the summer of 1794. The larger version of the apparatus was priced at 8 pounds 8 shillings with accessories available for a further 5 pounds. Users of the instrument of whom records survive all appear to have purchased them from the partners' factory at Soho. Perhaps partly as a marketing ploy, attempts were initially made to at-

41 Beddoes and Watt (1796), part I, p. 19*.
42 Beddoes and Watt (1794–96), part I, pp. 4–5.
43 Beddoes and Watt (1796), part II, pp. 181–182.
44 Beddoes and Watt (1796), part I, p. 5.

tract aristocratic users, notably those sympathetic to radical and opposition causes, such as the Duke of Bedford and Lord Daer. But a more reliable method of circulating the apparatus proved to be via networks of acquaintances. Beddoes persuaded several of his patients in Bristol and the surrounding area to order machines. The Manchester apothecary Thomas Henry was also enrolled in the effort; he ordered copies for himself and one of his sons and encouraged their use at the Manchester Infirmary.[45] Darwin was another important proselytizer for the instrument. In February 1794 he had asked Beddoes to publish a full account of the machine and its use, "which would put the experiment into many other hands." The following July he received an instrument from Watt and was soon trying it out on patients. By April 1795 he was ordering more apparatus and recommending it to the infirmaries in Nottingham and Shrewsbury.[46]

Darwin's correspondence with Watt also reveals some of the problems that arose as the machines were distributed in this manner. Written descriptions of procedures were sometimes found to be inadequate to allow patients to treat themselves unsupervised. Even when full instructions were given, patients like the troublesome Lord Daer could not always be relied upon to stick to them. Darwin pointed to a further fallacy in therapeutic methods – namely, the need to ensure that the patient inhaled the gases fully. There were also problems with the machine itself, such as maintaining airtight seals and ensuring the purity of the chemicals used in the reactions. Darwin found that manganese of low quality could produce oxygen that would be quite unbreathable.[47]

In addition to these technical problems, Beddoes was experiencing difficulties that stemmed from his political marginalization. In his 1793 *Letter to Erasmus Darwin,* he acknowledged the possibility that he would attract "ridicule and obloquy," and he went on, "Of course I must expect to be decried by some as a silly projector, and by others as a rapacious empiric." In 1794, Black wrote to Watt that Beddoes "does not do well to get into quarrels and disputes if they can possibly be avoided." Avoiding disputes, however, was not in Beddoes's character. He admitted in a letter to Watt at the end of 1795 that "I know well that my politics have been very injurious to the airs."[48]

The problems thrown in Beddoes's path by increasing political oppo-

45 Stansfield and Stansfield (1986), pp. 286–289, 294.
46 Darwin to Beddoes, 6 February 1794, and Darwin to Watt, 3 July and 17 August 1794, and 28 April 1795, in King-Hele (1981), pp. 242–244, 251–252, 258–269, 284–285.
47 Stansfield and Stansfield (1986), p. 290; Darwin to Watt, 1 September 1794, 2 April, 9 July and 21 November 1795, in King-Hele (1981), pp. 260–261, 283–284, 285–286, 288–289.
48 Beddoes [1794b], p. 4.; Black to Watt, 28 October 1794, in Robinson and McKie (1970), pp. 209–210; and Beddoes to Watt, 25 December 1795, quoted in Stansfield and Stansfield (1986), pp. 284–285.

sition indicate very clearly how his project was tied to a specific socio-political conjuncture. In advancing his program of a revived pneumatic medicine, Beddoes had communicated with his audience according to Enlightenment notions of how science should be constituted in the public realm. His style of writing, his cultivation of connections with progressive medical practitioners and among provincial elites, and his diffusion of instruments to render experiments widely replicable all recapitulated Priestley's methods of advancing science in his society. What Beddoes was experiencing, however, were difficulties stemming from the unreceptive political climate of the 1790s. These problems were to be perceived much more acutely and directly when his project moved on to investigate the properties of a new gas, nitrous oxide. The controversy surrounding experiments with this gas vividly demonstrated the obstacles confronting an attempt to constitute chemistry as a public science in the age of "the end of the Enlightenment."

Enthusiastic respirations: The nitrous oxide incident

Nitrous oxide became a concern of the Pneumatic Institution on the arrival of the young Cornishman, Humphry Davy (1778–1829), who was appointed assistant to Beddoes in October 1798. Davy had come to Beddoes's attention as the author of a crushing critique of a paper by Samuel Latham Mitchill, which had been published in part V of the *Considerations on Factitious Airs* in 1796. Mitchill's claim was that the gas he called "gaseous oxyd of azote" was an agent of putrefaction and was responsible for transmitting infectious diseases. "If a full respiration of the gaseous oxyd be made," he asserted, "there will be a sudden extinction of life; and this accordingly accounts for the fact . . . of many persons falling down dead suddenly, when struck with the contagion of the plague."[49] Davy's response to the paper was to begin a series of experiments that quickly refuted Mitchill's claims that the gas was poisonous. He developed a method of producing it in a relatively pure state by heating ammonium nitrate and, when he arrived at Bristol, set in train an investigation of its physiological effects and possible therapeutic benefits.

In April 1799, using a reaction vessel built by Watt that communicated with a silk bag held over his mouth, Davy recorded his first prolonged respiration of a sample of the pure gas. He noted "a feeling analogous to that produced in the first stage of intoxication." Further trials produced a thrilling in the nerves, an increased feeling of muscular power and "an

49 S.L. Mitchill, "Remarks on the gaseous oxyd of azote or of nitrogen," in Beddoes and Watt (1794–96), part V, pp. 41–69 (quotation on p. 56).

irresistible propensity to action."[50] Toward the end of 1799, Davy had
an air-tight breathing chamber constructed, in which he was confined on
one occasion for more than an hour, breathing a mixture of nitrous oxide
and air. When he emerged, he respired more nitrous oxide from a bag.
This produced the most extraordinary effects yet – a feeling of extreme
pleasure, a perceived loss of connection with the external world, and a
succession of novel images and perceptions. Davy described the experience in this way:

By degrees, as the pleasurable sensations increased, I lost all connection with
external things; trains of vivid visible images rapidly passed through my mind,
and were connected with words in such a manner, as to produce perceptions
perfectly novel. I existed in a world of newly connected and newly modified ideas.
I theorised – I imagined that I made discoveries. . . . My emotions were enthusiastic and sublime . . . with the most intense belief and prophetic manner, I exclaimed to Dr. Kinglake, "*Nothing exists but thoughts! – the universe is composed of impressions, ideas, pleasures and pains*".[51]

Predictably, Beddoes tried the effect of the gas on himself and proceeded to apply it to a number of patients at the Institution. He reported
that it seemed to provide relief in cases of paralysis but could provoke a
negative reaction in patients suffering from hysteria.[52] Perhaps more surprisingly, visitors to the Beddoes household were also offered nitrous
oxide to breathe, an experience that many of them found highly pleasurable. Friends of Beddoes who came to Bristol to inhale the gas included
some of those who had been prominent supporters of his research:
Edgeworth, James Watt, and his son Gregory, Thomas Wedgwood and
his brother Josiah.[53] The poets Robert Southey and Samuel Taylor Coleridge, the actor William Tobin, the chemist William Russell Notcutt, and
the lawyer Peter Mark Roget (1779–1869) also took part in the experiments. Their descriptions of what they experienced were published by
Beddoes in his *Notice of Some Observations Made at the Medical Pneumatic Institution* (1799) and by Davy in the fourth part of his *Researches, Chemical and Philosophical, Chiefly Concerning Nitrous Oxide* (1800).

Most of the subjects greatly enjoyed the experience; and they must also
have provided some amusement for the spectators. Tobin, for example,
recorded: "I threw myself into several theatrical attitudes, and traversed

50 H. Davy, "Researches, chemical and philosophical, chiefly concerning nitrous oxide . . .
   and its respiration," in Davy (1839–40), III, esp. pp. 270–271, 285–287; Beddoes
   (1799), pp. 7–8. See also Hoover (1978).
51 Davy (1839–40), III, pp. 289–290.
52 Davy (1839–40), III, pp. 320–324; Beddoes (1799), pp. 16–20.
53 Gregory Watt wrote to Davy that he was coming to visit Bristol in October 1800, and
   urged his host, "Get an air holder of gas prepared for I am determined to ascend the
   heavens" (Watt to Davy, 11 October 1800, in Davy MSS (RI), box 9, p. 85).

the laboratory with a quick step; my mind was elevated to a most sublime height." Edgeworth "capered about the room without having power to restrain himself." Thomas Wedgwood recorded, "I had a very strong inclination to make odd antic motions with my hands and feet. . . . I felt as if I were lighter than the atmosphere, and as if I was going to mount to the top of the room." Mrs Beddoes similarly, "could walk much better up Clifton Hill [and] frequently seemed to be ascending like a balloon."[54]

Such extraordinarily striking experiments posed immediate problems of interpretation and assimilation. Both Beddoes and Davy offered interpretations of the effects of nitrous oxide in terms of the Brunonian theory, though both seem to have felt that the action of the gas could not be fully explained by it. Beddoes described the gas as an excitant, and it seems to have been because of his commitment to this view that he failed to recognize its anaesthetic qualities. Davy's opinion was that the gas demonstrated the falsity of a Brunonian theory of excitation; his experiments had failed to show the debility that would be expected to follow excitation on the Brunonian view. Instead, nitrous oxide seemed to provide for a kind of "renovating excitability" – a heightening of mental and bodily activity without the corresponding after-effect of depression.[55]

Even more pressing than problems of theoretical interpretation were those surrounding assimilation of the experiments within the public culture of science to which Beddoes was committed. To establish the results as valid in this context meant satisfying a series of requirements. Skeptics would have to be convinced that real effects were being produced, and this could only be achieved by describing the subjects' experiences precisely and replicating the experiments more widely. As Beddoes and Davy both realized, skeptical outsiders would attribute the effects to the power of the imagination, not to the physiological properties of the gas.[56] To circumvent this objection, controls were introduced: The subjects were initially given pure atmospheric air to breathe, then switched to nitrous oxide but not told what effects to expect. By these "precaution[s] against the delusions of the imagination," Beddoes hoped to make his experimental results credible among the public at large.[57]

There remained, however, the problem of describing the experienced effects in terms that readers of the written accounts would find believable. In the case of this particular experimental phenomenon, Priestley's style of detailed and apparently objective description seemed quite impossible. Davy referred to the peculiar difficulties of recollecting and de-

54 Davy (1839–40), III, pp. 297, 308; Beddoes (1799), pp. 10, 13.
55 Davy (1839–40), III, pp. 276, 285–291, 330; Beddoes (1799), p. 27; Cartwright (1952), pp. 108, 112–114.
56 See Davy MSS (RI), box 20b, p. 163.   57 Beddoes (1799), p. 11.

scribing the experience of breathing nitrous oxide; to this extent he wondered whether verbal descriptions could ever be enough to win over skeptics.[58] Beddoes confirmed that, "it is impossible for the combined endeavours of the spectator and the subject of [the] experiment adequately to represent what was sometimes seen and felt."[59] Many of the subjects whose testimony was recorded by Davy and Beddoes referred to the difficulty of conveying their experience in ordinary language. Roget, the future compiler of the *Thesaurus,* was ironically one of those who noted his failure to find words to describe what he had felt. Another subject, James Thomson, expressed the difficulty well:

To be able at all to comprehend the effects of nitrous oxide, it is necessary to respire it, and after that, we must either invent new terms to express these new and particular sensations, or attach new ideas to old ones, before we can communicate intelligibly with each other on the operation of this extraordinary gas.[60]

Because the effects of nitrous oxide could not be adequately described, the only way the status of the experiments could be secured was for them to be widely replicated – everyone would have to breathe the gas. This was the expressed aim of Beddoes and Davy; they would pass the nitrous oxide experiments through the social networks of the enlightened community, describing the procedures in written form and circulating the apparatus required to reproduce them. Davy maintained in his *Researches* that he had described the procedures for producing the gas with sufficient detail that "if the pleasurable effects, or medical properties of the nitrous oxide, should ever make it an article of general request, it may be procured with much less time, labour, and expense, than most of the luxuries, or even necessaries of life."[61]

Unfortunately, the actuality does not seem to have accorded with the aspiration: Replication of the experiments elsewhere was not easily accomplished. Davy himself acknowledged the possibility that attempts to reproduce the effects would be "dangerous and inconclusive."[62] Apart from the difficulty of knowing from the written accounts precisely what physiological effects were being looked for, there were the familiar problems of managing the breathing apparatus. Beddoes's accounts referred to various sources of trouble: impurities or incorrect quantities of the reactant materials, leaks in the machine, incorrect proportions in the mixture of gas with air, insufficient time allowed for respiration, and so on. The anticipated difficulties of replication allowed Beddoes to specify in advance that, in the event that anyone should challenge the reported

58 Davy (1839–40), III, pp. 293–294; Davy MSS (RI), box 20b, p. 163.
59 Beddoes (1799), p. 8.   60 Davy (1839–40), III, pp. 303, 305.
61 Davy (1839–40), III, p. 74.
62 Draft letter from Davy to William Nicholson, undated, Davy MSS (RI), box 20a, p. 247.

effects of nitrous oxide, it would be possible for them to be shown to have used improper preparative techniques.[63]

Not only, then, were the experiences produced by the gas uncertain but replication of the experiments to the extent of producing any physiological effect was problematic. In November 1799, James and Gregory Watt were attempting to reproduce at Birmingham the effects reported from Bristol. Responding to a request for information from Black, James Watt admitted that preparation of the gas could be troublesome and that it had "very different effects on different people."[64] In the same month, Gregory Watt wrote to Davy for more specific instructions on preparative techniques. Neither he nor Robinson Boulton could perceive any physiological change on breathing the gas, even though their mode of preparation was apparently the same as Davy's, the gas seemed to have the same chemical properties, and they had remembered to hold their noses while breathing from the bag.[65]

It is not known what further information Davy sent to the Watts, but the problems of replication were apparently overcome. Significantly, Gregory Watt resorted to a religious metaphor to describe how replication of the phenomenon had eventually been achieved:

The Conversion of Robinson Boulton & myself has operated to the Conviction of some unbelievers here & damped the audacity of others. We have been most zealous apostles & have met with more credit as we were materially effected ourselves.[66]

Watt's metaphor also gives us a clue as to how the fate of the nitrous oxide experiments depended upon the wider context in which they were received. In a climate that was increasingly unfavorable to Beddoes's enlightened aspirations, the experiences of the Bristol group appeared to many as the shared delusions of an isolated religious sect. Reviving a seventeenth-century term, opponents of Beddoes's enterprise charged him with "enthusiasm," the pursuit of a private truth through unrestrained use of the imagination.[67] The epistemology articulated in Beddoes's *Observations on the Nature of Demonstrative Evidence* could give some plausibility to such a charge. There he had identified sensory experience with the unmediated knowledge of intuition, and had talked about the need to engage the fancy in scientific education, commenting "it is the enthusiasm she inspires, that has worked so many miracles in art and

63 Beddoes (1799), p. 31.
64 James Watt to Joseph Black, 6, 9, and 22 November, and 8 December 1799, in Robinson and McKie (1970), pp. 307–316.
65 Gregory Watt, postscript to letter by James Watt to Davy, 13 November 1799, Davy MSS (RI), box 9, pp. 71–75.
66 Gregory Watt to Davy, 24 March [1800], Davy MSS (RI), box 9, pp. 80–82.
67 On enthusiasm in the seventeenth-century context, see especially Heyd (1981).

science."[68] Beddoes was well aware that he was courting disparagement by his contemporaries as an enthusiast.

In this respect the controversy provoked by the nitrous oxide experiments can be understood as symptomatic of "the end of the Enlightenment." Like Mesmerism, nitrous oxide was understood by its proponents in terms derived from Enlightenment natural philosophy and was mobilized socially with overtly reformist or radical intent.[69] Hence, like Mesmerism, nitrous oxide came to be attacked by political conservatives as emblematic of the follies and sins of the Enlightenment as a whole. Its advocates were denounced and ridiculed as deluded enthusiasts, and the phenomenon itself was dissolved into the product of their overactive imaginations.

For Beddoes and Davy, the experiments with nitrous oxide were an extension of the normal practices of Enlightenment chemistry, which at the same time promised a much more fundamental and "sublime" knowledge of the phenomena of life. Beddoes saw the recent progress of chemical science as having brought gaseous fluids progressively within the range of instruments of investigation. The promise was that even more subtle fluids, such as electricity, light, and the nervous fluid, could also be made the subject of chemical research. Thus, he wrote of Lavoisier:

By contrivances calculated to convey the most distinct perceptions, he everywhere impresses a degree of conviction, which, not many years ago, would have been thought unattainable in the study of impalpable substances. The different kinds of air may perhaps be considered, with respect to our senses, as occupying a middle place between palpable bodies and the *aetherial* fluids, if any such exist. Perhaps another Lavoisier, by bringing these as much within the sphere of the senses, may exhibit almost mathematical evidence of the qualities of *fire, electricity* and *magnetism*.[70]

Beddoes and Davy sought to surpass Lavoisier's achievements by extending instrumental practices into the realm of more subtle forms of matter, thus bringing life itself within the reach of chemistry. As Davy explained, since the discovery of the role of oxygen in respiration, chemistry had begun to reveal knowledge of the phenomena of living matter. The experiments with nitrous oxide had shown that a chemical agent could have dramatic effects on actions and sensations, and hence offered the prospect of scientific control of human pleasure and happiness. Beddoes and Davy could see the way open to a "new philosophy" or "sublime chemistry" that would teach "the means of procuring pleasure &

68 Beddoes (1793a), pp. iii, viii–ix.   69 Darnton (1970).
70 Beddoes (1793a), pp. 110–112. Compare his remarks in [Beddoes], review of E. Peart, *On the Composition and Properties of Water*, in *Monthly Review*, 2nd ser., 23 (May–August 1797), 139–142, on p. 141. For similar views expressed by others, see Davy (1839–40), II, p. 32; and Bergman (1783a), pp. 17–19.

removing pain." The ultimate aim was (in Davy's words) "that the phenomena of life [might be] capable of chemical solution," or (according to Beddoes) that medicine might become "a matter of calculation."[71]

The philosophical importance of nitrous oxide was that it illuminated the relations between mind and body. Davy's notebooks from the time show that he interpreted the effects of the gas in a way that connected him directly with the tradition of Enlightenment materialism. He believed that his researches had provided a clear demonstration of the material basis of human actions, emotions, and perceptions. The materialist filiation of his views is made most explicit in an unpublished "Essay to Prove that the Thinking Powers Depend on the Organization of the Body," written in a notebook dating from the mid-1790s. There Davy developed the idea that the nervous system was activated by a material substance, a kind of ether, communicating between the brain and the muscles. He remarked, "The Phaenomena which were formerly attributed [to] *psyche* seem to be the effect of a peculiar action of fluids upon solids & solids upon fluids." "The Body is . . . a fine tuned Machine," he wrote, echoing the title of the most infamous of eighteenth-century materialist works, *L'Homme Machine* (1748) by Julien Offray de LaMettrie. Even more hazardously, he argued that the soul itself was a material entity and that the scriptures (properly interpreted) gave support to this view. "The Immaterialists maintain an impious Opinion," he claimed, "when they assert that God is unable to make matter think. [Despite] asserting his omnipotence, they limit his power." With this formulation, Davy connected himself with the "thinking matter" tradition of English materialism, which derived from John Locke via David Hartley and Joseph Priestley.[72]

Conservative thinkers, on the other hand, despised materialism and what they took to be an associated disposition to atheism. They blamed the Enlightenment as a whole for the appalling spectacle of the French Revolution. Several conservative writers of the 1790s seized on the nitrous oxide experiments as signs of the dangerous delusions encouraged by Enlightenment philosophy.[73] One group comprised George Canning, Charles Ellis, and Hookham Frere, editors of the *Anti-Jacobin Review* (founded in 1797). In 1798 they published "Loves of the Triangles," a satirical attack on the doctrines of Erasmus Darwin that also took aim

71 Davy MSS (RI), box 13h, pp. 15, 17; box 20a, pp. 20, 21, 266; Beddoes (1799), p. 34.
72 Davy MSS (RI), box 13f, pp. 33–47 (quotations on pp. 47, 41, 43). For background, see Yolton (1984).
73 Skepticism about the claims concerning nitrous oxide by John Ferriar, a former supporter of Beddoes's Pneumatic Institution, might have been motivated by his antipathy to materialism. See: [Ferriar], review of Beddoes (1799), in *Monthly Review*, 2nd ser., 30 (September–December 1799), 60–72, where the reviewer remarks that "to us the experiments do not convey demonstration . . . they rather create difficulties than furnish explanations." For Ferriar's anti-materialism, see: Ferriar (1793).

at pneumatic medicine and the contemporary interest in galvanism.[74] This was followed in 1800 by a poem, "The Pneumatic Revellers: An Eclogue," which satirized activities at the Pneumatic Institution. Beddoes and his friends were portrayed here as enthusiasts, who used the gases discovered by the satanic Dr Priestley to enjoy orgies of intoxication and sexual license. Beddoes was said to have been carried away by the force of his own imagination, deluded in his belief that nitrous oxide could be a material agent of enlightenment. A few lines of "The Pneumatic Revellers" convey the tone of the whole:

> When I tried it, at first, on a learned Society,
> Their giddiness seem'd to betray inebriety,
> Like grave Mandarins, their heads nodding together;
> But afterwards each was as light as a feather:
> And they, every one, cried, 'twas a pleasure extatic;
> To drink deeper draughts of the mighty pneumatic.[75]

The basic themes of the *Anti-Jacobin* offensive, and its condescendingly satirical tone, became common in other conservative attacks on pneumatic chemistry. One contemporary remarked that Beddoes had "converted the laboratory into a region of hilarity and relaxation."[76] The conservative Professor of Natural Philosophy at Edinburgh, John Robison, noted sardonically of the nitrous oxide experiments that "to those who are not hurt by the sight of folly, they are . . . very amusing."[77] Some of this rhetoric and imagery was repeated by the Carlisle surgeon Robert Harrington, for whom Davy and Beddoes were *"aerial flying* chemists" with their *"ecstatic, lunatic,* and *Laputatic* sensations."[78]

Further mocking criticism of the nitrous oxide experiments occurred in an anonymous work entitled *The Sceptic,* published in 1800. There pneumatic medicine was ridiculed together with Lavoisier's new theory of chemistry, galvanism, and Mesmerism. All these "wonders" were symptoms of an age of revolutionary upheaval, claimed the author, parodying the apocalyptic tone with which some enlightened philosophers had greeted the French Revolution. But, he went on, the supposed wonders were actually imaginary phantoms, not genuine phenomena at all. Of Lavoisier's antiphlogiston, for example, he wrote, "like a fog, [he] diffused himself all over Europe: and like a fog too, he is a mere vapour, which involves all objects in obscurity." The pneumatic chemists and

74 Reproduced in Rice-Oxley (1924), pp. 88–107, esp. pp. 88, 100 fn., 107 fn. For background to the *Anti-Jacobin* attack, see Garfinkle (1955).
75 *Anti-Jacobin Review and Magazine,* 4 (April–August 1800), pp. 109–118; see also the review of Beddoes (1799) in the same volume, pp. 424–428.
76 Joseph Cottle (1770–1853), author and bookseller, quoted in Cartwright (1952), p. 120.
77 Robison quoted in W.D.A. Smith (1982), p. 32.
78 Harrington (1804), pp. 72, 181.

physicians, or "the new set of empirics, the . . . quacks-pneumatic," had befuddled themselves and their patients in this all-embracing fog.[79]

In his *Considerations on Factitious Airs,* Beddoes seems to have anticipated ridicule of this kind. His introduction to part III of that work (published in 1796) included a spoof letter between two orthodox physicians who castigate what they see as the chaos and subversion being spread by new experimental therapies. One of the protagonists recounts how he dealt with patients of his who were tempted to try the new methods:

> . . . the consumptive daughter of Sir – –, who had been under my care for some weeks, was ordered to the Hotwells. Sir – asked if he should consult – [Beddoes], naming the man, out of whose hands, I am most decidedly of opinion, that all physicians, who have the state and stability of their profession at heart, should unite to keep patients. *"To be sure,"* I replied, "the disease is always fatal. It is, *in truth,* the very scourge of our island. *Would* to God some better means were devised – Researches for this purpose are *out of all question* laudable. I *most heartily* wish them success." Having thus tuned their minds to confidence with my key of candour, "but would you, Sir –," added I emphatically, "or *you,* my lady, have *experiments* made on your daughter?" My Lady shuddered at the question; she even started, nor could her countenance have expressed more horror, if she had actually beheld her daughter in the arms of an assassin.[80]

Beddoes's reputation never recovered from these attacks, of which (as Stansfield points out) he fully understood the political motivation.[81] Subsequent assessments of his career were strongly colored by the incident. Thus, when he died in 1808, the *Gentleman's Magazine* published an obituary letter, signed by "Amicus," which reiterated many of the charges leveled against him in the 1790s. Beddoes was said to have been of the school of atheists, "the doctrines of which have operated, with poisonous influence, on the great mass of society." The "great Empiric of Bristol"

79 [Anon.] (1800), pp. 1–11, 57–67. Davy was explicitly criticized for his theory of heat on p. 9, and experimental physicians were accused of sexual impropriety with their female patients on p. 10. Compare the description of oxygen as a "hocus pocus conjurer," in Harrington (1804), pp. 27, 146.

80 [Beddoes], letter supposedly from "Sir Jeremiah Morrison" to "Dr Daniel Lorimer Renshaw," June 1795, in Beddoes and Watt (1794–96), part III, pp. i–x (quotation on p. vii). The same introduction (p. ix) contained a poem by Beddoes, which anticipated the poetic form of the attacks on him by the *Anti-Jacobin*. This may have been a response to a satirical attack on Beddoes in the tract, [Anon.] (1794). From its mocking tone, this tract is clearly not by Darwin, as Levere (1984), pp. 192–193, explains. A possible identification of the pseudonymous author is given by a handwritten note at the end of the Cambridge University Library copy (shelfmark: Forster.a.22), which ascribes it to Richard Laurence, DD. The *D.N.B.* lists Laurence (1760–1838) as an Oxford contemporary of Beddoes (he matriculated at Corpus Christi College in 1778, and received his MA in 1785), but one who followed a very different career path, becoming Regius Professor of Hebrew, and subsequently Archbishop of Cashel in Ireland.

81 Stansfield (1984), p. 169.

could be seen to have been led astray by the force of his imagination and the "fervours of fanciful discovery."[82]

As we shall see in the next chapter, Davy had to work hard after he left Bristol in 1801 to rescue his reputation from the taint of association with Beddoes. His own subsequent comments on Beddoes's character were in terms similar to those just quoted.[83] Harrington for one thought that Davy was ashamed of his nitrous oxide "capers" and was doing his best to put them behind him.[84] He did not entirely succeed, however, because the incident was still brought up in an attack on him as late as 1824.[85]

As well as significantly shaping Beddoes's and Davy's subsequent careers, the nitrous oxide controversy arguably hindered the medical uses of the gas for the next forty years. There is evidence that other medical practitioners took note of the experiments at the time but their interest waned rather quickly. The effects of the gas were discussed in Thomas Charles Hope's lectures at Edinburgh in 1800 and in chemistry lectures at Guy's and St. Thomas's Hospitals in London in the early years of the following decade.[86] But in the three decades thereafter, nitrous oxide became the material of public entertainment, rather than of serious medical research. In the 1820s and 1830s it was employed in music hall stage acts in which volunteers would be intoxicated for the amusement of onlookers. In 1824, for example, a "Mr. Henry" was recorded as giving such performances at the Adelphi Theatre.[87]

Nitrous oxide thus passed into the realm of science-based entertainment, a competitor in that role to popular lectures on phrenology in the 1830s. This was the longterm effect of the ridicule of Davy's and Beddoes's initiative in the 1790s. The gas became lost to medical research and its anaesthetic properties were not fully revealed until the 1840s.[88] In the light of this, a wry remark made by Beddoes seems ironically prophetic. In his record of the observations at the Pneumatic Institution, he had pessimistically foreseen "that we might even prepare a happier aera for mankind, and yet earn from the mass of our contemporaries nothing better than the title of enthusiasts."[89] In the tormented times of the end of the Enlightenment, that was exactly what had occurred.

82 "Amicus," "Letter on the death of Beddoes," *Gentleman's Magazine*, 79(i) (February 1809), p. 120; reprinted in Cartwright (1952), pp. 158–159.
83 See Cartwright (1952), p. 75, where Davy is quoted on Beddoes's "wild and active imagination."
84 Harrington (1804), pp. 70, 208 fn.
85 "Anti-Dram," "Cheap drunkenness," *The Chemist*, 1 (March–September 1824), pp. 133–135.
86 W.D.A. Smith (1982), pp. 30–31.
87 W.D.A. Smith (1982), pp. 34–35, 39; *The Chemist*, 1 (1824), 133–135.
88 Cartwright (1952), pp. 265–6.
89 Beddoes (1799), pp. 4–5.

The end of Enlightenment science?

> . . . the period of his youth was one of peculiar excitement and innovation:
> the leaven of the French revolution was still fermenting; the mysterious
> phenomena of galvanism had recently been brought to light; . . . and pneu-
> matic chemistry had just then been called to the aid of medicine, with a
> confident expectation of wonderful effects, which deluded men of the
> soundest minds, and which could be corrected only by experience.
>
> John Davy, *Life of Humphry Davy* (1836)[90]

The episode of the Bristol nitrous oxide investigations shows the extreme
variety of interpretations of certain experimental phenomena during the
1790s. The gas, which for Beddoes and his allies was to be the material
agent of enlightenment and a means of release from physical suffering,
was seen by his conservative opponents as a comical delusion or a product of the dangerous enthusiasm of a radical sect. Given this polarity of
interpretations, the laughing-gas incident can be taken as emblematic of
the fate of the Enlightenment program for chemistry as a public science.
Reactions to the episode crystallized attitudes to a range of hotly disputed questions concerning the proper role of science in public culture.

These questions were of course implicated in the much wider controversy concerning matters of religion, politics, and the social order that
followed the outbreak of the French Revolution. Edmund Burke's *Reflections on the Revolution in France* (1790) is recognized as having pro-
foundly influenced conservative attacks on the ideals of the Revolution,
the values of its supporters in Britain, and their intellectual inheritance
from the Enlightenment. Burke's arguments and his extraordinary language and imagery also came to pervade subsequent reactionary attacks
on science. Taking their lead from him, a number of writers criticized the
pretensions and the mode of organization of Enlightenment science. They
charged natural philosophers with encouraging inappropriate dreams of
progress and spreading social subversion.

As Maurice Crosland has shown, Burke frequently used imagery derived from chemistry and sometimes referred to the role of Priestley and
other chemists in the revolutionary upheavals.[91] In the *Reflections* and in
the later *Letter to a Noble Lord* (1796), he attacked chemists and used
chemical metaphors for the chaos and destruction that their activities had
released. Thus he wrote of the Revolution, "The wild *gas*, the fixed air,
is plainly broke loose. . . ." Philosophical atheism was characterized as
an "unhallowed fire," a "hot spirit drawn out of the alembick of hell,"
which in France is now so furiously boiling." Of the revolutionary poli-

90 John Davy (1836), I, p. 79.   91 Crosland (1987).

ticians he wrote that "By their violent haste, and their defiance of the process of nature, they are delivered over blindly to every projector and adventurer, to every alchymist and empiric."[92] In the *Letter*, Burke made a more explicit attack on Priestley and the chemists who had worked for the new French government. He referred to the scheme of the French chemists J.H. Hassenfratz and Guyton de Morveau to extract saltpeter to make gunpowder from the mortar of demolished buildings. This project was, he suggested, symbolic of the destruction that awaited aristocratic property in England if the revolutionaries got their way. "There is nothing on which the leaders of the Republick . . . value themselves, more," he wrote, "than on the chymical operations by which, through science, they convert the pride of the Aristocracy, to an instrument of its own destruction."[93]

Crosland's discussion, though informative, can be extended further. His argument that Burke perceived science as a threat does not specify the reason for which Burke regarded certain sciences (or their practitioners) as a threat to social stability. Nor does Crosland's analysis relate Burke's use of metaphors drawn from chemistry to other aspects of his complex metaphorical repertoire, which has been discussed by several political historians and literary critics.[94]

Burke's imagery was structured by the contrast he perceived between the normal order of society, bound together by moral feeling and religious sentiment, and the subversion of this order that would follow from a rigid or over-extended application of metaphysical reason. He recruited metaphors from a variety of fields, including science, to illustrate this opposition. Chemical terms were among those used to represent the anarchy that resulted from subversion of the established order. Thus, ". . . the revolution in France is the grand ingredient in the cauldron"; its effects would be to tear society apart into a "chaos of elementary principles."[95] A properly ordered society, on the other hand, was portrayed as part of the harmonious system of the universe revealed by natural philosophy:

By a constitutional policy, working after the pattern of nature, we receive, we hold, we transmit our government and our privileges . . . Our political system is

92 Burke (1968), pp. 90, 187, 282.
93 Burke (1796), 27–28 (quotation on p. 28 fn.). As Crosland points out, Burke was making a stab at Priestley's "gunpowder sermon" of 1787, in which he had talked about "laying gunpowder, grain by grain, under the old building of error and superstition" (Crosland (1987), p. 285).
94 The following discussions of Burke's political aims and writing style have been useful to me: Blakemore (1988); Boulton (1963); Goodwin (1968); Hughes (1976); and Paulsen (1983), pp. 1–87.
95 Burke (1968), pp. 93, 195.

placed in a just correspondence and symmetry with the order of the world . . .
Thus, by preserving the method of nature in the conduct of the state, in what we
improve we are never wholly new.[96]

Burke believed that philosophical rationality, if unchecked, would draw
men away from their natural moral passions, which included reverence
for God and earthly rulers.[97] Philosophers who tried to strip away the
natural mystery and sublimity of human and divine power risked un-
leashing the basest animal instincts among their followers. Hence Burke
repeatedly referred to the appalling conjunction of metaphysical reasoning and brute sensuality that he perceived among the supporters of the
Revolution; or, in his own words, that "unfashioned, indelicate, sour,
gloomy, ferocious medley of pedantry and lewdness – of metaphysical
speculations blended with the coarsest sensuality."[98] Oxymorons like this
were deployed powerfully in his onslaught on the intellectuals who backed
the Revolution in the *Letter to a Noble Lord*. There Burke wrote of "can-
nibal philosophers," "a set of literary men converted into a gang of rob-
bers and assassins," "a den of bravoes and banditti, assum[ing] the garb
and tone of an academy of philosophers."[99]

Burke saw the readiness of scientists to engage in experiments, whether
chemical, economic or social, as a sign of the distortion of proper moral
feeling that arose from unrestrained indulgence of philosophical reason.
"These philosophers are fanatics," he warned, ". . . they are carried with
such an headlong rage towards every desparate trial, that they would
sacrifice the whole human race to the slightest of their experiments."[100]
In a clear allusion to Priestley, he cited pneumatic experiments with mice
as an indication of the danger to humanity if experimental philosophers
were to gain control of society:

The geometricians, and the chymists bring, the one from the dry bones of their
diagrams, and the other from the soot of their furnaces, dispositions which make
them worse than indifferent about those feelings and habitudes, which are the
supports of the moral world. . . . These philosophers consider men in their experiments, no more than they do mice in an air pump, or in a recipient of mephitic gas.[101]

The proposed Republican constitution of France was just such an in-
humane experiment, in Burke's view, and demonstrated the lack of moral

96 Burke (1968), p. 120.
97 On the natural respect for authority, see Burke (1968), p. 182: "We fear God; we
    look up with awe to magistrates; with reverence to priests; and with respect to nobil-
    ity. Why? Because when such ideas are brought before our minds, it is *natural* to be
    affected . . ."
98 Quoted from Hughes (1976), p. 38.   99 Burke (1796), pp. 24–25.
100 Burke (1796), p. 26.   101 Burke (1796), p. 27.

restraint characteristic of the intellectually arrogant. The project to extract "revolutionary gunpowder" from ruined buildings further demonstrated the great dangers of allowing such an "experimental philosophy" to hold sway over human affairs.

We can detect in these passages from Burke one theme of the rhetoric used against Beddoes's nitrous oxide experiments later in the decade. Burke had prepared the way for a portrayal of experimentation – particularly when human subjects were used and the goal was a distant prospect of general social improvement – as an immoral enterprise. He also contributed a further element to the attacks on Beddoes, – namely, the revival of the seventeenth-century satirical vocabulary in which enthusiasm had been identified and ridiculed. Burke saw revolutionary fervor as an almost material entity transmitted from agitators to their audiences. Here again, chemical metaphors were used. Of Priestley's friend Richard Price, for example, it was said that "his enthusiasm kindles as he advances; and when he arrives at his peroration, it is in a full blaze." Price's audience were affected by "the fumes of his oracular tripod," so that they emerged "reeking from the effect of the sermon."[102]

In this Burkean imagery we can locate the antecedents of the savage satires against Beddoes in the *Anti-Jacobin, The Sceptic,* and the writings of Robert Harrington. Whereas Burke likened revolutionary fanaticism to the breathing of dangerous fumes, the *Anti-Jacobin* reversed the metaphor and asserted that Beddoes's gas-breathing experiments partook of the enthusiasm of naive believers in enlightenment and revolution. For Harrington, and also for the author of *The Sceptic,* gases such as nitrous oxide and oxygen were metaphorically grouped with revolutionary enthusiasm itself as wild, uncontrollable spirits. Thus Burke set the tone for the conservative satirical attack on Enlightenment chemistry as a kind of revived alchemy – a deluded attempt to conjure with potentially dangerous spirits.

The influence of Burke's metaphors was not confined to written culture. He also stimulated the imagination of several cartoonists, whose work was being systematically used by the political establishment to shape popular consciousness, and who played a significant part in turning British political debate against the cause of the revolutionaries.[103] Priestley was frequently represented in caricature, paying the price for his high-profile identification with the supporters of the Revolution. Fitzpatrick has shown that from 1790 until his emigration in 1794, he appeared in at least twenty-seven cartoons, all critical of his radical sympathies and

102 Burke (1968), pp. 93, 157, 159.
103 For the caricaturists and Burke's influence on them, see Paulsen (1983), chh. 5, 6; and Butler (1984), p. 50.

Published as the Act directs by Bentley & Cᵒ July 1ˢᵗ 1791.   *Annibal Scratch fecit.*

DOCTER PHLOGISTON,
The **PRIESTLEY** *politcian or the*
Political Priest.

Figure 4. Anonymous caricature of Priestley as "Docter Phlogiston." From the British Museum, Department of Prints and Drawings (DG no. 7887). [Reproduced by permission of the Trustees of the British Museum.]

his activities as an agitator.[104] Some of these used chemical imagery similar to Burke's to suggest the confusion and chaos that Priestley's irresponsible oratory had stirred up. An anonymous print of 1791 showed him as "Dr. Phlogiston," brandishing his political and religious tracts, which were belching out an insidious black smoke (Figure 4). The same year the caricaturist W. Dent published a print, "Revolution Anniversary or Patriotic Incantations," showing Priestley in the company of Charles James Fox and other liberal Whig politicians, dancing around a boiling cauldron from which a cloud labeled "French spirits" was emerging. The drawing was published on 12 July 1791 in anticipation of radical celebrations of the forthcoming second anniversary of the fall of the Bastille (Figure 5). The influence of Burke's description of Price's sermon – with the "cauldron," the "oracular tripod," and the "fumes" of enthusiasm – is clear. In this striking and widely circulated form, Priestley was again branded as a conjurer of destructive and anarchic spirits. The following year a similar theme was portrayed in a drawing by James Sayers, which showed Priestley's son William appearing before the French National Assembly (Figure 6). William was presented as a surrogate for his father, holding a firebrand and an electrical conductor connected to a jar marked "Phlogiston from Hackney College." Priestley's phlogiston was again being invoked as a symbol of revolutionary fanaticism and its dangerous communicability.[105]

The pervasiveness that Burke's images achieved through their reproduction in the works of the caricaturists was an indicator of their extraordinary imaginative potency. Progressive writers whom Burke had attacked, such as Priestley, Thomas Paine, and Jeremy Bentham, resented his use of figurative allusions and metaphors and feared their efficacy with a popular audience.[106] According to Priestley, Burke was guilty of deploying eloquence without wisdom and of stooping to play the part of a "satirist" and "declaimer."[107] Bentham's view was that "the power [Burke] trusted to was *oratory* . . . the art of misdirecting the judgment by agitating and inflaming the passions."[108] But apart from protesting at Burke's rhetoric and reasserting their faith in the eventual victory of reason over superstition and servitude, there was little that the reformists could do to counter his campaign. Even the use of terms such as "en-

---

104 Fitzpatrick (1984) provides a very helpful discussion of these caricatures. They are all listed with details of publication in British Museum (1978). The ones discussed here are nos. 7887 (p. 806), 8108 (pp. 918–919), and 7890 (pp. 808–809).
105 Fitzpatrick (1984), pp. 347, 357–358, 362.
106 Boulton (1963), pp. 75–133; Garrett (1975), pp. 133–134; Fruchtman (1982).
107 Priestley (1791b). The quotations from Burke's own words on the title page are clearly intended by Priestley to be applied to Burke himself.
108 Bentham quoted in Boulton (1963), pp. 122–123.

Figure 5. "Revolution Anniversary," by W. Dent. From the British Museum, Department of Prints and Drawings (DG no. 7890). [Reproduced by permission of the Trustees of the British Museum.]

Figure 6. "M. Francois introduces Mr. Pr***tly," by James Sayers. From the British Museum, Department of Prints and Drawings (DG no. 8108). [Reproduced by permission of the Trustees of the British Museum.]

lightened" had been given a sarcastic intonation in Burke's writings.[109] He had fundamentally undermined the whole vocabulary of enlightened debate on intellectual culture and its political and social functions.

This was also the case with respect to ideas about the social organization of intellectual life. Priestley, Beddoes, and their allies had relied on an Enlightenment tradition in which the formation of networks of acquaintances and the constitution of informal societies were seen as the most desirable means of encouraging innovation. Burke's writings initiated a conservative counterattack on these assumptions. One of the strongest messages of the *Letter to a Noble Lord* was a critique of aristocratic patronage of radical intellectuals, from which Priestley had himself benefited. Aristocrats like the Duke of Bedford, who were gullible enough to encourage such people, were likely to find themselves in the unenviable situation of a mouse cornered by a cat or an ox awaiting the butcher, Burke suggested, when the revolution they supported came to affect their own domestic arrangements.[110]

No less dangerous than aristocratic patronage of enlightened philosophers was the "clubability" that had been such a prevalent feature of their lives. Burke portrayed reformist political clubs as hotbeds of revolutionary enthusiasm, where natural moral restraints were shrugged off in the excitement of democratic discussion and egalitarian social intercourse. In 1790 he referred to the "licentious and giddy coffee-houses" of Paris, and went on to assert that the policies of the French government were distorted by being subjected to discussion in public meetings, open to all comers: "In these meetings of all sorts, every counsel, in proportion as it is daring and violent, and perfidious, is taken for the mark of superior genius."[111] In 1796 he repeated the warning against groups of intellectuals who met together to encourage one another in throwing off the fear of God and earthly rulers: ". . . when in that state they come to understand one another, and to act in corps, a more dreadful calamity cannot arise out of Hell to scourge mankind."[112]

In subsequent conservative discourse the critique of Enlightenment patterns of socialization was amplified and extended. "Association," which for Priestley and his followers had been a basic means of generating and transmitting scientific knowledge, and by which they had asserted their independence from government patronage, was undermined. The most notorious work in this class was the *Proofs of a Conspiracy against all the Religions and Governments of Europe* (1797) by the Edinburgh Professor of Natural Philosophy, John Robison. Robison's paranoid attack on a supposedly European-wide conspiracy tied responsibility for the

109 For example, in Burke (1968), pp. 157, 166–168.   110 Burke (1796), pp. 27, 29–30.
111 Burke (1968), p. 160.   112 Burke (1796), p. 26.

Revolution and all the disorder that followed from it to a network of secret "associations" known as the "Illuminati." The conspiracy was said to have embraced all the leading philosophes and even some misguided rulers who had encouraged them. According to Robison, many local philosophical societies and subscription libraries were covertly engaged in subversive activities for the Illuminati.[113]

Robison's account of the Enlightenment and the French Revolution as a large-scale secret conspiracy proved popular, going through four editions in just over a year, and it chimed with the mood of a British government increasingly concerned about all forms of possible subversion. In 1795, Prime Minister Pitt introduced the first of a series of parliamentary measures against "seditious meetings," which required all groups and societies for intellectual debate to be licensed and forbade them to discuss religion or politics. In 1800, further fuel was added to establishment paranoia by William Hamilton Reid's *The Rise and Dissolution of the Infidel Societies of this Metropolis,* an account of several free-thinking political and religious groups that had sprung up in London in the previous decade. With all the zeal of a former devotee turned informer, Reid set out to expose the subversive and irreligious aims of these groups and to trace their inspiration in the doctrines of Voltaire, Rousseau, Paine, Godwin, and Priestley. Atheism and materialism, alchemy and astrology were said to be rife in these clubs, and their debates were described in terms that clearly owed much to Burke's idea that social interaction had the effect of loosening moral restraints. The antireligious views expressed by some members were, Reid noted, "by no means natural, but forced from the hot-beds of the clubs." On the belief that uninhibited debate was a means of arriving at the truth, as Priestley had claimed, Reid was scathing:

The supposition of a candid discussion in these assemblies is a mere farce; for in proportion as party-spirit enlarges its sphere of action, candour is uniformly jostled out of its place. The number of hands held up, for or against a question, is always more attended to, than the weight of arguments in its favour.[114]

In this climate of establishment paranoia, Priestley and his allies had good reason to perceive that an attack was being made on their rights to association and free-ranging philosophical discussion. In his *Letters to . . . Edmund Burke* (1791), Priestley insisted upon the right of free discussion and reasserted its value as a means of achieving philosophical truth.[115] The following year his friend and disciple Thomas Cooper re-

---

113 Robison (1797). See also Morrell (1971b); Brown (1987).
114 Reid (1800), pp. 4, 8, 16, 24, 89–91 (quotation on pp. 30–31). Reid acknowledged the influence on his work of Robison and of the Abbé Barruel, whose writings had echoed the conspiracy theme.
115 Priestley (1791b), p. 146.

plied to personal criticism that Burke had made of him and the younger James Watt in the House of Commons. He tackled Burke's rationale for recent suppression of political reform societies, restating the Enlightenment justification for association as a condition of social progress:

The God of Nature has given to Man, not merely hands to labour, but a head to think: he has given him the capability of obtaining knowledge, of mental Improvement, and of social Intercourse. One great use of Society is to bring these Capabilities into Action, that not only each Individual, but each Community of Individuals, and finally the Human Race, by means of mutual Communication, may proceed in the glorious Career of mutual and progressive Improvement.[116]

Even members of supposedly respectable scientific societies felt the pressure of the reactionary criticism of associations. Dr Hugh Downman, delivering his Presidential Address to the new "Society of Gentlemen" at Exeter in June 1792, felt obliged to refer to the contemporary debate about the dangers of a general diffusion of knowledge. Although he endorsed the ideals of free intellectual enquiry, he made it clear that such freedom should be confined to the polite and virtuous classes: ". . . the extension of knowledge beyond certain limits is forbidden by that state of society to which it owes its very existence," he wrote. When knowledge mingled with ignorance, "political mischief" could be the result, due to an upsurge of "that restless turbulence of spirit which murmurs at, and endeavours to subvert the gentlest and best constituted authority."[117] By 1798, even a member of the liberal Manchester Literary and Philosophical Society was voicing similar fears. The Rev. Thomas Gisborne addressed the Society on "The benefits and duties resulting from the institution of societies for the advancement of literature and philosophy." He stressed the dangers of acrimony, factionalism, and intellectual pride. He also referred to external perceptions that the pursuit of science was likely to lead to infidelity. Although Gisborne believed that the charge could be countered by pointing to the pious figure of Newton, whose life demonstrated the humility that a truly profound study of nature induced, it is clear that conservative anxieties about the subversive impact of public science had been felt even here.[118]

Burke's onslaught on Enlightenment philosophy and its subversive public influence was clearly very persuasive in Britain in the 1790s. His barely controlled rage powered a forceful argumentative style, which appealed to intellectuals like Robison, and his extraordinary aptitude for vivid imagery caught the attention of cartoonists like Gillray, Dent, and Sayers. Burke provided the conceptual ammunition and the visual metaphors by which opinion was rallied against radical Enlightenment science. In this

116 Cooper (1792), p. 65.   117 [Anon.] [1796], p. 4.
118 Gisborne (1798). For Gisborne's subsequent writings as a social moralist, see the D.N.B.

context the incident of the nitrous oxide experiments appears as a symptomatic episode, if not exactly a typical one. The laughing-gas case concentrated the attention of conservative critics of Priestleian public science. They saw an opportunity to heap ridicule on the radical aspirations of Beddoes and his allies and, by implication, to damn the whole campaign to make the universal diffusion of scientific knowledge into a means of moral and material progress.

We have seen some of the reasons why Beddoes's enterprise was vulnerable to this attack. His ambitious schemes for preventive therapy and the spread of happiness were easily portrayed as the dreams of a deluded "projector." The antics of those who came to the Bristol Pneumatic Institution to breathe nitrous oxide required little distortion to bring out their comic and ridiculous aspect. Furthermore, Beddoes's manner of organizing and presenting his experiments left him without effective means of establishing the efficacy of the gas in the face of conservative ridicule and ostracism. Beddoes regarded experiment as a simple matter of recorded experience and (as we saw) identified the immediately demonstrative quality of experience with "intuition."[119] This was readily converted by his opponents into a willing indulgence of "speculation" and even "enthusiasm." In addition, Beddoes's manner of assembling acquaintances to share the breathing experience and recruiting support and transmitting knowledge of techniques via the networks of the provincial intelligentsia proved equally vulnerable. The reactionary climate of the 1790s weakened the prestige of such groups and loosened the ties of common political and religious objectives that bound them. As Beddoes came to realize, pneumatic therapy could make little headway against such strong opposing forces.

We shall see how Humphry Davy responded to this impasse by radically changing the situation, presentation, and subject of his work. Leaving Bristol in 1801, he moved to the Royal Institution in London at the focus of the scientific world. There he assembled a substantial public and specialist audience, whose attention he focused upon his experimental researches. Particularly in the field of electrochemistry, he developed new instrumental resources, which he deployed in public demonstrations to manifest their power. And he learned the value of a tightly organized, precisely focused mode of discourse as a means of persuasion. In all these respects, Davy utilized resources that had originally been applied to chemistry by Lavoisier, resources that Beddoes, committed as he was to Priestley's vision of the moral order of public science, had ignored.

119 See the comments recorded in Stansfield (1984), p. 176.

7

Humphry Davy: The public face of genius

> The love of discovery, and of intellectual acquirement, may indeed be called an almost instinctive faculty of the mind; but still a combination of circumstances is required to call it into useful exertion. The truly insulated individual can effect little or nothing by his unassisted efforts. It is from minds nourishing their strength in solitude, and exerting that strength in society, that the most important truths have proceeded.
>
> Humphry Davy, extract from an unpublished lecture[1]

Through the work of Sir Humphry Davy (1778–1829), chemistry was given an entirely new form as a public science. In the first ten years of the nineteenth century it emerged from the crisis of the previous decade with greatly enhanced esteem and respectability. Davy's career was emblematic of this transformation, and indeed was substantially responsible for it. By moving in February 1801 from Thomas Beddoes's Pneumatic Institution in Bristol to the very different setting of the Royal Institution (the RI) in London, Davy extricated himself from the confusion of the "end of the Enlightenment" and placed himself in a more central position from which he could address a wider public. He was translated, as a friend wrote, from "the stormy Parnassus" to "the mild and unvarying temperature of the central grotto of science."[2]

From then on, Davy assembled around himself a large and diverse audience for his work. In his lectures at the RI between 1801 and 1812, he addressed hundreds of members of London's fashionable elite. His reputation as a lecturer stimulated efforts to extend the benefits of scientific instruction to an even larger population through the periodical press and popular textbooks. Gradually Davy also became a powerful voice within the emerging community of specialist chemists. His research papers, presented mainly under the auspices of the Royal Society, commanded respect from many other chemists and set new standards for their work.

Davy's relations with the different components of his audience, from

---

1 Davy (1839–40), VIII, p. 321.
2 Letter from Gregory Watt to Davy, 7 February 1801, in Davy MSS (RI), box 9, p. 87.

middle-class public to specialist chemists, were interlocking and mutually reinforcing. Specialists joined the general public in attendance at his RI lectures and among the readers of periodicals in which his work was reported. On a number of occasions, he appealed to his public audiences to adjudicate questions that were in dispute between specialist chemists. He created a situation in the lecture theater in which his audience assumed a passive, acquiescent role in the face of his forceful presentation; and he used this effectively to induce other specialists to accept his claims.

In this chapter we shall see how Davy exploited the relationship with his audience to consolidate and extend his scientific achievements. After specifying relevant aspects of his context and his lecturing practice at the RI, we shall consider the techniques (experimental, rhetorical, and social) that he used in his research career. In the work that culminated in his first two Bakerian Lectures, delivered to the Royal Society in 1806 and 1807, Davy will be seen to have converted the voltaic pile (a type of electrical battery) from an experiment into an instrument, with which new discoveries could be made.[3]

In discussing this incident, I aim to recover the social dimension of Davy's achievement. I shall show how he had to counter a series of alternative views of the pile's mode of action in order to establish its efficacy as an analytical instrument. He did this by gaining authority over his audience and mobilizing certain rhetorical forms to give his experimental claims persuasive power. Davy had to engage in a strenuous defense of the analytical effect of the voltaic pile, before his discoveries with it could win recognition. Having established its potency, he could use the pile to make his famous discoveries of the elements sodium and potassium, announced in 1807. These discoveries gained wide acceptance, with the result that the discipline of electrochemistry consolidated around the use of the voltaic pile as an instrument of analysis.

Davy's subsequent assertion of the elementary nature of chlorine, formerly known as oxymuriatic acid, also called for the mobilization of his full repertoire of persuasive resources. In the prolonged debates that followed the announcement of his claim in 1810, it was subjected to severe scrutiny by other members of the chemical community and his experimental evidence was seriously questioned. Attempting to win his argu-

---

3 The notion that scientific instruments are constructed artifacts, with a history that can be analyzed in social and rhetorical terms, arises from recent work in the sociology and philosophy of science. See, for example, Pinch (1986), esp. pp. 212–214; Hacking (1983), ch. 11. My treatment of Davy also owes much to David Gooding's work on Michael Faraday. Gooding discusses how Faraday worked to transfer experimental effects from the private space of the RI laboratory to the public space of its lecture theater, magnifying the phenomena and masking the labor required to produce them, in order that nature would appear to speak directly to his audience. See in particular, Gooding (1985a, 1985b).

ment, he again tried to enlist the support of nonspecialist audiences. But in this case the strategy had only limited success. Some of Davy's opponents remained stubbornly unconvinced, and most specialists were only won over by the later discovery by French chemists of what appeared to be an analogous element – iodine.

By examining the processes of communication and persuasion by which Davy's scientific discoveries were established, we can gain a greater understanding of the achievements for which his career has been remembered among historians of chemistry.[4] These can be shown to have relied upon the construction of a particular pattern of relations with his audience. A phrase that Davy himself used, "the lever of experiment," seems a pertinent one to describe this situation.[5] He was referring to the power of certain significant experiments to stir nature from its position, and thereby to force it to reveal itself. But experiments also operate socially as a lever – they move opinions and secure acceptance of beliefs. They can do this, however, only if embodied in effective linguistic forms, and communicated via functioning social structures. In other words, the lever of experiment turns on the fulcrum of the social relations between experimenter and audience. Davy's success lay in placing that fulcrum in a new institutional context, and thereupon moving the lever of experiment with unprecedented effect.

Davy's career: The creation of a public audience

The first consequences of his [Davy's] success in the line of mere exhibition were unfavourable, and threatened to be fatal; for he was led away by the plaudits of fashion, and must needs join its frothy, feeble current. For a while he is remarked to have shown the incongruous combination of science and fashion, which form a most imperfect union, and produce a compound of no valuable qualities, somewhat resembling the nitrous gas on which he experimented earlier in life, having an intoxicating effect on the party tasting it, and a ludicrous one on all beholders.

Henry Brougham, "Davy" in *Lives of the Philosophers*[6]

Mr. Davy is one of those philosophers whose imaginary flights and credulity are equally shewn in his *ecstatic sensations* upon his breathing the gaseous oxyde. Educated under that vague theorist, Dr. Beddoes, and now placed in the Institute,

---

4 A traditional appraisal of Davy's achievement is given by Partington (1961–70), IV, pp. 29–76. The most comprehensive and helpful study of Davy's chemical theory is Ziemacki (1974). For other perspectives on Davy's natural philosophy, see: Levere (1971), ch. 2; and Knight (1978), chh. 2, 3. Further helpful studies are included in Forgan (1980). Biographies include: Hartley (1972); and Treneer (1963). I have also found the numerous articles of June Z. Fullmer particularly informative on Davy's life and career; in particular Fullmer (1962, 1964, 1980). Recently, an interesting overview has been given by Lawrence (1990).
5 Davy (1839–40), VIII, p. 318.   6 Brougham (1872), p. 110.

he must be influenced by his patrons. But I must candidly acknowledge that he displays great ingenuity in his experiments; and if he was not under the dominion of the present absurd theory, and his genius properly applied, he might be an ornament to science. . .

   Robert Harrington, *Death-Warrant of the French Theory of Chemistry*[7]

When Davy took up his appointment at the Royal Institution, he entered an institution that was not yet two years old. The circumstances in which it was formed have been excellently described in Morris Berman's social history of its early years.[8] The RI was the creation of a group of aristo-crats and wealthy gentry, centered on Sir Joseph Banks, the President of the Royal Society. Many of those who took a leading role in the new institution had previously invested their energies in the Board of Agricul-ture and the Society for Bettering the Condition of the Poor. They viewed the new organization as an outlet for their philanthropic concerns and a means of serving their own (mainly agricultural) economic interests. Their plans were crystallized by the initiative of the ambitious adventurer Count Rumford, who published a set of proposals for a scientific philanthropic and educational institution in 1796 and subsequently lobbied for its foundation. After the institution was formally launched, in a meeting at Banks's house in March 1799, Rumford's involvement lessened, and it ceased altogether in 1803.[9]

Thereafter, the direction taken by the RI was determined by the eco-nomic and intellectual interests of the Proprietors, and was governed on a regular basis by a committee of Managers. The Proprietors, 280 of whom were recruited in the first year, subscribed a substantial sum (ini-tially set at fifty guineas) to acquire life-long rights to attend lectures and exhibitions, and to elect the Managers. The latter, acting on behalf of the Proprietors and other classes of subscribers, shaped the programs of teaching and research.

Chemistry was immediately established as a focus of research and lec-tures at the RI. Initially, Davy was appointed assistant chemical lecturer to Dr. Thomas Garnett, who had been recruited from Anderson's Insti-tution in Glasgow.[10] He was promoted to Professor of Chemistry on Garnett's resignation in May 1801 (Figure 7). Although much of Davy's early work on applied chemistry was undertaken at the direction of the Managers, his successes in research and teaching also gave him a large measure of autonomy. His achievements reoriented the institution's pro-gram of public education toward a fashionable middle- and upper-class audience and helped alleviate its endangered financial situation.

7 Harrington (1804), p. 208 fn. 8 Berman (1978), esp. chh. 1, 2.
9 Berman (1978), pp. 28–31. Cf. Jones (1871), chh. 2, 3; and Brown (1979), chh. 20, 21.
10 For a comprehensive account of Garnett's career, see Lythe (1984).

*Figure 7. Portrait of Humphry Davy by H. Howard. [Reproduced
by permission of the National Portrait Gallery, London.]*

The degree to which Davy retained his independence in his new institutional setting was a matter of particular significance for him and for
his former friends in the English provinces. In taking up his position,
Davy had broken with the custom among his enlightened colleagues of
maintaining independence from government patronage and metropolitan
institutions. The RI was unquestionably a creature of the metropolitan
scientific establishment, and to take a job there was arguably to surrender
one's independence to its landowning Proprietors and to the cronies of
Sir Joseph Banks.

A degree of anxiety about this aspect of his move was reflected in a
letter Davy wrote to his mother at the time: "You will all, I dare say, be
glad to see me getting amongst the Royalists, but I will accept of no
appointment except upon the sacred terms of independence." His friend
S.T. Coleridge expressed the worry that however strong Davy's attach-

ment to the principle of independence, he would inevitably be corrupted. Seven years later, the Whig reformer Henry Brougham judged that this had not happened; he remained confident that Davy's "political sentiments are as free and as manly as if he had never inhaled the atmosphere of the Royal Institution." Davy's talents, Brougham concluded, "have escaped unimpaired from the enervating influence of the Royal Institution; and indeed [have] grown prodigiously in that thick medium of fashionable philosophy."[11] Davy's translation to a metropolitan institutional base was nonetheless viewed with some anxiety in progressive and enlightened circles. At best, he was surrendering part of his personal autonomy; at worst, running the risk of moral corruption by "fashionable philosophy."

Davy himself however had no qualms about abandoning the political aims of his former associates. After his move to the RI, he adopted a uniformly conservative political stance. He condemned the "devouring flame of anarchy" released by the French Revolution, and echoed Edmund Burke in his denunciations of the revolutionaries as "a nation of cannibals" and "savages." In private reflections, he recorded his hopes for "gradual reform and progressive change," but publicly he stressed the value of social hierarchy as an encouragement to emulation and economic activity. In his religion he was never known to be anything other than a conforming Anglican.[12]

From Davy's point of view, the move to the RI was a rational and even a necessary one, if an appropriate audience for science were to be found and won over. The political polarization of the 1790s had marginalized the proponents of Enlightenment science to the extent that it was difficult for them to get their message across. As Davy was well aware, Beddoes and Priestley had suffered personal ridicule and vituperation, and their writings had been deprived of respect in society at large. He was clearly determined to avoid their fate by establishing his credentials with a wider audience and in a more secure manner. In one of his notebooks at the time he considered how science might be advanced in the current state of society. Books and lectures remained important, he recognized, but elite patronage must also be secured: "The rich & privileged orders . . . are ultimately the guardians of refinement & civilisation & even of science." The aims of enlightenment should be maintained, and might still be expected to be realized, but only if the enlightened ceased to talk only to themselves. The problem hitherto had been that different social groups

---

11 Letter from Davy to his mother, 31 January 1801, quoted in Davy (1839–40), I, pp. 84–85. Coleridge's relations with Davy are well treated in Levere (1981b), esp. pp. 27–31. Henry Brougham's comments are in *Edinburgh Review*, 11 (1807–8), no. 22, p.390; and 12 (1808), no. 24, p.399.

12 Davy (1839–40), I, pp. 213–214, 231.

"have been too much insulated in small circles of self interest."[13] The remedy was to find a way of addressing a wider public.

This was what Davy achieved through the RI. His lectures there attracted large audiences of men and women drawn from the affluent classes, and they were widely reported in cultural periodicals and the general press. A whole genre of popular texts sprang up to relay his doctrines to readers further afield.[14] A commentator on his first lecture series on galvanism described those attending as "Men of the first rank and talent, ... blue-stockings and women of fashion, the old and the young." Another noted that the lectures were "attended not only by men of science but by numbers of people of rank and fashion; a proof that this Institution bids fair to promote a taste for philosophical pursuits among those whose wealth has but too often fostered the idea that such subjects were beneath the[ir] notice." Thomas Carlyle aptly joked that Davy had made the RI into "a kind of sublime Mechanics' Institute for the upper classes."[15]

The prominence of women among his audience was a feature that many commentators noted, and that Davy himself welcomed as a sign of social refinement and progress. "The standard of the consideration and importance of females in society is, I believe, likewise the standard of civilisation," he wrote.[16] His view of the position of women in intellectual life gave them an important but limited role – they were expected to absorb and transmit science as part of a general cultural education, but not to attempt to penetrate into the realm of the specialist. This view had been proposed before, by scientific educators such as Priestley and Adam Walker, and it was subsequently to inspire Jane Marcet to compose her *Conversations on Chemistry* after attending one of Davy's courses.[17]

That Davy's RI lecturers were masculine science, presented to an audience that was to a significant degree feminine, was one important dimension of their rhetorical construction. But male and female auditors alike found themselves seduced by his mode of delivery, and captivated by the forceful presentation of his own personality and of the powers of nature under his control. His carefully cultivated oratorical talents and attractive personal appearance were mobilized to produce this effect, as was his choice of convenient and spectacular experimental demonstra-

13 Davy MSS (RI), box 13c, pp. 54–70, esp. pp. 56, 59.
14 The audience at Davy's lectures is discussed in Foote (1952); Berman (1978); and Fullmer (1980). Press reporting, and Davy's attitudes to it, are mentioned in Fullmer (1962), pp. 150–152. Reviews of Davy's published (and some unpublished) lectures are listed in the helpful bibliography: Fullmer (1969).
15 Anonymous observers quoted in Davy (1839–40), I, p. 88; and in *Philosophical Magazine*, 10 (1801), pp. 86–87. Carlyle quoted in Foote (1952), p. 7.
16 Davy (1839–40), VIII, pp. 353–355.
17 Priestley (1778), pp. 137–138; Walker (1807), p. v; [Marcet] (1817), I, pp. vii, ix, 3–4.

tions.[18] Through these means, Davy appeared in the guise of the scientific genius, in line with the prevalent Romantic notions that he had reflected upon in his early notebooks.[19] Most observers of his lectures would have agreed with the historian Harriet Martineau, that Davy "presented most strongly to the popular observation the attributes of genius."[20]

The most sophisticated among his listeners realized that these performances were artificial and highly crafted. At the RI lectures, in contrast to the Bristol nitrous oxide experiments, "inspiration" was a rhetorical feint. The French traveler Louis Simond recorded of Davy that:

> ... he knows what nature has given him and what it has withheld, and husbands his means accordingly. You may foresee by a certain tuning or pitching of the organ of speech to a graver key, thrusting his chin into his neck, and even pulling out his cravat, when Mr. Davy is going to be eloquent, – for he rarely yields to the inspiration till he is duly prepared.[21]

It was this contrived oratorical style that Davy used to present chemistry in his lectures. A prime example of his rhetoric on behalf of the subject was his "Discourse Introductory to a Course of Lectures on Chemistry," published a few months after it was first read in the RI lecture theater on 21 January 1802.[22] In the "Discourse," Davy echoed many of the traditional legitimatory themes, but added a very contemporary emphasis. He mobilized significant new discursive resources to support his claims for the importance and relevance of chemistry.

The "Discourse" opened with grandiose, but not particularly novel, claims for the range of chemistry's concerns. Davy's audience was told that chemistry encompassed all material substances in their changes and interactions with one another. It therefore occupied a central position in relation to the other sciences, which depended upon it for knowledge of the specific properties of bodies. This was the source of the widespread utility of the subject, a wider knowledge of which would yield improvements in many of the practical arts. Among these, Davy mentioned agriculture, metallurgy, bleaching, dyeing, tanning, and the manufacture of porcelain and glass.[23]

Davy articulated the relationship of chemistry to its dependent arts in a way that owed much to his eighteenth-century predecessors. Improve-

18 On Davy's cultivated oratorical style, see: Davy (1839–40), I, pp. 91–93; and on his coaching of John Dalton in lecturing technique, Hartley (1972), p. 46; Treneer (1963), pp. 88–89. For remarks on his appearance: Fullmer (1962), p. 158; Treneer (1963), p. 86; *Philosophical Magazine*, 9 (1801), 281–282. For examples of spectacular demonstrations: Foote (1952), p. 11; Levere (1981b), p. 30; Davy (1839–40), II, pp. 188–209, 211–213.

19 For example in Davy MSS (RI), box 13c, p. 32. Davy's notions of genius are discussed in Levere (1980); and Lawrence (1990).

20 Martineau (1849–50), I, p. 594.   21 Quoted in Fullmer (1962), p. 157.

22 Davy (1839–40), II, pp. 311–326.   23 Davy (1839–40), II, pp. 311–312, 315–318.

ments in all the various fields of technological practice would flow from theoretical knowledge of the basic science, which in turn derived from "a systematic arrangement of facts" and a philosophical overview of their interconnections. If problems were to arise in application, then these must derive from some external hindrance to improvement and not from difficulties in theory. For example, obstacles to progress in tanning methods, which Davy had investigated, were said to be due to "the difficulty occurring in inducing workmen to form new habits," not to "any defect in the general theory of the art as laid down by chemical philosophers, and demonstrated by their experiments."[24]

If given free reign, chemistry could contribute substantially to the material and moral progress of society, which in turn further stimulated the cultivation of science. Hence the history of chemistry was in effect the history of social development:

To be able indeed to form an accurate estimate of the effects of chemical philosophy, and the arts and sciences connected with it, upon the human mind, we ought to examine the history of society, to trace the progress of improvement, or more immediately to compare the uncultivated savage with the being of science and civilisation.[25]

Although Davy was willing to go this far, he could not follow through with a full-scale articulation of the Enlightenment vision. He stopped well short, for example, of Priestley's image of scientific improvement as the engine of social progress and political emancipation. The theater of the RI was clearly not the place for such an avowedly radical perspective. On the contrary, Davy deliberately distanced himself from the aspirations of his enlightened mentors, Priestley and Beddoes, whose ideals had recently undergone social opprobrium. His vision was presented as a realistic and pragmatic one, which accepted the necessity of social inequality and inescapable limits on material progress:

The unequal division of property and of labour, the difference of rank and condition amongst mankind, are the sources of power in civilized life, its moving causes, and even its very soul. . . . In this view we do not look to distant ages, or amuse ourselves with brilliant, though delusive dreams concerning the infinite improveability of man, the annihilation of labour, disease, and even death. But we reason by analogy from simple facts. We consider only a state of human progression arising out of its present condition. We look only for that time that we may reasonable expect, for a bright day of which we already behold the dawn.[26]

It was in terms of this conservative reformulation of the Enlightenment ideal, in which progress would not disrupt the existing hierarchy, that Davy described the social function of chemistry. His concluding words

24 Davy (1839–40), II, p. 317. For Davy's work on tanning, see Berman (1978), pp. 49–53, 85–86.
25 Davy (1839–40), II, p. 318.   26 Davy (1839–40), II, p. 323.

in the lecture linked the maintainance of social order with the aesthetic appreciation of the natural world that science could engender. A wider knowledge of experimental philosophy, he suggested, would incline men to prefer stability and harmony in the natural and the social worlds:

The man who has been accustomed to study natural objects philosophically, . . . perceiving in all the phenomena of the universe the designs of a perfect intelligence, . . . will be averse to the turbulence and passion of hasty innovations, and will uniformly appear as the friend of tranquillity and order.[27]

Thus, social order was to rest upon natural theology – the perception "in all the phenomena of the universe" of "the designs of a perfect intelligence." In introducing the arguments of natural theology into chemistry, Davy was forging a link between two subjects whose relations had until then been tenuous and intermittent. Chemistry, inherently a human art, could only with difficulty be used to illustrate the wonders of God's work in the world of nature.[28] Priestley had been one of the few writers previously to make the connection, basically because his determinist view of human action made man's art appear as a link in the uninterrupted chain of causation leading back to God. Human achievements in chemistry were thus seen as part of a divinely determined natural order.[29]

Davy made the link between chemistry and natural theology in a different way, having rejected Priestley's determinism along with much of the rest of his metaphysics. He saw the divine order as manifested in an ultimate unification of the fundamental laws governing the powers of nature.[30] Confident that these basic laws would be progressively revealed to him, the philosopher could devoutly pursue the study of chemistry, a science that, "must be always more or less connected with the love of the beautiful and the sublime; . . . [and] is eminently calculated to gratify and keep alive the more powerful passions and ambitions of the soul."[31] Apparently giving way to these passions, Davy introduced florid rhetorical excursuses on natural theology into his lectures, to the great applause of his audiences who were thereby reassured that chemistry would endorse, and not subvert, religious orthodoxy.[32]

In the 1802 "Discourse," Davy was providing his audience with a re-

27 Davy (1839–40), II, p. 326.
28 Even writers who had interests in both natural theology and chemistry, such as Robert Boyle, seem to have made little direct connection between the two. The first systematic work on the subject was William Prout's Bridgewater Treatise (1834). See also Brock (1985), ch. 4.
29 McEvoy and McGuire (1975); Brooke (1984).
30 Levere (1971), ch. 2, esp. pp. 42, 45, 49.   31 Davy (1839–40), II, p. 325.
32 Louis Simond recorded the applause that greeted Davy's digressions into natural theology (quoted in Fullmer (1962), p. 157). John Davy records that his brother's religious eloquence was so impressive that he was urged by clerical friends to enter the church (Davy (1839–40), I, p. 127).

assuringly sanitized version of Enlightenment hopes for scientific progress. The perspective of natural theology suggested that science would reveal harmony and stability in the natural world, and encourage the same qualities in society. Progress would occur without disruption to the social order, stimulated by advances in philosophical theory that would be passed down the cognitive and social hierarchy to the practitioners of the dependent arts.

This image of improvement was one Davy repeated elsewhere, and which he attempted to convert into actuality in his work on the practical arts. Consider, for example, his lectures on agricultural chemistry, originally delivered to the Board of Agriculture between 1802 and 1812, and published as the *Elements of Agricultural Chemistry* in 1813. There Davy developed his view of progress in the chemical arts into what might be described as an aristocratic or paternalistic model of agricultural improvement. He emphasized the dependence of agriculture on the fundamental principles supplied by chemistry, and combined this with an image of progress stimulated by the upper classes in society:

It is from the higher classes of the community, from the proprietors of land; those who are fitted by their education to form enlightened plans, and by their fortunes to carry such plans into execution; it is from these that the principles of improvement must flow to the labouring classes of the community.[33]

One can imagine that this was very much what Davy's audience wanted to hear. The mainly aristocratic members of the Board of Agriculture were no doubt flattered to be described as distinguished and enlightened; they readily saw themselves as bearing the prime responsibility for improvement by disciplining their workforce to adopt more productive habits.[34] Davy's approach was rhetorically in line with the paternalistic tradition established in agricultural chemistry by Scottish writers in the mid-eighteenth century. Davy's "philosophical farmer" mirrored the outlook of Lord Kames's "gentleman farmer," and his text was also close in orientation to the *Treatise on Agriculture and Chemistry* by Archibald Cochrane (Lord Dundonald) (1795). Cochrane had argued strenuously for the need to apply chemistry to agriculture, a science he saw as "morally and politically conducing to the true happiness of man, . . . whence flow health, social order and obedience to lawful authority."[35] Davy did not quite share Cochrane's degree of paranoia about the pernicious political influence of industrialization and the ideas of the French Revolution, but his work partook of the same conservative spirit.

In many of its recommendations for practice, Davy's text also ad-

33 Davy (1813); reprinted in Davy (1839–40), VII, pp. 169–391; VIII, pp. 1–152 (quotation on VII, p. 197).
34 Davy (1839–40), VII, pp. 177–178, 197.   35 Cochrane (1795), pp. 3–4.

vanced little beyond eighteenth-century precedent. His main claim to originality lay in his suggested method for soil analysis – a standard procedure for testing any soil sample that involved evaporation, various titrations and precipitations, and the careful weighing of reaction products. The method was more accurate and comprehensive than any proposed before, but it also made greater demands on the skill and resources of the operator. The reader was told to purchase a sensitive balance (capable of weighing to an accuracy of one grain), a sieve, an Argand lamp, a collection of glassware, "Hessian crucibles; porcelain . . . evaporating basins; a Wedgwood pestle and mortar," filter papers, "and an apparatus for collecting and measuring aëriform fluids." Although Davy claimed that a "small closet is sufficient for containing all the materials required," it is clear that a substantial investment was expected from the philosophical farmer.[36] Davy was not recommending that the landowner seek specialist help with soil analysis; he was still assuming the farmer could apply the methods himself.

This failure to endorse specialist expertise was indicative of Davy's distance from the aims of many other chemists of his time. The *Edinburgh Review* seems to have accurately anticipated the reception of his text when it remarked that Davy's method was "suited to the views of the agriculturalist," but that a properly comprehensive analysis would require "a complicated apparatus, much time and labour, and all the resources of the analytical chemist."[37] Davy's procedure proved a popular one, being reprinted many times in periodicals and repeated in the works of a number of other chemists,[38] but analytical techniques were beginning to become much more the prerogative of the specialist chemist than Davy seemed willing to allow. The author of an anonymous *Treatise on Soils and Manures* pointed out in 1818 that what "may be a very simple process for an expert chemist . . . is neither so easy to describe, nor so cheap to practice in occasional experiments." In the face of this growing gap between expert and amateur, the notion that arguments "drawn from the depths of philosophy" could be imposed magisterially upon the practitioner of the arts, seemed quite inappropriate to this author.[39]

Returning to the 1802 "Discourse," we can see that contemporary changes in the discipline were not entirely absent from Davy's account,

36 Davy (1839–40), VII, pp. 198, 304–305.
37 Review of Davy, *Elements of Agricultural Chemistry*, in *Edinburgh Review* 22 (Oct 1813–Jan 1814), 251–281 (quotation on p. 267). The review is ascribed to John Gordon, an Edinburgh physician, in Houghton, ed. (1966–87), I, p. 451; IV, p. 782.
38 For a survey of aspects of the reception of Davy's work, see Miles (1961). The soil-analysis procedure was first printed in Davy (1805); it can also be found in Brande (1819a), pp. 538–560; and Accum (1820), pp. 407–439.
39 [Anon.] (1818), pp. iii, 23 fn.

despite its emphasis on the theme of ordered and harmonious progress in chemistry. Another, potentially more disturbing, theme was also present: the emergence of dramatic new experimental methods that promised to reshape the science quite radically. These "instruments" were assigned considerable prominence in the lecture, and displayed as further evidence of the successes chemistry had achieved. "New instruments and powers of investigation" were invoked to show that chemistry, as opposed to those sciences that only studied surface appearances, could penetrate into matter itself. The science was contrasted with natural history, which was satisfied with contemplation of the external forms of animals, plants, and minerals. It would be to chemistry therefore that natural philosophers would have to turn – for example, to establish a rigorous system of mineralogical classification, or to understand the fundamental internal processes that distinguished living beings.[40]

This last was a subject in which Davy showed significant interest in the course of his career; and it was one in which he saw imminent prospects of results from the new instrumental powers now at his control. Precisely because they offered the promise of knowledge of the fundamental processes of life, including human life, these new powers gave a god-like status to the man who wielded them. They enabled man, Davy explained, "to interrogate nature with power, not simply as a scholar, passive and seeking only to understand her operations, but rather as a master, active with his own instruments."[41] Specifying what these powers were, he reviewed some recent discoveries:

The composition of the atmosphere, and the properties of the gases, have been ascertained; the phaenomena of electricity have been developed; the lightnings have been taken from the clouds; and lastly a new influence [galvanism] has been discovered, which has enabled men to produce from combinations of dead matter effects which were formerly occasioned only by animal organs.[42]

Many of these powers were shown by Davy in the course of his lectures at the RI. He demonstrated how they could be, as he put it, "according to circumstances, instruments of comfort and enjoyment, or of terror and destruction."[43] In the lecture theater they were also shown as instruments of investigation – that is, as powers that could be mastered by the experimenter to yield useful knowledge. The rhetorical force of such demonstrations could be considerable, and their effect was to establish acceptance among his audience that the powers of nature were indeed Davy's to command. He had long recognized the peculiar ability of experiments to convey this message, having written in one of his early notebooks that

40 Davy (1839–40), II, pp. 312–314.   41 Davy (1839–40), II, p. 319
42 Davy (1839–40), II, p. 321.   43 Davy (1839–40), II, p. 318.

"the language of exp[erimen]t is universally intelligible & the truths it conveys can never be forgotten."[44]

To teach this lesson, however, it was essential to choose appropriate experiments – ones that would enhance the apparent mastery of the experimenter, and that did not produce disorder or confusion. Davy made more than one attempt in his early days at the RI to demonstrate the effects of nitrous oxide, but it proved difficult to reproduce the Bristol results in a striking and controlled way. After an attempt to do so in a lecture in June 1801, Davy reported enthusiastically to a former colleague at the Pneumatic Institution:

My last lecture was on Saturday evening. About 500 persons attended. . . . There was respiration, nitrous oxide and unbounded applause – Amen. To-morrow a party of philosophers meet at the Institution, to inhale the joy-inspiring gas. It has produced a great sensation – çà ira.[45]

Press comments on the same incident, however, painted a rather different picture. The *Philosophical Magazine* made the occasion seem quite comical and anarchic. A Mr Underwood was said to have "lost all sense to everything else, and the breathing-bag could only be taken from him at last by force."[46] The journal implied that it was in order to avoid a repetition of these ludicrous scenes that the experiments were continued a few days later at a private meeting.[47] Respiration of nitrous oxide appears to have dropped out of the repertoire of public experiments after this, a change that apparently reflected a desire to avoid ridicule and disorder. The final straw may have been the publication of James Gillray's popular satirical cartoon, "Scientific Researches! – New Discoveries in PNEUMATICKS!," in May 1802 (Figure 8). The cartoon depicted the chaotic and distasteful results of a respiration experiment in unforgettable imagery, and may well have been a factor in curtailing experiments of this kind in public lectures.[48]

Ridicule of the lecturer was certainly not the outcome Davy sought from his performances. His aim was rather to control the situation in the theater, to give a striking demonstration of the power of his instruments, and to coopt that power to enhance his personal persuasiveness as an

44 Davy MSS (RI), box 13c, p. 94.
45 Davy to J. King, undated letter [apparently June 1801], in J. Davy, ed. (1858), pp. 64–65.
46 Report of lecture on 20 June 1801, in *Philosophical Magazine*, 10 (1801), 86–87 (quotation on p. 86).
47 The "select party," which met to continue the experiments on 23 June 1801, comprised Count Rumford, Sir Charles Blagden, W.H. Wollaston and others.
48 Gillray's cartoon is discussed in Wright and Evans (1851), pp. 468–469; and British Museum (1978). It has been identified as referring to a previous incident in March 1800, when Davy assisted Garnett in demonstrating nitrous oxide at the RI, in a letter by June Fullmer in *Scientific American*, 203 (no. 2) (August 1960), 12–14.

Figure 8. "Scientific Researches," by James Gillray. From the British Museum, Department of Prints and Drawings (DG no. 9923). [Reproduced by permission of the Trustees of the British Museum.]

experimenter. Galvanic phenomena produced by the voltaic pile yielded a good fund of demonstrations of this kind, as Davy showed in his early lectures on this subject. The pile could produce shocks, sparks ("of a dazzling brightness"), and loud noises. The sparks would be different colors (white, purple, yellow, red, or blue), according to which metal composed the electrodes. And party tricks could be shown: A little gunpowder and mercury fulminate could be ignited by the galvanic fluid, or letters and figures could be drawn on gold leaf by burning off the metal by electricity.[49]

The effect of these spectacular demonstrations was both to show the efficacy of natural forces such as galvanism, and simultaneously to demonstrate Davy's power to command them through the instruments at his disposal. In other words, his possession of instruments such as the voltaic pile, by which natural forces could reliably be controlled, was projected

---

49 See especially, "An account of some experiments on galvanic electricity made in the theatre of the Royal Institution," in Davy (1839–40), II, pp. 211–213.

as an integral part of his presentation of himself as an experimental philosopher. In the rhetoric of Davy's lectures, the power of the instruments and that of the experimenter buttressed one another.

In the lecture theater of the RI, Davy had established his identity in the eyes of his audience as an experimental genius, and that identity was integral with the perceived strength and reliability of his instruments. It was this authority over his audience and over nature that Davy attempted to transfer to other settings, such as his Bakerian Lectures to the Royal Society and his published research papers. In these other contexts, different rhetorical techniques had to be used to establish a similar relationship between experimenter and audience. Davy used techniques of literary representation and argument to give his experimental accounts credibility, and hence to convey an impression of the masterful power of the experimenter. Again, it was his possession of effective and reliable instruments of investigation that constituted the status of the experimenter as controller of nature.

The voltaic pile: The making of an instrument

Nothing tends so much to the advancement of knowledge as the application of a new instrument. The native intellectual powers of men in different times are not so much the causes of the different successes of their labours, as the peculiar nature of the means and artificial resources in their possession.

Davy, *Elements of Chemical Philosophy* (1812)[50]

In pliant obsequiousness to theory, the cathode jar was almost full of gas; the anode jar was half full: I brought this to Enrico's attention, giving myself as much importance as I could, and trying to awaken the suspicion that, I won't say electrolysis, but its application as the confirmation of the law of definite proportions, was my invention, the fruit of patient experiments conducted secretly in my room. But Enrico was in a bad mood and doubted everything. "Who says that it's actually hydrogen and oxygen?" he said to me rudely. "And what if there's chlorine? Didn't you put in salt?"

Primo Levi, *The Periodic Table*[51]

When Davy came to give his first Bakerian Lecture to the Royal Society in 1806, the control of the voltaic pile he had already acquired in the eyes of his audiences at the RI was a substantial asset to him. Of course, the audience at the Society's meeting was different from that which had attended the lectures at the RI; but there was an overlap – many of the leading chemists in the Society had already seen him perform in the lecture theater. In addition, the Bakerian Lecture, like Davy's previous per-

50 Davy (1812), p. 54.   51 Levi (1985), p. 27.

formances, reached a wide public audience, who read reports in newspapers and general periodicals.[52]

The Bakerian also differed from Davy's previous lectures in that no accompanying experimental demonstrations were presented. Rather than invoking the persuasive force of a display of natural powers, Davy had to use literary techniques to convey a convincing account of a series of experimental manipulations. To his work in the laboratory he had to add the rhetorical work of literary composition in order to assert the efficacy and security of his prime instrument of research, the voltaic pile. His success in doing this enabled him to consolidate a community of users around his own interpretation of the pile as an analytical instrument. He was aided in this task by the dramatic discoveries of new metallic substances that he extracted from the alkali earths.

In order to transform the voltaic pile from an experimental device into an instrument, Davy had first to wrest it from its original context in eighteenth-century natural philosophy. Galvanism, like Mesmerism and the breathing of nitrous oxide, was initially subject to the politically polarized interpretations of the end of the Enlightenment. On the one hand, galvanic phenomena were seen by some radicals as instances of the vital action of imponderable fluids. The galvanic fluid was thought of as similar, if not identical, to ether, the underlying cause of heat, light, electricity, and nervous action. Thomas Beddoes hailed the discovery of the voltaic pile as significant for this reason. It would, he wrote, form "the origin of a new and more subtle chemistry; in which etherial fluids ... will make the principle figure."[53] But from the other side of the political spectrum the anonymous author of The Sceptic (1800) grouped galvanism with oxygen, Mesmerism, and revolutionary hysteria itself, as phantom spirits that caused chaos and delusion. The ghostly spirit, this author noted, was supposed to have "fled into Italy, and entered into the body of a Frog, which professor Galvani was dressing for his Wife's supper." The phantom was destined to remain invisible and completely useless.[54]

Davy's early notebooks show his interest in galvanism to have originated in hopes like those expressed by Beddoes. Galvanic phenomena promised to open a window into the realm of the subtle fluids that animated living matter and activated the nervous system. Galvanism seemed to him to connect the organic and inorganic, the material and mental,

---

52 The Bakerian Lectures were delivered at meetings of the Royal Society, and subsequently published in the *Philosophical Transactions*. Reports and reviews appeared in journals such as Tilloch's *Philosophical Magazine*, *Nicholson's Journal*, *Edinburgh Review*, *Monthly Review*, and others. Fullmer (1969), records many of these reports.

53 Beddoes in *Monthly Review* (1801), quoted in Stansfield (1984), p. 110.

54 [Anon.] (1800), p. 6.

realms.[55] Hence in July 1800, Robert Southey wrote to Davy that the voltaic pile was important because it implied that, "the galvanic fluid[,] stimulated to motion, . . . is the same as the nervous fluid, & your systems will prove true at last."[56]

For this reason, Davy shared the enthusiasm for galvanic investigations that swept Europe in the wake of Luigi Galvani's description of his original experiment in 1791. As Naum Kipnis has shown, Galvani's claims to have produced motion in the limb of a frog by applying a metallic couple to connect nerve and muscle generated intense controversy. Dispute focused on the questions of the origin and the nature of the galvanic fluid: Did it originate in the frog's nerves, or in the metals applied to the limb? And was the fluid the same as the electricity commonly produced from electrical machines, or was it a different entity altogether? As Kipnis has also shown, the discovery of the voltaic pile in the spring of 1800 refocused these disputes, but did not conclusively resolve them.[57] The pile showed that galvanic fluid could be produced from combinations of inorganic materials, but it did not kill off research into animal sources of electricity – Davy himself continued to take an interest in this field throughout his career.[58] Nor did the pile decide the question of the identity of the galvanic fluid with common electricity. Few in fact accepted this identity, even in 1800.[59]

The importance of the voltaic pile for Davy hinged upon its use by William Nicholson and Anthony Carlisle in July 1800 to effect what they claimed was the electrical decomposition of water.[60] They described how they had inserted wires from the opposite poles of the battery into water. This apparently produced hydrogen gas from one electrode, and oxygen, which formed a compound with the metal of the other electrode. Davy immediately accepted this result, realizing that it implied the possibility

---

55 Davy MSS (RI), box 20b, p. 108; box 22b, pp. 36–37, 137–138, 155–156.
56 Southey to Davy, 26 July 1800, Davy MSS (RI), box 9, p. 109.
57 Kipnis (1987), esp. pp. 135–137.
58 On animal electricity, see Davy (1839–40), II, pp. 221–228; and his last published paper (of 1829), which returned to the subject of the electric eel or torpedo, ibid., VI, pp. 359–364.
59 Kipnis (1987), pp. 125–131, 136. For instances of Davy's own hesitation in affirming the identity between the two "electricities," see: Davy (1839–40), II, pp. 201, 205.
60 Volta described the construction of his pile in a letter to Sir Joseph Banks, read to the Royal Society on 26 June 1800, and subsequently published as: Volta (1800). Nicholson and Carlisle's experiment was described in: Nicholson (1800–1801), published in July 1800. The speed with which Volta's apparatus was applied to water, and the readiness among some researchers to interpret the effect as a decomposition, owes much to the previous attempts to decompose water using frictional electricity. See, for example, Pearson (1797–98). This possibility is discussed in Sudduth (1980). Note that the term "electrode," which I have used for ease of recognition throughout, is anachronistic. Contemporary investigators tended to refer to wires, points, or poles.

of using the voltaic pile as an analytical instrument for hitherto undecompounded chemical substances. On 6 August 1800 he wrote in his notebook that Nicholson and Carlisle's result had "placed such [a] wonderful & important instrument of Analysis in my power."[61]

The subsequent deployment of the voltaic pile in the Royal Institution lectures was directed at demonstrating this instrumental efficacy and Davy's control over it. Garnett had already shown the voltaic decomposition of water in the RI theater as early as May 1800.[62] Following his example, Davy showed its many spectacular and dramatic effects. He told his listeners that "whatever is brilliant or impressive in the experiments made in this course of lectures, will be owing to the agencies of this instrument."[63] The audience was also advised that the visible powers of the apparatus were as nothing compared with those that could not be seen: "It is not by flashes, by explosions, and sparks, but by quiet, gradual, and almost unperceived operations, that it produces its greatest effects."[64] Those who attended the lectures were thus given forceful visual and verbal evidence to accept Davy's assertion that Volta had provided experimenters with a powerful analytical instrument – "a key which promises to lay open some of the most mysterious recesses of nature."[65]

In order that it should realize this potential, however, it was necessary for the pile to become widely accepted as effective and reliable. In sociologists' terms, it had to be "black-boxed" – that is, transformed from a subject of experimental investigation into an unproblematic tool for further research.[66] The problem was that the integrity and efficacy of the pile as an analytical instrument depended crucially upon the acceptance among chemical researchers of Nicholson and Carlisle's result. Unless it was agreed that the voltaic pile really did decompose water into *just* hydrogen and oxygen, it could not function as an instrument of analysis at all. In this respect, Davy faced an uphill task, for in 1800 there was no universal acceptance of the claimed electrical decomposition of water.

One reason for the confusion about the mode of action of the pile on water was the fact that the apparatus was relatively easily replicated. To construct a pile, all that one needed was metal wires, pieces of damp cloth or cardboard, and metal discs. Nicholson gave easy-to-follow instructions for reproducing his apparatus, recommending the use of silver half-crowns and discs of zinc, and giving the address of a supplier from

61 Davy MSS (RI), box 20b, p. 2.   62 Sudduth (1980), p. 27.
63 Davy (1839–40), VIII, p. 282.   64 Davy (1839–40), VIII, p. 171.
65 Davy (1839–40), VIII, p. 282.
66 On the "black-boxing" of instruments, whereby they become accepted as devices for producing new knowledge whose workings are not brought into question, see Pinch (1986); Collins (1985), esp. ch. 6; Latour (1987), esp. pp. 2–3, 80–83, 131.

whom the latter could be obtained.[67] In Edinburgh, John Thomson (the future biographer of William Cullen) constructed a pile from silverware provided by his patron, Lord Lauderdale.[68] Replication, to the extent of producing some kind of action upon water, was easily and rapidly achieved. The result was a plethora of papers in periodicals such as *Nicholson's Journal* and Tilloch's *Philosophical Magazine,* suggesting a variety of interpretations of the pile's action. Contributions to the debate appeared from a wide range of geographical and social locations: William Cruickshank and Colonel Henry Haldane from London, William Henry from Manchester, Charles Sylvester from Sheffield, D. Gardner (Lecturer in Chemistry at the City Dispensary, London), Joseph Priestley from Pennsylvania, and other experimenters from Scotland, France, Italy, and Germany.

Some of these authors directly attacked the assertion that water was decomposed by the action of the pile. In March 1802, Priestley continued his rearguard attack on Lavoisier's theory of chemistry by denying the voltaic decomposition of water. The dephlogisticated air produced at the zinc wire came from air dissolved in the water, Priestley asserted; it was not evolved if the water was covered with a layer of oil, or if it had been exhausted of its dissolved air. Properly interpreted, the voltaic pile actually demolished the chimerical hypothesis of Lavoisier.[69]

The fact that the supposedly constituent gases were produced at different electrodes, which might be widely separated, impeded acceptance of Nicholson and Carlisle's claim. William Cruickshank, writing in September 1800, basically accepted their result, but noted hesitantly, ". . . how this can be effected, is by no means so easily explained."[70] For others, the absence of a plausible explanation of the phenomenon was an insuperable obstacle to agreement that the pile acted analytically. An anonymous writer in *Nicholson's Journal* in January 1801 asked how it was

67 See Nicholson (1800–1801). For examples of other investigators following his recommendations, see: Henry (1800–1801); Haldane (1800–1801).
68 See the letter from John Thomson to John Allen, dated 12 June 1801: "Lord Lauderdale and I made the galvanic experiment last week, and I exhibited it to the [Chemical] Society on Saturday. We are getting tubes with gold wires and glass stoppers to try its effects on caustic liquids, and we are getting a very broad plate of zinc made, to try whether the increase in power be in proportion to the increase of surface. In that case, his Lordship's whole service of plate will be converted into a galvanic battery." (Quoted in "Biographical notice of John Thomson," in Thomson (1832–59), II, p. 17.) According to the *D.N.B.,* Thomson was conducting a Chemical Society at Lord Lauderdale's house in Edinburgh at this time. His previous interest (and that of other Edinburgh philosophers) in galvanism is revealed in Fowler (1793), esp. pp. 10–12, 75, 80, 169–176.
69 Priestley (1802). For similar views, see also: "Experimentalist" (1801); and "H.B.K." (1806).
70 Cruickshank (1800–1801) (quotation on p. 257).

possible for a component particle, "insensibly to hurry through the water for a distance of six inches or more, and there to make its appearance in the character of gas." He concluded:

It appears to me therefore that the separation of the two gases from the connection with the pile of Volta, is ... unexplained according to the doctrine of the decomposition of water. ... I do not see how, in strict philosophy, we are warranted in saying more of the experiment with the pile, than that one kind of electricity and water produce one kind of gas, and the other another gas. Is it not an assumption in this experiment to say, that the bases of the gases are the component parts of water?[71]

The alternative explanation for the pile's action to which this writer alluded gained greater credence during 1801, when the work of Johann Wilhelm Ritter was reported from Germany. Ritter took the separate production of hydrogen and oxygen as conclusive proof that they were not the components of water. A more likely explanation for what was occurring, he proposed, was that oxygen was produced from water plus positive electricity, and hydrogen from water plus negative electricity. In Ritter's view, water was thus shown to be what he had always thought it was – an element.[72]

The voltaic decomposition of water did have its defenders. In November 1804, Charles Sylvester mounted a strong attack on those (like Ritter) who claimed that there were two types of galvanic electricity. To make this assumption was more arbitrary, Sylvester insisted, than to accept that the two components of water were produced separately.[73] On the other hand, Davy's papers from this period evaded the problem of understanding the effect of the pile upon water; his investigations of its mode of action simply assumed its efficacy as an analytical instrument.[74] This was a perilous assumption in view of what was clearly not a closed debate. Davy's research could be undercut by renewed challenges to his understanding of the way the pile operated. In 1805 and 1806 it looked as if this were just what would happen. From a number of sources came repeated claims that acids (or occasionally alkalis) were produced de novo from the action of the voltaic pile on water. If these claims had been accepted, the effect would have been fundamentally to undermine the status of the pile as an instrument of chemical analysis.

71 "Experimentalist" (1800–1801) (quotation on p. 515).
72 Ritter's experiments were reported in a letter from an unidentified "Dr. G.M." to William Babington ("G.M." (1800–1801)). See also Sudduth (1980), p. 31. For background on Ritter's approach to galvanism, see Wetzels (1990).
73 Sylvester (1804).
74 Davy's twelve early papers on the voltaic pile were published in Nicholson's Journal (1800–1801), in the Philosophical Transactions (1801), and in the Journal of the Royal Institution (1801–1802). They are collected in Davy (1839–40), II, pp. 139–228. For commentary and analysis, see Russell (1959–63).

In 1805, a correspondent calling himself "William Peel" wrote from Cambridge to Alexander Tilloch, editor of the *Philosophical Magazine*. Peel claimed that he had produced muriates (i.e., chlorides) of soda and potash, by applying a voltaic current to pure water.[75] Tilloch expressed reservations about Peel's techniques, but his results did seem to confirm those reported by the Italian experimenter Pacchiani, who had described producing muriatic and oxymuriatic acids (i.e., hydrochloric acid and chlorine) from galvanic action upon water.[76] Other investigators (Cuthbertson in London and Brugnatelli and Alemani in Italy) confirmed that muriatic acid was indeed produced, no matter how stringent the precautions taken to use only distilled water and to avoid any contamination by organic substances.[77] Defenders of the analytical power of the voltaic pile were therefore given cause to realize by the end of 1806 that the efficacy of their instrument was severely threatened. If these results were correct, the pile was not acting simply to analyze water into oxygen and hydrogen. Somehow, acids (muriatic, oxymuriatic, and nitric) were being produced by its action, perhaps as Sylvester speculated by oxidizing hydrogen to various degrees.[78]

This was the situation that Davy confronted as he delivered his first Bakerian Lecture to the Royal Society on 20 November 1806. To establish the power of his instrument, he had to reduce this confusion to order and to convince a new audience that the voltaic pile was his to command. He had to muster his arguments to reduce the freedom for alternative interpretations and to marginalize those who upheld them. To do this, Davy introduced a series of refinements of experimental technique that had the rhetorical effect of blocking off various alternative views of the pile's action. They also had the social function of making replication of the experiments more difficult, thereby securing Davy's claims from challenge by those who did not share his resources of equipment and skill.[79]

The first section of the lecture is the crucial one, from this point of

75 The letters, dated 23 April and 4 June, are in the *Philosophical Magazine*, 21 (1805), 279–280; 22 (1805), 152–155. In his 1818 account of the history of galvanism, John Bostock recorded that "no such individual as Mr. Peel could be found in Cambridge, so that the letter bearing his name is a complete fabrication." (Bostock (1818), pp. 62–63.) There is indeed no mention of such a person as Peel in the *Alumni Cantabrigienses*.
76 See Tilloch's comments, *Philosophical Magazine*, 22 (1805), 152–154.
77 Letter from John Cuthbertson, 22 March 1806, *Philosophical Magazine*, 24 (1806), 170–171. Other reports concerning the production of acids and alkalis are in *Philosophical Magazine*, 9 (1801), 181–185; 24 (1806); 176–180, 185; 25 (1806), 57–66, 130–142, 364–365; 27 (1807), 260–261, 338–339, 339–343.
78 Sylvester (1806).
79 The lecture is in Davy (1839–40), V, pp. 1–56. The technical refinements he introduced were not presented by Davy as tactical moves in a debate, but they did serve that function. Like other technical innovations, such as increasing the precision of measurement, making experiments harder to perform can be a way of establishing knowledge by reducing the freedom for alternative experimental claims.

view. There, in a series of experiments, Davy sought retrospectively to secure the basis of his electrochemical work: the belief that the voltaic pile was an instrument of analysis that decomposed water into hydrogen and oxygen only. The order in which the experiments were described was probably not that in which they had originally been performed, but one deliberately constructed to persuade the reader to accept the integrity of Davy's instrument. This section was a remarkably persuasive piece of writing. Davy surveyed the alternative claims about the action of the pile, including the latest experiments by Charles Sylvester (reported in *Nicholson's Journal* only the previous August), and gave the impression of a field of enquiry beset with a confused variety of interpretations. He announced what his own conclusion would be: that all the acids and alkalis reported to have been generated in electrolysis were artifacts produced by contamination. And then he described experiments to support his view. His refinements of technique consisted in using a large battery, tiny agate cups, pure platinum electrodes, repeatedly distilled water, and a piece of amianthus (purified asbestos) to connect the two separate cups in which the electrodes were placed.[80]

With this arrangement he told how at first he produced both acid and alkali, from (respectively) the positive and the negative electrodes. However, after repeated operations in the same agate cups, he succeeded in eliminating the production of acid and greatly reducing that of alkali. He concluded that saline matter in the composition of the agate cups or in "intimate adhesion in its pores" was responsible for producing the acid and probably some of the alkali. Attention was then directed at what Davy said was the only other possible source for the alkali: the water itself. He described experiments conducted using pure gold cups, which initially yielded small quantities of alkali at the negative electrode. This quantity was not increased, however long the operation was continued. By subjecting the distilled water he was using to a lengthy process of evaporation, Davy established that saline matter had been carried over in the original distillation. This saline matter could be eliminated by especially slow distillation of the water, and no alkali would then be produced under electrolysis.[81]

Further experiments were then described, which showed that alkalis (as well as acids) could be produced from the vessels in which electrolysis was conducted – wax cups produced potash and soda, resin cups yielded potash, and so on. Glass itself yielded soda, as could be seen if a small piece of glass were added to a gold cup in which a negative electrode was operating. Finally Davy pointed to the presence of dissolved nitrogen in the water as a plausible source for the appearance of acids and alkalis.

80 Davy (1839–40), V, pp. 2–4.   81 Davy (1839–40), V, pp. 5–8.

He cited Priestley as authority for the view that nascent oxygen would coexist in solution at the positive pole with nitrogen, forming an acid, whereas nascent hydrogen would expel any dissolved nitrogen from the water around the negative pole. This suggestion was apparently confirmed by experiments performed under reduced atmospheric pressure, or with another gas replacing the atmosphere over the cups of water. In these circumstances, Davy claimed, neither acid nor alkali was produced in detectable quantities.[82]

All these experimental reports were mustered to support Davy's overall conclusion:

To detail any more operations of this kind will be unnecessary; all the facts prove that the fixed alkali [and the acid are] not generated, but evolved, either from the solid materials employed, or from saline matter in the water.[83]

By driving a wedge between generation and evolution of products, Davy categorized the acid and alkali as artifacts produced by contaminants, whereas oxygen and hydrogen were genuine products of chemical analysis. Having thus established the fundamental efficacy of the voltaic pile as an analytical instrument, he went on to use it as such and to investigate its mode of action. These later parts of the Bakerian Lecture have absorbed the attention of several historians, who have puzzled over the discrepancy between the metallic-contact theory of the pile's action that Davy advanced there and the chemical theory proposed in his earlier papers.[84] This discussion, however, overlooks the importance of the first section of the lecture, which was the linchpin about which the whole turned. If Davy had not responded vigorously and resourcefully to those who were threatening to undermine the status of the pile as an instrument of analysis, he would not have been able to gain acceptance of its use as such, and an investigation of its mode of action would have been meaningless. The instrumental power of the pile was not purely natural; it had to be constructed by Davy through resourceful experiment and effective argumentation.

In written form, the Bakerian Lecture exhibits a series of strategic moves toward solving the problem of the sources of the reported acids and alkalis. Davy appears to encircle the troublesome substances, progressively cutting off their potential hiding places. The contaminants are never removed entirely, and each claimed identification of their sources is potentially open to challenge, but the reader is induced to acquiesce in Davy's assurance that they could be eliminated. Hence, one is led to accept his conclusion: "It seems evident then that water, chemically pure, is decom-

82 Davy (1839–40), V, pp. 9–12. 83 Davy (1839–40), V, p. 9.
84 See especially Russell (1959–63); Ziemacki (1974), esp. pp. 235–245.

posed by electricity into gaseous matter alone, into oxygen and hydrogen."

Davy's text operates in a similarly strategic manner against the potential skepticism of its readers. His apparently exhaustive elimination of possibilities operates as a kind of buried dialogue, anticipating and answering the objections of a skeptical reader. The length of the procedure as described and the repeated emphasis on refinements of technique and the sensitivity of tests convey an impression of Davy as an extraordinarily careful and skillful experimenter. The technical refinements also made experiments in this field harder, by raising the stakes in terms of the necessary level of equipment and skills for making contributions in future. Investigators who lacked these resources would no longer need to be taken seriously.

Most reviewers of the Bakerian Lecture readily accepted its conclusions. Some also commented on its rhetorical effectiveness. Even those who confessed they could not follow the details of the argument said they were convinced by its overall drift.[85] Henry Brougham, writing in the *Edinburgh Review,* pointed to the way the detailed argument of the first section of the lecture supported Davy's other experimental claims: "The examination of [this enquiry] gives us an irresistible disposition to confide in all the other processes of the author, which he passes over more hastily, or only refers to in general terms."[86]

As Brougham noted, Davy had augmented his future credibility by virtue of the persuasiveness of his arguments on this occasion. By establishing the efficacy of the voltaic pile as an instrument of analysis, he had armed himself with the means of producing new knowledge of the chemical composition of matter. Thus the foundation was laid for the discoveries of two new metals, announced by Davy in his second Bakerian Lecture in November 1807.[87] He described there how his expectation that the pile could be used to analyze previously undecompounded substances had borne fruit. A current passed through a sample of fused potash or soda yielded globules of the new metals, which were named "potassium" and "sodium." The pile was taken to have produced these metallic elements by analyzing their oxides, potash and soda. The startling new discoveries were very widely accepted; some British reviewers hailed them as the greatest contributions to science in recent times.[88]

The discovery of two entirely new metals lent considerable support to Davy's claims for the analytical power of the voltaic pile. Most research-

85 See: reviewer in *Philosophical Magazine*, 26 (1806–1807), 181, 266–267, 269; and John Bostock in *Monthly Review*, 2nd ser., 54 (1807), 1–4.
86 Brougham (1807–1808), p. 398.   87 Davy (1839–40), V, pp. 57–101.
88 *Philosophical Magazine*, 29 (1807–1808), 180–183; *Monthly Review*, 2nd ser., 57 (1808), 225–227; Brougham (1808).

ers accepted that the pile had acted to decompound the metallic oxides, and hence could be expected to act generally as an instrument of analysis. Davy had largely succeeded in establishing the pile as an analytical tool, and it is clear that the production of new metallic substances around which opinion rapidly consolidated was an important part of this accomplishment.

It is worth noting, however, that agreement with Davy's claims was not universal, even after the discovery of the alkali metals. Some investigators continued to hold theories like Ritter's, which denied that water had been decomposed. Robert Harrington of Carlisle was one of these. A similar view was taken by Ezekiel Walker, of Lynn in Norfolk, who published a lengthy series of articles in the *Philosophical Magazine* through the 1810s, which purported to give "New Outlines of Chemical Philosophy." Walker's trenchantly expressed view was that ". . . the theory of the compound nature of water does not rest upon common experience, nor upon any known truth, but upon a bare assertion without any proof whatever." He claimed that hydrogen and oxygen gases were composed of water combined with (respectively) negative and positive electricity (or "photogen" and "thermogen"). The gases were indeed released when electricity was applied to water, but the process was one of combination not decomposition.[89]

Another writer who took this kind of view was George Smith Gibbes, a physician and chemical lecturer in Bath who perversely saw Davy's discoveries as authorizing a revival of the phlogiston theory. Explicitly connecting his opinion with that of Ritter, Gibbes claimed it had been confirmed by the "late splendid discoveries of Mr. Davy." Water was again seen as an element and fire (electricity) as a compound of a "positive galvanic principle" with a "negative galvanic principle." In Gibbes's view, the voltaic pile was "an instrument, by the operation of which fire is instantly decomposed, and its two component parts separated. Thus in water, the one side produces vital air, and the other side inflammable air."[90]

Though a quite consistent point of view, this was not of course Davy's own opinion. Although he sometimes indicated that he saw the possibility of reviving a version of the phlogiston theory, he always insisted upon the decomposition of water as the basis of his account of the workings of the voltaic pile. Most other chemists agreed with him, and some even

89 Walker's articles were published in the *Philosophical Magazine,* and his views summarized in E. Walker (1823). The passage quoted is in *Philosophical Magazine* 43 (1814), p. 26.

90 Gibbes (1809), p. 20. Thus vital air is positive galvanic principle + water, inflammable air is negative galvanic principle + water. This author's earlier work (Gibbes, 1799) had given tentative support to the antiphlogistic theory.

used the fact that water was analyzed by the pile to argue against Davy's account of the formation of the alkali metals. Charles Sylvester voiced this argument in his *Elementary Treatise on Chemistry* (1809). Discussing the supposed analysis of potash, Sylvester remarked that the water that was part of its composition would be expected to be itself decomposed into hydrogen and oxygen. Davy had noted the production of oxygen, but had said nothing about the hydrogen. Sylvester suggested that potassium could actually be a compound of potash with this nascent hydrogen.[91] The possibility was taken up as a challenge to Davy's work by the French chemists Gay-Lussac and Thenard. The implication was that the efficacy of the voltaic pile as an instrument of analysis was secure, but that the status of the alkali metals as elements was questionable.

By mapping the diversity of opinion that remained after Davy's first two Bakerian Lectures, we gain an impression of the social dimension of his achievement. As sociologists of science have shown, the "blackboxing" of a working instrument is coincident with the consolidation of a community of practitioners around its routine use. Davy had indeed succeeded in constructing such a community of users of the voltaic pile. Excluded were those who refused to accept that the pile acted to analyze water into its component parts. They tended to write from marginal provincial locations, either (like Harrington) performing no experiments or working (like Walker and Gibbes) in a tradition of small-scale experimentation on the boundaries of chemistry, optics, and electricity – a tradition that was rapidly coming to look quite ineffective and outdated. Whereas before 1806 the voltaic pile was the focus of a free-for-all debate of conflicting interpretations, Davy had succeeded in imposing a single dominant interpretation and a concomitant definition of experimental competence. Recognized expert chemists did not thereafter step out of line. Davy recruited allies among the leading London chemists to replicate and extend his analyses with the pile. W.H. Pepys and J.G. Children built their own versions of the apparatus and collaborated with Davy on subsequent work.

The direction of this work was shaped by a competitive rivalry into which Davy was drawn with the French chemists Joseph Louis Gay-Lussac (1778–1850) and his collaborator Louis-Jacques Thenard (1777–1857). The issues in this dispute showed both Davy's continuing vulnerability and the magnitude of his recognized achievement. Gay-Lussac and Thenard mounted a challenge to the status of the alkali metals as elements, but they did so on the basis of agreement on the analytical action of the pile. Immediately after the 1807 lecture the French chemists set about constructing their own apparatus, in order to confront Davy's dis-

91 Sylvester (1809), pp. 119–135.

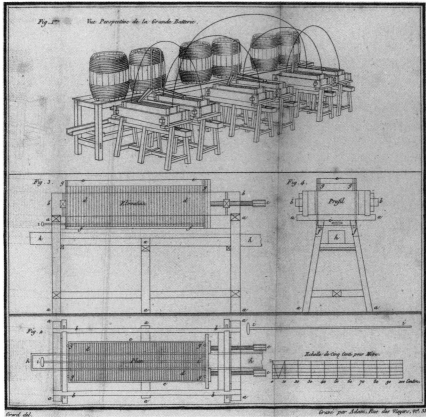

Figure 9. The voltaic pile of Gay-Lussac and Thenard. From J.L. Gay-Lussac and L.-J. Thenard, Recherches Physico-chimiques, *2 vols. (Paris: Detervile, 1811), vol. II, plate I. [Reproduced by permission of the Syndics of the Cambridge University Library.]*

coveries on his own ground. The 1806 lecture had already been widely hailed in France, and its author awarded a Napoleonic medal for his endorsement of the metallic-contact theory of the pile's action, which was popular among French physicists. The following year an order was given to construct a new voltaic pile of unprecedented power, to compete with that under Davy's command (Figure 9).[92]

Davy's response was in turn to initiate a subscription fund among the patrons of the RI to support the construction of a new and more power-

92 For the French background, see Sutton (1981). The Paris pile is described, with detailed plates, in Gay-Lussac and Thenard (1811), I, pp. x-xi, 1–7.

ful battery there. In his petition, presented to the Managers on 11 July 1808, he stressed that national priority in this area of science was in danger of being lost, an outcome that would be "dishonourable to a nation so great, so powerful, and so rich." The Proprietors and other subscribers rose to the challenge, collecting more than 1,000 pounds. A pile of 2,000 double plates, each of 8 inches square, was rapidly built.[93] Electrochemistry entered a new phase: No longer a widely dispersed activity with easily replicated apparatus, it was now the subject of an international battle involving powerful and expensive equipment and refined experimental procedures.

Before resuming our account of this rivalry in the following section, let us recapitulate how Davy had achieved this transformation.[94] He had built and exploited a solid relationship with his audience, initially at the RI and later at the Royal Society. He had beaten off a series of challenges to his interpretation of the basic phenomenon of the voltaic pile, the electrolysis of water. Without general agreement that this produced only hydrogen and oxygen, the pile could not have been used as an instrument of analysis at all. If its workings had remained problematic, it would have been useless as a tool of research. What was necessary was for the apparatus to be "black-boxed," with the formation of a community of researchers who agreed on the way in which it was to be used. Davy's success came from mobilizing phenomeno-technics and rhetorical and social resources to construct a persuasive demonstration of the power of his instrument. He showed, as Faraday was also to do later, that "he who proves, discovers."[95]

Compare this situation with the ridicule that greeted the Bristol nitrous oxide experiments. Those experiments were widely denounced as visionary, enthusiastic, and speculative. Skepticism and derision severely inhibited their acceptance as genuine knowledge, adding to the difficulties that were experienced in replicating them. In the face of such antagonism, Davy had to disguise the roots of his own interest in topics such as galvanism and the voltaic pile.[96] The success that he subsequently achieved with the pile flowed in part from having abandoned Priestley's model of the social organization of public science. Priestley and Beddoes had aimed

93 Davy quoted in Jones (1871), pp. 355–357 (quotation on p. 356). See also Davy (1839–40), IV, pp. 110–111; V, pp. 282–283.
94 As a measure of the importance of Davy's work in transforming this field, compare the account of the uncertainties prevalent in 1806, in Parkinson (1807), pp. 338–347, with the whiggish story told in Bostock (1818).
95 Gooding (1985b).
96 Thus, in a lecture given in 1810, Davy denounced speculation that sought to connect Galvani's discoveries with the causes of animal motion—speculation of the kind he had himself indulged in just over a decade before. Davy (1839–40), VIII, p. 269: ". . . a spirit of generalisation was indulged in, romantic, and far removed from that of sound reason and unprejudiced investigation."

to diffuse knowledge and techniques through descriptive writings and via networks of acquaintances, while preserving an independence from central institutions. Davy on the other hand moved into and exploited to the full central institutions such as the RI and the Royal Society; he formed around himself a new kind of public audience for science, and consolidated in this context the rhetorical and social resources that helped him establish the power of his instruments.

In fact, so completely did Davy's audience accept the instrumental efficacy of the voltaic pile, that his discoveries with it were widely said to have been expected, given the promise of the device that he now had at his command. It was as if mere possession of the instrument guaranteed discoveries. Henry Brougham put the point with characteristic vigor: Davy, he wrote, owed much to "the powerful instrument which former discoveries had put into his hands." He went on, "Any man possessed of his habits of labour, and the excellent apparatus of the Royal Institution, could have almost ensured himself a plentiful harvest of discovery."[97] Davy had so successfully convinced his audience of the power of the voltaic pile that his own agency could easily be lost sight of – the instrument now seemed to yield discoveries entirely by itself.

Paradoxically, Brougham's comment also drew attention to the unprecedented concentration of resources that Davy had deployed. His remark hints that the discoveries might only be producible by the uniquely powerful battery of the RI. This kind of charge, with its insinuation that his results could not be replicated by other investigators, worried Davy. He and his brother repeatedly assured readers that the discoveries had all been achieved using a battery of quite modest size, and that they could very readily be replicated.[98] Although many among his audiences heeded these reassurances, some did not. As we shall see in the next section, allegations that his experiments were not readily replicable continued to hinder the acceptance of Davy's work well after the Bakerian Lecture of 1807.

Underlying these tensions were political differences about the proper form of public science. Brougham wrote as a liberal Scottish Whig, in the leading journal of that group, the *Edinburgh Review*. He did not hesitate to express his distrust of the metropolitan scientific establishment and its institutions.[99] Brougham's judgment was that Davy had not succumbed to the corrupting atmosphere of the RI, and had maintained his "manly" independence. But other observers were not so sure. Harrington described Davy as a self-styled "Hercules . . . seated on the shoulders of Sir

---

97 Brougham (1808), p. 395. Compare the similar comments in his essay on Davy in Brougham (1872), I, 107–122, esp. p. 115.
98 Davy (1839–40), V, pp. 62–63 fn., 106–107 fn.; VIII, p. 355 fn.
99 See, for example, Brougham (1810–11), p. 403.

Joseph Banks."[100] It was clear to everyone that Davy had abandoned his roots among provincial enlightened intellectuals and assumed a position in the central institutions of the scientific establishment. Those who were politically antipathetic to the world in which he now moved could point to the concentration of resources and power that he had in his hands. For them, Davy's work had made the voltaic pile less democratically available, and more like a private means of discovery. Davy may have convinced many of his critics of the analytical efficacy of the pile, but he continued to face objections to other aspects of his work from disgruntled opponents.

Chlorine and "the lever of experiment"

CAROLINE: But what was Sir H. Davy's reason for adopting an opinion so contrary to that which had hitherto prevailed?
MRS. B.: There are many circumstances which are favourable to the new doctrine; but the clearest and simplest fact in its support is, that if hydrogen gas and oxy-muriatic gas be mixed together, both these gases disappear, and muriatic acid gas is formed.
EMILY: That seems to be a complete proof; is it not considered as perfectly conclusive?
MRS. B.: Not so decisive as it appears at first sight; because it is argued by those who still incline to the old doctrine, that muriatic acid gas, however dry it may be, always contains a certain quantity of water, which is supposed essential to its formation. So that, in the experiment just mentioned, this water is supplied by the union of the hydrogen gas with the oxygen of the oxymuriatic acid; and therefore the mixture resolves itself into the base of muriatic acid and water, that is, muriatic acid.

Jane Marcet, *Conversations on Chemistry* (1817)[101]

When the RI's new battery was finished in 1809, Davy gave a special lecture to introduce it to his audience and to thank the subscribers who had made it possible. He described the building of the apparatus as a fitting response to the threat of French experimenters taking over the lead in electrochemistry. Individual munificence, spurred by emulation, had shown itself capable of matching anything the Napoleonic regime could do. He expressed his faith that "our sciences will always flow from the voluntary efforts of individuals, from whom the support will be an honour – to whom it will be honourable."[102]

In these terms, Davy connected the construction of the new pile with the traditions of English Enlightenment science – a voluntary activity, dependent on individual patronage, stimulated by emulation and the quest

---

100 Harrington (1819), p. 1.   101 [Marcet] (1817), II, p. 148.
102 Davy (1839–40), VIII, pp. 355–356.

for honor. But at the same time he acknowledged that a change had taken place: His science now drew its strength from a single powerful experimental apparatus which required the widest public support. Widespread curiosity about science would have to be fostered if material support was to continue, and should be precisely focused on the leading institutions of research. Some coherent direction of the diverse forms of local scientific culture was required because "To divide and to separate the sources of scientific interest, is to destroy all their just effect."[103] Public interest had to be oriented toward a central locus of scientific activity, which could then exploit the material support the public would deliver.

In the years following this lecture, Davy worked to mobilize British public opinion in his support, as he found himself embroiled in sustained rivalry with Gay-Lussac and Thenard.[104] As John Davy noted, his brother and the French chemists were like contending generals, maneuvering their weapons – the voltaic batteries in London and Paris – to attack one another just as the real generals of England and France were doing in the continuing war between the two nations.[105]

The techniques Davy used to defend himself in these prolonged controversies were simultaneously experimental, rhetorical, and social. He tried to win his arguments by doing more experiments than his opponents, exploring more possible claims and counterclaims; he replicated his opponents' work assiduously to demonstrate their errors; and he attempted to sustain a higher degree of precision in all aspects of measurement. The facts produced by experiment would have been impotent without the persuasive force they were given by the use of these techniques. In addition, Davy manipulated chemical language by making several suggestions for changes in nomenclature that would embody his empirical claims. Finally, he buttressed his assertions as to the results of specific experiments by mobilizing various kinds of audiences as witnesses. Davy took advantage of his relations with the general public and many specialist chemists by invoking their authority as witnesses to his experiments. He succeeded in turning "the lever of experiment" by placing its fulcrum on the base of the relationship with his audience that he had so successfully constructed.

To do this, he literally brought the lecture theater into the laboratory, assembling at the site of his experiments an audience of expert chemists and prominent patrons (Figure 10). In 1804 the basement laboratory at

103 Davy (1839–40), VIII, p. 357.
104 The rivalry is discussed in Crosland (1978), pp. 71–91; and idem. (1980).
105 Quoted in Crosland (1978), p. 76. The background of the Napoleonic wars is a pertinent one. In his lectures, Davy described his debates with Gay-Lussac and Thenard in terms of a rivalry between English and French schools of chemistry; and, as Fullmer has shown, he also had to defuse criticism that he was lacking in patriotism for having accepted the award of the Napoleonic medal (Fullmer (1962), pp. 150–151).

*Figure 10. Interior view of the laboratory at the Royal Institution,
with the arch leading through to the basement lecture
theater at the right. From W.T. Brande,* A Manual of
Chemistry *(London: John Murray, 1819), plate II. [Re-
produced by permission of the Syndics of the Cambridge
University Library.]*

the RI had been enlarged by knocking down a wall dividing it from an
adjacent cellar (Figure 11). In the newly cleared space a bank of seats was
built, capable of accommodating up to 150 spectators, each of whom
would have a clear view of the laboratory bench. The declared intention
was to create "a theatre for those who attend the experiments of research."[106] The audience invited here was more select than that which
attended in the public lecture theater upstairs. Members of the RI Committee of Chemistry were apparently among those who were invited to

106 Jones (1871), p. 258; quoting from the RI Managers' Minutes (in Greenaway (1971–
75), III, pp. 172–173). For a brief description of the modified laboratory, see: Chilton
and Coley (1980), p. 176 and plates I, II.

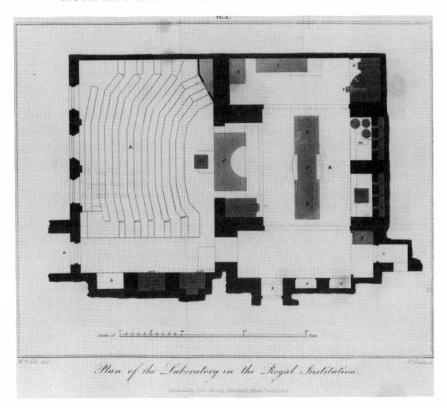

Figure 11. *Plan of the laboratory of the Royal Institution, with the basement lecture theater at the left. From W.T. Brande,* A Manual of Chemistry *(London: John Murray, 1819), plate I. [Reproduced by permission of the Syndics of the Cambridge University Library.]*

sessions in the basement.[107] By invoking this audience as witnesses in his published papers, sometimes in general terms, sometimes as named individuals, Davy brought their authority to bear in support of his experimental claims. For example, the chemist W.H. Pepys and "a numerous assembly" were invoked to contradict a French claim that potassium was

107 Their presence is mentioned at experiments reported in the paper, "On a combination of oxymuriatic gas and oxygen gas," read to the Royal Society on 21 February 1811 (Davy (1839–40), V, pp. 349–357). Although John Davy denied that the basement laboratory was fitted up specifically for display, he admitted that his brother received friends there and conversed with them about what he was doing. He also noted that the space was used to accommodate students for the practical chemistry course (J. Davy (1836), I, p. 254).

capable of forming a compound with hydrogen. In response to another reported observation by Gay-Lussac and Thenard, Davy recorded:

I have repeated the experiment ... more than twenty times, and often in the presence of some of the most distinguished chemists in this country, from whose acuteness of observation, I hoped no source of error could escape.[108]

Citing respectable witnesses to his experiments was one way in which Davy mobilized his audience in specialist controversy. Using this and other tactics, he defended his identification of potassium as a metallic element. Challenges to this claim continued for several years after the substance was discovered. The Manchester chemist John Dalton had to be taken to task by the Davy brothers, when the second part of his *New System of Chemistry* (1810) identified potassium as a compound.[109] In 1811 the Edinburgh chemical lecturer John Murray (d. 1820), writing in *Nicholson's Journal*, took up a claim that he ascribed to Claude-Louis Berthollet that potash contained combined water.[110] If this were the case, then its decomposition by the voltaic pile could not be as Davy had described: The oxygen he had observed could have come from decomposed water, and potassium could be a compound of hydrogen with some unknown basis. Responding to Murray, John Davy curtly cited his brother's own determination of the water content of potash, and referred his opponent also to a brief statement by Gay-Lussac and Thenard in which (he claimed) they accepted the elementary status of potassium[111] Gay-Lussac and Thenard themselves, however, in their *Recherches Physico-Chimiques* (1811), by no means took Davy's side. They insisted that potash and soda did contain water, which would be decomposed by the operation of the pile. It was thus possible that potassium and sodium should continue to be regarded as compounds of the unknown bases of potash and soda with hydrogen.[112]

In mounting his defense of potassium and sodium, Davy was pushed into a dilemma, as he tried to reconcile the defense of his discoveries of two new elements with an explicit methodological commitment to reducing the number of laws and principles in nature. In the June 1808 paper in which he criticized the French chemists' view that potassium was a compound of potash with hydrogen, he was prepared to speculate that (like all metals) it might be formed by a union of hydrogen with some yet-unknown basis.[113] As the assault on his discoveries continued and as the analytical efficacy of the voltaic pile was brought into question again,

108 Davy (1839–40), V, pp. 158 fn., 207.
109 Dalton (1808–27), I (i), pp. 260–263; I (ii), pp. 53–54. For Davy's response, see: Davy (1839–40), V, pp. 322–323 fn.; and J. Davy (1811a), pp. 205–206.
110 Murray (1811b); Berthollet (1809).   111 J. Davy (1811b).
112 Gay-Lussac and Thenard (1811), II, pp. 215–258. See also idem. (1810).
113 Davy (1839–40), V, pp. 131–132.

Davy dropped such speculation and insisted upon adherence to a rigorous definition of an element as the actually attained limit of chemical analysis.[114]

This definition, which was defended coextensively with Davy's actual discoveries, tied them to his claim to have a special authority over the instruments of analysis. If one accepted that Davy had a particular knowledge and command of the most appropriate and powerful analytical instruments, one would be inclined to accept his assertions as to what substances were elements. But conversely, if one doubted the efficacy of his instruments one would be likely to dispute some of his identifications of elements and might also conclude that such a definition was unhelpful. Despite its overt epistemological modesty, the definition had to be enforced quite firmly on those who challenged Davy's view of where the limits of chemical analysis lay. This was particularly the case in the controversy that followed his announcement in 1810 of the elementary status of chlorine.

Of all the elements Davy discovered, chlorine was the one on which his claim was most open to challenge. The opinion that what was known as oxymuriatic acid might not be an oxide of muriatic acid but a simple substance was first advanced tentatively by Gay-Lussac and Thenard in 1809.[115] To them, this was a hypothetical alternative to the traditional view, equally consistent with experimental facts but on balance less simple to reconcile with the remainder of chemical theory. Davy converted French opinion into English fact, and in two papers published in July and November 1810 committed himself to the assertion that oxymuriatic acid was an element.[116]

To defend this claim, Davy brought forward no dramatically new experimental evidence. There was no new substance to be exhibited, because the basic phenomenal properties of oxymuriatic acid had been familiar since the description of its preparation by Carl Wilhelm Scheele in 1774.[117] All Davy was able to do was to assert that available techniques had failed to decompose this substance, that it should therefore be considered an element, and that chemical theory could be revised and made consistent with this result. Each step of the reasoning was potentially contentious. In taking the first step, Davy exploited the accepted power of his analytical instruments. He confirmed that oxymuriatic acid could not be reduced by heating with charcoal, and added that neither could it

114 For an invocation of this definition, see Davy (1839–40), V, p. 527 fn.; and J. Davy (1811a), p. 193.
115 Gay-Lussac and Thenard (1809).
116 "Researches on the oxymuriatic acid, its nature and combinations," and "On some combinations of oxymuriatic gas and oxygen," in Davy (1839–40), V, pp. 284–311, 312–348.
117 Scheele (1897).

be decomposed by the most potent voltaic battery at his command.[118] The implication that it should therefore be regarded as an element required acceptance of his definition of what an element was. He reminded his readers, "names should express things and not opinions; and till a body is decompounded, it should be considered as simple."[119] The conclusion was rendered more plausible by a sketch of how chemical theory could be reconstructed to incorporate the new element. It was portrayed as electronegative and closely analogous to oxygen; it was capable of combining vigorously with metals and combustible substances, and of displacing oxygen from oxides.[120]

These arguments were supported with the same persuasive tactics that Davy had deployed in his previous papers. Witnesses to the experiments were cited. He insisted that what he was reporting were facts, and not in the least hypothetical. Above all, he gave the entity a new name – "chlorine" – which implied its elementary nature and obliterated the suggestion that it was an oxide.[121] The renaming strategy clearly owed something to Lavoisier, who had shown very well how adherents could be recruited by getting people to use a new language. But, presumably mindful of this precedent, Davy was careful to validate his change of nomenclature by citing the approval of the specialist chemical community. The change of name was advisable, he proposed, "to assist the progress of discussion, and to diffuse just ideas on the subject"; he had adopted the term "chlorine" only after "consulting some of the most eminent chemical philosophers in the country."[122]

The employment of these tactics did not guarantee acceptance of Davy's claims. Gay-Lussac and Thenard remained unconvinced, at least at first. In their *Recherches,* they described the notion that oxymuriatic acid was an element as no more than a plausible hypothesis that was not easily reconciled with Lavoisier's theory of acidity.[123] The Swedish chemist Jöns Jakob Berzelius (1779–1848) also resisted Davy's assertion for a further ten years.[124] Brougham, assessing Davy's papers on oxymuriatic acid, judged that he had not entirely proved his point. He lamented that "the experiments are not detailed with such minuteness as is essentially necessary, when any new point of doctrine is to be established," and he resented the resort to new nomenclature, complaining, "We wish Mr.

---

118 Davy (1839–40), V, p. 292–293.   119 Davy (1839–40), V, p. 347 fn.
120 Davy (1839–40), V, pp. 292, 296–297, 341–345.
121 Davy (1839–40), V, pp. 291, 345, 348 fn.
122 Davy (1839–40), V, pp. 345, 347–348.
123 Gay-Lussac and Thenard (1811), II, pp. 155–166, 171–176. According to Lavoisier's theory, muriatic and oxymuriatic acids had to contain oxygen, the principle of acidity (Crosland (1973); LeGrand (1972)).
124 Melhado (1981), pp. 216–217, 217 fn.

Davy would attempt to confine himself within the common limits of the language."[125]

The most sustained challenge to Davy's assertion that oxymuriatic acid was an element was made by John Murray, a private lecturer on chemistry, materia medica, and pharmacy in Edinburgh, and the author of two well-known chemical textbooks.[126] In a series of papers in *Nicholson's Journal* between February 1811 and April 1813, Murray mounted a consistent and plausible attack on Davy's view. John Davy sustained a tenacious defense on his brother's behalf. The controversy reveals much about the rhetoric of argumentation and the difficulties of securing agreement over the interpretation of experiments. It also shows the uses that could be made of public audiences in attempting to resolve specialist disputes. In addition to being a skilled chemist, Murray was experienced as a public lecturer and demonstrator; he waged his campaign on this battlefield as well as in the scientific press.[127]

Murray's challenge led off from the supposition that acids such as muriatic acid required combined water in order to exist as independent entities. On this basis, the experiments on which Davy rested his claims for chlorine could be given an alternative explanation. For example, oxymuriatic acid combined with hydrogen to form muriatic acid because its oxygen formed the water that was part of the final product. Charcoal would not be expected to reduce oxymuriatic acid because charcoal could not provide the water that was needed to constitute muriatic acid in its gaseous form.[128] John Davy in reply tied the defense of chlorine to that of his brother's definition of an element, a definition he described as "a fundamental principle of modern chemistry." He denounced Murray's muriatic acid with combined water as "an imaginary body, that I confess I am altogether ignorant of."[129]

One of the experiments discussed was that in which Murray claimed to have used oxymuriatic acid to provide the oxygen to convert carbonic oxide (carbon monoxide) to carbonic acid (carbon dioxide). According to Davy's account, in which oxymuriatic acid was actually the element chlorine, this could not be possible. Not surprisingly, when the Davy

---

125 Brougham (1810–11), p. 406; idem. (1811), p. 472.

126 This John Murray (d. 1820) should not be confused with his son, John Murray (1798–1873), or his namesake, John Murray (1786?–1851). All three are listed in the *D.N.B.* The John Murray who took part in this controversy published Murray (1801, and 1806); his private lectures in Edinburgh are mentioned in Morrell (1972), p. 13 and fn.

127 There is a little information on Murray's experimental skills and lecturing in Thomson (1830–31), II, pp. 268–269; and in Murray's obituary in the *Gentleman's Magazine*, 90 (ii) (1820), 185–186.

128 Murray (1811a).   129 J. Davy (1811a), pp. 193, 199.

brothers repeated Murray's experiment with what they considered appropriate precautions to purify and dry the gases, they found no carbonic acid was formed. In John Davy's view, Murray's gases must have been contaminated with water and/or atmospheric oxygen. Replications of the experiments, with more rigorous procedures, had supposedly produced results consistent with Davy's new theory.[130]

The Davys' strategy of replicating their opponent's experiments in order to expose his errors was one Murray found difficult to handle. He could only respond by repeating his own experiments, with "every attention . . . to ensure accuracy," and reiterating that the original results stood.[131] Murray's position was that he was denying, not the facts, but the Davys' inferences from them. He sought to show that the traditional account of oxymuriatic acid was preferable because of its coherence with established chemical theory. Hence, he was committed to portraying both accounts as equally hypothetical constructs. Murray insisted:

... the very possibility of the proposition being called in question without any doubt being expressed of the accuracy of the experiment on which it rests is a sufficient proof, that it is not a simple expression of the fact, as he [John Davy] and his brother suppose, but an inference from the fact.[132]

John Davy drew a different line between facts (which included deductions from experiments) and hypotheses.[133] He defended his brother's discovery of chlorine as an experimental fact and hence was obliged to deny the factuality of many of Murray's experimental claims, displaying them as mistaken inferences from improperly conducted experiments. For example, on Murray's assertion that muriatic acid, when combined with potash, released water, Davy commented: "If the above is a simple expression of facts, the theory which expresses those facts must be correct. But I have not been able to witness such facts."[134]

The early stages of the debate thus proceeded in a manner that both protagonists realized was likely to be inconclusive. While John Davy complained that Murray was denying plain experimental facts, Murray continued to uphold his own results and the validity of his inferences from them. He complained of Davy's dogmatic style of expression, while presenting himself as open-minded and tolerant of alternative theories. The Davys mobilized the preponderance of prestige witnesses, (including such members of the RI Chemical Committee as Charles Hatchett and William Thomas Brande), as against Murray's assistant Mr. Ellis and "some other friends."[135] But the debate could not be resolved by these means.

130 J. Davy (1811a), pp. 201–202.   131 Murray (1811c).
132 Murray (1811d), p. 195.
133 J. Davy (1811a), pp. 194, 205–206; idem. (1811c), pp. 39–40, 40–41.
134 J. Davy (1811c), p. 40.   135 J. Davy (1811c), p. 42; Murray (1811d), p. 190.

The stalemate was only broken when John Davy recruited powerful new allies to support his case. In September 1811 he declared triumphantly:

I have now to announce the existence of a new acid gas, which operated in Mr. Murray's experiment, without his knowledge of its existence, and was the cause of those phenomena which he erroneously attributed to the formation of carbonic acid gas.[136]

John Davy's act of producing a new gas (subsequently named "phosgene") and asserting its (unknown) presence in his opponent's experiments proved a powerful move in the dispute. He had previously drawn Murray's attention to the possible complications caused by a new compound of oxymuriatic acid and oxygen, recently described by Humphry Davy in a paper read to the Royal Society on 21 February 1811. Murray had however denied that that gas, subsequently named "euchlorine," had been present in his experiments.[137] Now he was faced with another one, "a new and peculiar compound of carbonic oxide and oximuriatic gas," which behaved in a way liable to be confused with the carbonic acid that Murray claimed a mixture of those two gases produced.[138] Murray's response was tentative and defensive. He recorded that he could not himself produce the new gas in the way John Davy had described; but, pending the publication of a fuller description of its preparation, he had to accept that there might be "some peculiarity necessary to the success of Mr. J. Davy's experiment." He went on, "I know sufficiently the disadvantage to which any experimentalist is subjected, who undertakes the examination of experiments of which only a general account is given."[139]

Murray retreated to consideration of another experiment, which seemed to offer the possibility of a decisive resolution of the whole debate: the reaction of muriatic acid gas with ammonia. According to Davy's theory, neither of these gases contained water, so nor could the salt that was the product of their combination. Murray claimed, however, that muriatic acid did contain water and that this could be established by extracting water from the product salt. He wrote, "This experiment then has the advantage of being conclusive on the subject of the present discussion; the state of the fact only requires to be ascertained, and with due precaution this is not difficult of attainment."[140]

Murray spoke too soon. Concentrating attention on the reaction of muriatic acid with ammonia focused the debate on what promised to be "an *experimentum crucis*" but did not in the event resolve it. John Davy rose to the challenge and announced that the water that Murray had

136 J. Davy (1811d), p. 30.   137 Murray (1811d), pp. 200–201.
138 J. Davy (1811d), p. 30.   139 Murray (1811e), p. 227.
140 Murray (1812a), p. 131.

found in the product (muriate of ammonia) had been absorbed from the atmosphere or from insufficiently dried reactant gases. The salt, he claimed, was deliquescent.[141] Murray in reply denied this, pointing out that the salt neither became damp nor showed any gain in weight on exposure to the air. He also drew support from a successful replication of his experiment, reported from the Liverpool Literary and Philosophical Society, by the physicians John Bostock and Thomas Stewart Traill.[142]

Bostock lent further support to Murray's cause by objecting strongly when the first volume of Humphry Davy's *Elements of Chemical Philosophy* (1812) presented the question of chlorine in an entirely one-sided way. In what was proposed as an authoritative textbook, Davy stated that the traditional view of the composition of the gas was "now universally given up." He ascribed the intransigence of certain (unnamed) Scots and Frenchmen to their belief in a fictitious entity called dry muriatic acid.[143] Writing in the *Monthly Review,* Bostock expressed resentment at the dogmatic tone Davy had adopted in defense of his fanciful beliefs, and particularly criticized the construction of a nomenclature based on the chlorine theory:

. . . to employ such [terms] as are at present the subject of controversy, and even to make them the basis of a nomenclature, can only tend to impede the progress of science, and lead us to suppose that we have acquired knowledge when in fact we have only learnt a new language.[144]

Apart from producing a textbook, the Davy brothers were also using public experimental demonstrations to try to bring the controversy to a favorable resolution. In January 1813, John published "An Account of an Experiment made in the College Laboratory, Edinburgh," which described what was clearly meant to be a decisive refutation of Murray's arguments: a version of the "crucial experiment" performed by Humphry in Murray's own city, with an assembly of illustrious witnesses including Thomas Charles Hope (Professor of Chemistry at Edinburgh University), Sir George Mackenzie, John Playfair and "some other gentlemen." Ammonia and muriatic acid gases were dried for approximately sixteen hours and then allowed to mix. The product salt was heated and a tiny amount of dew was given off. This was judged far too small a quantity of water to support Murray's case and was ascribed to "uncombined moisture derived from various sources," such as the mercury over which the gases were mixed. John described the experiment curtly and

141 J. Davy (1812).   142 Murray (1812b); Bostock (1812).
143 Davy (1839–40), IV, p. 177.
144 [Bostock] (1813), p. 152. Comparable remarks were made by other reviewers in *Annals of Philosophy, 1* (1813), 371–377; *Quarterly Review, 8* (1812), 65–86; and *Philosophical Magazine, 40* (1812), 145–151, 297–307, 434–444.

factually, and with a lofty air of finality refused to reply to any further objections by Murray.[145]

Murray, however, was not to be beaten in this manner. A few months later, he hit back with his objections to the Davys' experiment and (having taken a leaf from his opponents' book) a description of his own replication of it. The Davys had employed an inappropriate mode of heating the salt, he complained, which diminished the chances of extracting what would only be expected to be a small quantity of water. When he had repeated the experiment, with some of the same witnesses present, "The quantity of water, Dr. Hope was satisfied, appeared considerably larger than in Sir Humphry's experiment." Calculations about the expected quantity were beside the point, Murray suggested, because the reaction of muriatic acid with ammonia was not the one best suited to reveal how much combined water the acid contained. The essential point was that water *was* found: "The production of any water is incompatible with Sir Humphry's hypothesis, and therefore refutes it."[146] True to his promise, John Davy declined to reply.

The controversy between Murray and the Davy brothers invites analysis in terms derived from the sociology of modern science. The inconclusive nature of the experiments that were performed provides an excellent illustration of what H.M. Collins has called "the experimenters' regress."[147] Rather than being astonished, as contemporaries apparently were, that the available experimental facts seemed equally compatible with beliefs on both sides of the debate, we should note Collins's observation that a sufficiently determined and resourceful opponent can always find ways to challenge any empirical claim. Experiments by themselves cannot convince someone who refuses to assent to them, and such a person can always stipulate grounds for their refusal. Collins also argues that it is impossible to specify a priori what constitutes a replication. Because the author of a written description cannot control the actions of another experimenter, both actors retain space in which to argue that their experiments differ in some important respect. Thus there will always remain grounds for disagreement about whether a confirmation or a falsification has actually been achieved, or whether the experiment has simply not been properly replicated. Designations of the meanings of experiments are not forced by nature; they emerge from social and rhetorical processes of negotiation and persuasion.

Thus, although we can see that the Davys' strategy of supposedly replicating Murray's experiments in order to show where he went wrong might in other circumstances have been an effective one, we can also see

145 J. Davy (1813).   146 Murray (1813), pp. 266, 270.
147 On the experimenters' regress, see Collins (1985); and, for a historical case study, Shapin and Schaffer (1985), ch. 6.

why in this instance it failed to produce a satisfactory resolution of the controversy. To call Murray's bluff by accepting his suggestion for a crucial experiment and performing it publicly in his own city was a bold move by the Davys, one that in their opinion was crowned with success. But Murray retained room for maneuver and exploited it resourcefully: He denied that the experiment had been properly conducted, and repeated it with an outcome more favorable to his point of view.

There was therefore no agreement between the parties that the dispute had been concluded. Later accounts by the Davy brothers presented as accepted fact that Murray had been misled by the unknown presence of euchlorine and phosgene in his early experiments and that the crucial experiment with muriatic acid and ammonia had decisively demonstrated his error.[148] Murray never accepted that the dispute had ended in this way. He continued to reject the discovery of the two gases as irrelevant to an understanding of his experiments and to insist that the claimed outcome of the crucial experiment had subsequently been reversed. In the third edition of his *Elements of Chemistry* (1814) he still gave an even-handed description of the two hypotheses regarding oxymuriatic acid and refused to accept that the Davys had proved their case.[149] He maintained this position through to the fourth edition of his *System of Chemistry* (1819), published the year before his death.[150] In 1818 he had also reported further experiments to show that muriatic acid gas contained combined water, thus reiterating the evidence for judging the chlorine theory unproven.[151] In Murray's view, the controversy had not been resolved at all.

In the light of this, it is worth asking, who *was* convinced about the chlorine theory and how were they persuaded? The Davys repeatedly performed public experimental demonstrations to convince audiences of the truth of their claims. Humphry Davy's command of his public and specialist audiences was an important resource, and one to which Murray had no adequate counterpart. Hence the brothers' enrollment of expert witnesses, a contest in which (as we have seen) they gained the advantage. Appeal was also made to a wider public. In February 1812, Humphry showed an audience at the RI what was wrong with Murray's experiment with muriatic acid and ammonia, a display that was described in a number of monthly journals:

... the Professor clearly proved, that the presence of water was owing to the hygrometric qualities of the salt, which, when exposed to the atmosphere for an instant, absorbs moisture directly; and he showed an experiment, in which, when

148 Davy (1839–40), I, p. 123; V, pp. 347–348 fn.
149 Murray (1814), I, pp. 484–500.   150 Murray (1819), II, pp. 416–478.
151 Murray (1818).

muriatic gas and ammonia were combined out of the atmosphere and heated, not an atom of water could be procured from them.[152]

In terms of a distinction that Collins has recently made in a discussion of public science, this exhibition would have to be classified as a demonstration rather than an experiment.[153] It is hard to see how a skeptical and informed onlooker could have been *shown* that "not an atom of water" was produced. Less well informed and less committed observers could however be expected to be persuaded, and it was presumably with this aim that Davy presented such demonstrations. Throughout the controversy, he described his view of chlorine in his RI lectures as an established fact supported with conclusive experimental manipulations. Skating over the areas of dispute, he showed his public audiences demonstrations that supposedly gave unambiguous support to his doctrines. He exploited their chauvinism also, presenting chlorine as an alternative to a pernicious and erroneous French system of chemistry.[154] Evidence of the effectiveness of these displays comes from Michael Faraday, who attended Davy's lectures in March and April 1812. The young Faraday was readily won over to the view of chlorine as an element and declared himself satisfied with the demonstration of the absence of water in muriatic acid. An uncharitable view might number Faraday among those whom Harrington described as Davy's "devoted amateurs, who can see, or not see, any thing as he *dictates or decrees.*"[155]

Specialists could find grounds for dismissing such demonstrations as inconclusive. Murray himself, who found ways of criticizing the Davys' designated crucial experiment in Edinburgh, would certainly not have been persuaded by anything done in the public lecture theater of the RI. Thomas Thomson, who was much less committed than Murray to opposing the chlorine theory, likewise hesitated to judge the "intricate" matter of the outcome of the Edinburgh demonstration.[156] T.C. Hope, a specialist witness at the Davys' crucial experiment, later endorsed the alternative outcome of Murray's replication. Those with significant knowledge of the issues at stake were clearly not as easily persuaded by experimental demonstrations as were members of the general public.

It seems that most specialist chemists remained unconvinced by the chlorine theory until well after the dispute between the Davys and Mur-

---

152 Description quoted from the *Monthly Magazine*, 2nd ser., 33 (1812), 259–263, on p. 260. Identical reports appeared in the *Philosophical Magazine*, 39 (1812), 132–140; and in *Nicholson's Journal*, 2nd ser., 31 (1812), 236–237.

153 Collins (1988).

154 E.g. Davy (1839–40), VIII, pp. 311–319. The strongly anti-French tone of Davy's lectures of 1811 is recorded in newspaper clippings preserved in the MSS of John Davy (RI), box 2, item 5.

155 Williams (1960); Harrington (1819), p. 36.

156 *Annals of Philosophy*, 3 (1814), p. 14.

ray had been abandoned. Acceptance does not seem to have been wide-spread among chemists until about 1816-18.[157] Three factors can be identified that swung support behind Davy at that time. First, the discovery of iodine, announced triumphantly by Gay-Lussac and Thenard in August 1814. As the properties of this surprising new substance were explored, it began to appear that it was an element that shared many of the attributes of chlorine. The plausibility of the view that chlorine was itself an element was thus increased.[158] Second, following from this, was the apparent shift of position by the leading French chemists, who gave their allegiance to the chlorine theory and even pressed their claim to have originated it.[159] Third, Davy himself continued to build support for his view. Although overtly dismissive of continuing challenges to the chlorine theory, he tacitly acknowledged that the debate was still open by publishing a series of papers on iodine and fluorine compounds that helped to strengthen perceived analogies between the newly identified elements and chlorine.[160]

The movement of opinion in the specialist community in favor of Davy's view was chronicled by Thomson in the annual reviews of scientific developments published in his journal. In 1814, Thomson was still expressing serious reservations about Davy's theory. The following year he reported the recent work on iodine and fluorine compounds, but it was only in 1816 that he related these developments to the chlorine theory which he now described as "pretty generally admitted." Accordingly, in 1817 the fifth edition of Thomson's *System of Chemistry* was largely rewritten to incorporate what he described as the revolutionary new doctrine.[161] Other authors were more cautious. In the fifth edition of her *Conversations on Chemistry* (1817), Jane Marcet was still recommending readers to wait for a more decisive judgment by specialist chemists. The pupils "Caroline" and "Emily" in Marcet's dialogues were told by their mentor "Mrs. B." that, "the new doctrine has certainly gained ground very rapidly, and may be considered as nearly established; but several competent judges still refuse their assent to it."[162] The eighth edition of Samuel Parkes's *Chemical Catechism* (1818) was revised to incorporate Davy's theory and to include a discussion of the properties of iodine.[163]

---

157 Compare LeGrand (1974), esp. pp. 224–225.
158 On the discovery of iodine, see Crosland (1978), pp. 80–87. Murray himself acknowledged that this discovery had persuaded many others of the elementary status of chlorine (Murray (1819), II, p. 426).
159 See *Annals of Philosophy*, 7 (1816), p. 28.
160 H. Davy (1814a, 1814b, 1814c, 1815).
161 *Annals of Philosophy*, 3 (1814), 13–14; 5 (1815), 12–14; 7 (1816), 27–30; Thomson (1817), I, pp. vii, 184–185.
162 [Marcet] (1817), II, pp. 148–150.
163 Parkes (1818), pp. 167, 485–488. Compare the seventh edition (1816), pp. 173–174 fn., 181–182 fn.

The same year, William Henry's *Elements of Experimental Chemistry* endorsed the chlorine theory, noting that it had been rendered more plausible by the discovery of the apparently analogous new element. Privately, however, Henry and his Manchester colleague John Dalton continued to express their doubts.[164]

By 1818, Davy could have confidence that his theory was winning a large measure of acceptance among specialist chemists. In a paper read to the Royal Society he was able magisterially to dismiss experiments reported by Murray and Andrew Ure to the Royal Society of Edinburgh. They claimed to have shown experimentally that water could be produced by passing muriatic acid gas over heated iron, implying that chlorine had to be a compound containing oxygen. Davy was dismissive of these claims: "To take up the time of the society by long experimental details and theoretical speculations on such an occasion, will be unnecessary; I shall therefore only transiently mention the sources of error. . . ." The errors Davy mentioned had a familiar ring: He asserted that the oxygen to form the water was derived from lead oxide or alkali in the flint glass of the apparatus, or from atmospheric air remaining in the reaction tube.[165]

Davy's credit with his audience was apparently high enough at this point for his assurances to be accepted without much question. William Henry and Thomas Thomson both thought his response was adequate to put the matter beyond doubt. A writer in the *Monthly Review,* while noting that the controversy regarding the nature of chlorine was not yet closed, also found Davy's reply to the Murray-Ure experiments convincing.[166] Clearly the climate of informed opinion was much more favorably disposed toward his view of chlorine as an element than it had been only three or four years previously.

Davy's tactics in the debate with Murray parallel those he had employed successfully against Gay-Lussac and Thenard and against the other early investigators of the voltaic pile. In those disputes he had used forceful experimental and rhetorical techniques, forged in the context of appeal to a public audience, and had also invoked the authority of that audience itself to support his descriptions of experiments. By these methods his experiments had gained meaning as supports for his theoretical

164 Henry (1818), I, pp. v, 419–421, 425. Compare the account of a private conversation with Dalton and Henry in Manchester on 10 May 1818 by the American chemist John Griscom. Griscom recorded that "They are neither of them entire converts to the new theory of chlorine" (quoted in Farrar, Farrar and Scott (1974), p. 195).

165 Murray (1818), esp. pp. 291–301, 316; H. Davy, "On the fallacy of the experiment, in which water is said to have been formed by the decomposition of chlorine," in Davy (1839–40), V, pp. 524–527 (quotation on p. 524).

166 Henry (1818), I, p. 425; [Thomson] in *Annals of Philosophy, 11* (1818), 220; *12* (1818), 450; *Monthly Review,* 2nd ser., 87 (1818), 189.

claims: The voltaic pile was established as a powerful analytical instrument and sodium and potassium were defended as new metallic elements. In this way, Davy turned "the lever of experiment" on the fulcrum of his relations with a public audience. In the case of chlorine, however, such tactics were not successful, at least not until well after the dispute with Murray. The Davys' experiments, even those showing purportedly conclusive results before public audiences, were not accepted by their opponent or by others in the specialist chemical community.

Collins's analysis provides a way of understanding how Murray retained freedom to argue his position. Public demonstrations may be distinguished from true experiments and are likely to be regarded by informed observers as less convincing. Skillful and knowledgeable chemists were only likely to be persuaded by a dialogue within the experimental community. Investigators who were perceived as retaining privileged access to resources, or who made what seemed like illegitimate appeals to the judgment of public audiences, would be distrusted by their peers.

There were reasons for such distrust of Davy, deriving from the social and political context in which he forged his career. He had left his roots in the English provincial Enlightenment to move to the fashionable culture of the capital and the heart of the metropolitan scientific world. He could be seen as having turned his back on his scientific mentors, Beddoes and Priestley, on their progressive "Country" politics, and on the democratic form of public culture of science that they had espoused. Some of Brougham's comments on Davy's achievements seem to arise from such suspicions. When Davy launched his chlorine theory, Brougham protested at his dogmatic tone and at the perversion of a common chemical vocabulary into the service of an unjustified hypothesis. Similar reservations expressed about aspects of Davy's work by Bostock, Henry, Dalton, and Thomson may also have stemmed from distrust among liberal provincial intellectuals of Davy's moves to entrench himself at the apex of the scientific establishment.

On the other hand, Davy's considerable scientific accomplishments cannot be denied, and the role in them of the social context in which he worked should not be ignored. The centralization of resources through focusing public support on the RI yielded remarkable dividends. Davy's voltaic battery was constructed and acceptance of its results consolidated. Public experimental demonstrations clearly had an important role in this process. And in international competition with the French chemists, Davy's strategy brought victories comparable with those being gained by English generals on the field of battle. If Davy were a general of science, planning grand strategy, mobilizing the heavy artillery, consolidating the lines of supply, and pressing on regardless of mutterings in the ranks, then that was apparently what was required.

Looking at Davy in this way lends support to, and helps to extend, Ludwik Fleck's analysis of "the general epistemological significance of popular science." Fleck proposed that "certainty, simplicity, vividness originate in popular knowledge;" that they are generated by processes of "exoteric" communication with a public audience.[167] We have seen Davy enhancing the vividness of his experiments in order to present them in his lectures, simplifying complex and controversial subjects, and thereby consolidating the certainty of disputed facts. This simplified and more certain knowledge could then be fed back into esoteric discourse among specialists. The resolution of specialist disputes could be effected by the continual circulation of discourse from the esoteric to the exoteric realms and back again. To a significant degree, Davy's accomplishments as a chemist confirm Fleck's claim; they show how appeal to a public audience can consolidate and strengthen knowledge in the expert realm.

The case of Davy also demonstrates how the character of this process might be affected by historical circumstances. Faced with the failure of certain Enlightenment forms of public science, Davy mobilized a new audience in support of a newly reconstructed and more resource-intensive discipline of chemistry. His concentration of instrumental power and his presumption of public support for his experimental claims were frequently resisted, particularly by those who recalled the more egalitarian and democratic ideals of public science that Priestley had upheld. The Enlightenment, however, was over. Davy had anticipated that nineteenth-century chemistry was to depend upon substantially increased resources of apparatus and skills, and he saw how the audience of chemistry could be expanded to deliver the support chemists would now require. That support took the form of financial contributions and faith in the authority of the scientist and his knowledge. Davy had shown how a public audience could be wielded as a weapon in the battles through which a new scientific discipline was established.

167 Fleck (1979), pp. 106, 113–115. See also Whitley (1985).

8

Analysis, education, and the chemical community

"The ancient teachers of this science," said he, "promised impossibilities and performed nothing. The modern masters promise very little; they know that metals cannot be transmuted and that the elixir of life is a chimera. But these philosophers, whose hands seem only made to dabble in dirt, and their eyes to pore over the microscope or crucible, have indeed performed miracles. . . . They have acquired new and almost unlimited powers; they can command the thunders of heaven, mimic the earthquake, and even mock the invisible world with its own shadows."

Mary Shelley, *Frankenstein* (1818)[1]

The chemist, indeed, is flattered more than any one else with the hopes of discovering in what the essence of matter consists; and Nature, while she keeps the astronomer and the mechanician at a great distance, seems to admit him to a more intimate acquaintance with her secrets. The vast powers which he has acquired over matter, the astonishing transformations which he effects, his success in analysing almost all bodies, and in reproducing so many, seem to promise that he shall one day discover the essence of a substance which he has so thoroughly subdued.

John Playfair, "Biographical account of Hutton" (1805)[2]

As the words of the Scottish physicist John Playfair confirm, Mary Shelley's fictional Professor Waldman was giving voice to a view that was common in her time. Chemistry indeed appeared capable of working miracles. It was widely believed to have proved the most spectacularly successful science of recent times. The new discoveries had placed unprecedented powers in the hands of the chemist and promised revelations of the inmost secrets of matter. It was mainly to chemistry, therefore, that hopes for further scientific progress turned in the first two decades of the nineteenth century.

In the previous chapters we have seen some of the reasons why people believed that chemists now had virtually miraculous powers at their dis-

1 Shelley (1968), p. 307.   2 Playfair (1805), p. 74.

posal. Priestley's discoveries and their exhibition in public lectures had made known the dramatic properties of the new airs. Lavoisier's chemical revolution had shown how gases were intimately involved in the most common and diverse chemical processes. Their therapeutic application, by Beddoes and others, had produced startling and unexpected effects. And finally, Davy's spectacular use of the voltaic pile wielded the force of electricity to produce entirely new substances by analyzing species of earth.

Davy's role was undoubtedly crucial in spreading public awareness of the extraordinary achievements of chemistry and its continuing potential. But he was not solely responsible for this widespread enthusiasm, any more than public knowledge of pneumatic medicine had been created single-handedly by Priestley. Also critically important was a community of specialist chemists, particularly in London, that was growing significantly in size and influence at this time. In this chapter we move the spotlight onto them. We shall find that Davy was central to this community but quite atypical. His work was widely commented upon and respected but his methods of research and the way he built his career offered no model for other chemists to emulate. There was in fact some degree of resentment at the way in which he profited from aristocratic patronage and rather ruthlessly exploited the resources of the RI and the Royal Society in his research. Other chemists, lacking such resources, were forced to develop their own techniques and build their careers independently of Davy's example.

These chemists worked in a variety of fields of applied chemistry and in a growing market for scientific education, especially in the capital. Many of them retained an attachment to the enlightened values that had informed eighteenth-century public science. Thus, even while they devoted themselves to improving the techniques and instruments of chemistry, they voiced a commitment to making the science accessible to as wide an audience as possible. Education and communication seemed to them a natural complement to the continuing development of specialized apparatus and skills. This conjunction of attitudes was a characteristic one at the beginning of the nineteenth century. Whereas in the previous century expertise and complex equipment had sometimes been seen as threats to the ideal of science as a publicly accessible enterprise, such concerns were now heard less often. The commitment to widespread public diffusion of science was however still strongly held. Hence the exploration of new textual forms and new institutions for public education coexisted with the progressive enhancement of specialist expertise.

After surveying the context in which these developments occurred, we shall turn to the field of chemical analysis (particularly the analysis of minerals), to examine in more detail the relationship between increasing specialist skill and public science. In mineralogy, Davy's spectacular deployment of the voltaic pile proved less relevant than more humble methods, such as solution analysis and the use of the blowpipe. These techniques could be more widely and readily applied by the majority of chemists. The picture is not, however, one of steady and unproblematic improvement in methods. The demand that techniques should be accessible to nonexpert practitioners continued to be made and constrained the form of development that occurred. We shall see these questions being raised in the debates surrounding different analytical approaches. Thus, the general issues of expertise and public access to science were played out in arguments within the field of mineralogical chemistry.

In this chapter the first two decades of the nineteenth century are viewed from the perspective of eighteenth-century ideals of public science. The virtue of this approach is that it enables us to see how the influence of Enlightenment values persisted. The overall implication is that it was a specific conjunction of the commitment to science as a public culture and the dramatic advances in techniques that made this such a climactic period in the history of chemistry.

Specialist careers in the London chemical community

The aristocracy of chemistry would fain confine the science to their own class, on the same principles as the Royal Society accepts title or fortune as a qualification for the fellowship, when knowledge happens to be wanting. If a man cannot afford to expend five hundred pounds in an apparatus, let him stick to his work-shop. Philosophy in a vinegar cruet! science in a salt cellar! forbid it peers, princes, and prelates – forbid it ministers and secretaries. . . . Fortunately, however, the spirit of the age does not accord with the views of the dandy philosophers; they may black-ball at Somerset House, segregate from Albemarle-street, or shut themselves up in the atheneum; . . . but they will only . . . have the mortification of seeing that the world goes on better without them.

Correspondent in *The Chemist* (1824)[3]

Davy's extraordinary scientific accomplishments were, as we have seen, bound up with his success in assembling and marshalling a diverse audience for his work. The middle-class public who attended his lectures at the Royal Institution and the specialists who met in his basement laboratory or at the meetings of the Royal Society were persuaded to accept his construction of a powerful voltaic pile and the numerous refinements of technique that it entailed. Those who resisted his claims to control

3 "A constant reader" to the Editor, in *The Chemist*, 2 (1824–25), 46–47 (quotation on p. 47).

over the instruments of analysis were progressively marginalized in the scientific community. Beyond the walls of these institutions his work was also reported in scientific and general periodicals and purveyed by the authors of popular textbooks. Jane Marcet's *Conversations on Chemistry* (1806) and Jeremiah Joyce's *Dialogues in Chemistry* (1807) both offered themselves as popular expositions of Davy's doctrines, designed to complement his public lectures at the RI.[4] By the time of its fourth edition in 1807, James Parkinson's more practically oriented *Chemical Pocket Book* (originally published in 1799) was also recommending that readers attend Davy's lectures for demonstrations of the new discoveries it described.[5]

Extravagant praise for Davy and his achievements became commonplace in the increasing number of chemical textbooks. Even Charles Sylvester, whose *Elementary Treatise on Chemistry* (1809) took issue with Davy's assertion that the alkali metals were new elements, praised him as a "truly great and indefatigable chemist."[6] Some writers and lecturers relied much more closely on his researches. William Babington, Alexander Marcet, and William Allen, the chemical lecturers at Guy's Hospital, were giving a sympathetic account of Davy's theory of the elemental status of chlorine as early as 1811, when it was still highly controversial.[7] In the same year, John Webster, a schoolteacher, also welcomed the new theory, recognizing that it was destined to produce "a remarkable change in the whole theory of chemistry." Webster even retailed Davy's explicitly speculative notions as established facts, recording that sulfur and phosphorus had been shown to be compounds.[8] Slightly more cautiously, the chemical manufacturer Samuel Parkes (1761-1825) remarked in his *Rudiments of Chemistry* (1810) that since Davy's discoveries with the voltaic pile, ". . . it is scarcely possible to say what substances are not compound bodies."[9]

With even his speculations likely to be taken as demonstrated facts, and with a host of writers hanging upon his every announced result, Davy found his work enshrined at the center of a growing body of didactic literature. He also had a substantial influence on the style of chemical discourse through the rhetorical innovations he had pioneered, particularly in the 1802 "Discourse Introductory to a Course of Lectures on Chemistry." We can summarize these innovations under three headings: natural theology, "politeness," and political conservatism.

As has been noted already, Davy forced chemistry into a closer rela-

4 [Marcet] (1806), I, pp. iii–vi; Joyce (1807), I, dedication to Davy (unnumb. pp.).
5 Parkinson (1807), pp. v–vii.   6 Sylvester (1809), p. viii.
7 Babington, Marcet and Allen (1811), pp. 33–34.
8 Webster (1811), pp. xix, 91, 95, and unnumb. pp. at end of volume.
9 Parkes (1810), p. 152.

tionship with natural theology than it had hitherto enjoyed, by proposing that the contemplation of God's laws was the ultimate aim of the science. Coincidentally, Thomas Thomson's *System of Chemistry,* published in the same year as Davy's "Discourse," noted that chemistry was the science best suited to imparting the noblest ideas of God's handiwork. Thomson wrote, "No study can give us more exalted ideas of the wisdom and goodness of the Great First Cause than this, which shews us everywhere the most astonishing effects produced by the most simple though adequate means. . . ."[10] By means of such pronouncements, chemistry could be distanced from the suspicions of materialism that had surrounded it in the previous decade.

Following the lead of Davy and Thomson, many writers included similar comments in their works. The Manchester physician and chemist William Henry (1744-1836), whose earlier *Epitome of Chemistry* (1801) had shown no concern for the theological credentials of the subject, made a point in the enlarged *Elements of Experimental Chemistry* (1810) of insisting that this science was just as appropriate for revealing the design of creation as astronomy: "In these minuter changes, we shall find, there is not less excellence of contrivance, than in the stupendous movements of the planetary system."[11] For a Unitarian like Henry, such remarks might have appealed as a way of defusing Anglican suspicion and antipathy. Brief comments along the same lines were made by Webster and by Richard Reece in their textbooks.[12] More substantially, Parkes (another Unitarian) explained that his *Chemical Catechism* (1806) had been compiled with the aim of giving "in a popular form, a body of incontrovertible evidence of the wisdom and beneficence of the Deity." Although some readers might think such considerations irrelevant, "no writer, as a parent, could lose sight of the necessity of embracing every favourable opportunity of infusing such principles into the youthful mind, as might defend it against immorality, irreligion, and scepticism."[13] By 1834, William Jones showed just how formulaic such pronouncements had become by reducing them to execrable poetry in his *Chemical Science in Verse:*

> Let not the Christian think 'twill undermine
> Religion – or give power to infidels;
> No! every law speaks of the Power Divine,
> And every wondrous change a story tells
> Of power omnipotent, and every line
> Upon the page of Chemistry, but swells
> With wonder, and proclaims the praise of Him
> Who form'd the whole from Chaos, dark and dim.[14]

10 Thomson (1802), I, pp. 3–4.   11 Henry (1801); idem. (1826), I, p. xiv.
12 Webster (1811), p. xiii; Reece (1814), p. vi.
13 Parkes (1818), p. iv. See also Knight (1986).   14 W. Jones (1834), p. 8.

Bad poetry aside, another of Davy's legacies to the language of chemical popularization was the presentation of the subject as a component of polite learning. Davy had shown how chemistry could be exhibited as a part of general middle-class culture, and his example in this respect was widely imitated. A phrase of A.F. de Fourcroy, to the effect that chemistry should be considered an indispensable part of a liberal education, was much quoted by writers and lecturers to make this point.[15] Friedrich Christian Accum (1769-1838), who cited Fourcroy in his lectures at the Surrey Institution in 1811, was just one chemist who exploited the market for polite education in the subject. Ten years later, Andrew Ure perceived general readers as one of the audiences (along with manufacturers and medical students) for his revised version of William Nicholson's *Dictionary of Chemistry*.[16] Even those whose didactic efforts were primarily aimed at medical students made clear the broadly educative virtues of the discipline. Babington and Allen at Guy's Hospital noted that the subject "has become in some degree necessary in the general system of education." And Reece told his auditors at the Chemical and Medical Hall in Piccadilly that chemistry was an essential acquirement to distinguish civilized man from the untutored savage.[17]

Davy had also taken the position of a number of writers of the previous century in remarking that the presence of women at scientific lectures was a sign of a civilized culture. Their continued attendance at many chemical lectures was recorded by contemporaries as confirmation that the aim of scientific education was public enlightenment. In 1819 at the London Institution, a city replica of the RI, William Thomas Brande (1788-1866) echoed Davy's call for women to attend. They were invited:

... to partake of that healthy and refined amusement which results from a perception of the variety, order and harmony, existing in all the kingdoms of nature; and to encourage the study of those more elegant departments of science, which at once tend to exalt the understanding and purify the heart.[18]

The view that chemical education could properly include women among its audience underpinned the acceptance of Mrs. Marcet's *Conversations*. The author, who disclosed her sex though not her identity in the publication, admitted that she could not claim the title of a chemist. The subject was nonetheless (she insisted) a proper one for female education, and polite conversations were the appropriate means of imparting it. Marcet's fictional instructress "Mrs. B." told her young pupils that chemistry was not to be regarded solely as a practical concern:

15 For example, in Sylvester (1809), p. v.; Weldon [1825?], p. iii.
16 Accum (1810), title page (verso); Ure (1821), p. ix.
17 Babington and Allen (1802); p. v; Reece (1814), pp. v-vi.
18 Brande (1819b), p. 36–37.

Nature also has her laboratory, which is the universe, and there she is incessantly employed in chemical operations. You are surprized, Caroline; but I assure you that the most wonderful and the most interesting phenomena of nature are almost all of them produced by chemical powers. Without entering therefore into the minute details of practical chemistry, a woman may obtain such a knowledge of the science, as will not only throw an interest on the common occurrences of life, but will enlarge the sphere of her ideas, and render the contemplation of nature a source of delightful instruction.[19]

The image of chemical knowledge as a component of general enlightenment also inspired its communication at lower levels of society. As the subject became identified with middle-class gentility, it came to be an object of emulative appropriation by aspirant working-class individuals. The short-lived London Chemical Society of 1824, the creation of a group of artisans and political radicals, seems to have been an expression of this. When the physician and leading light of the Mechanics' Institute movement, George Birkbeck, addressed the society, he told them that the Enlightenment dream of unceasing and limitless diffusion of knowledge was still a valid aim. Working men, and even women, could expect to share in this process, he announced:

It may not be out of place here to state, that chemistry is not only not intended to be confined to *learned* men but not even to *men* exclusively. Hitherto, ladies have conferred the honour of their presence upon all our public proceedings; and we are exceedingly desirous . . . that they should hereafter become participators also, as members.[20]

Although the rhetoric of polite knowledge provided a resource with which chemists could continue to capture a would-be enlightened audience, many of them had to be careful to soften the political edge given to the Enlightenment vision by men like Priestley and Beddoes. This was Davy's third contribution to the language of chemistry as a public science. Middle-class audiences clearly appreciated his assurances that the progress promised by chemistry posed no threat to the social order. In similar terms, Brande assured the wealthy and prominent patrons of the London Institution that their support for science would enhance the splendor of the commercial class in the eyes of the rest of society. Progress and enlightenment were indeed to be striven for, but respect for morality and religion and "the general veneration of the Constitution" would prevent any social upheaval from ensuing. Brande called down God's blessing on the Institution to ensure that "commerce and science may here be entwined in perpetual friendship, uniting their strength for the

19 [Marcet] (1806), I, pp. iii, 2.
20 Report of Birkbeck's lecture, 25 November 1824, in *The Chemist*, 2, 162–168 (quotation on p. 164). On the society, see: Brock (1967); and Russell (1983), pp. 139–146.

glory of the empire, the stability of the throne, the perpetuity of our glorious constitution, and the prosperity of the people at large."[21]

The conservative political values reflected in Davy's rhetoric gave it appeal in many quarters in the 1800s and 1810s, during the lengthy wars with Napoleonic France. Like Davy, many aspirant scientists and intellectuals suppressed the memory of the radical enthusiasm of the 1790s. The chemist and mineralogist Arthur Aikin (1773-1854), who was the son of an outspoken Dissenting physician and had been closely involved in the Hackney Dissenting community while Priestley was living there, left the Unitarian ministry in 1795 and engaged in no political activity after that time.[22] Jeremiah Joyce (1763-1816) had been a prominent member of the radical Society for Constitutional Information in the 1790s, and spent twenty-nine weeks in the Tower of London awaiting trial on treason charges in 1794.[23] After 1800 he turned his attention to writing popular scientific texts and encyclopedias. The surgeon James Parkinson (1755-1824) wrote radical pamphlets in the 1790s and was examined by the Privy Council in 1794 when he came forward to exonerate friends accused of involvement in a plot to assassinate the king. After this episode, he published only on medicine, geology, and paleontology.[24] P.M. Roget, a more patrician medical practitioner and chemist, quickly dropped his association with Beddoes after the nitrous oxide fiasco, when he perceived that his reputation was in danger. He went on to make an outstandingly successful career in the medical and scientific institutions of the capital.[25]

In these three respects – theologically, culturally, and politically – Davy provided elements of a rhetoric that could be used to package chemistry as a public science. Although talk of politeness and enlightenment could have been turned in progressive or even radical directions, Davy had given it a distinctly conservative twist. In its theological and political implications, his discourse was unmistakably antireformist, reflecting the aristocratic and wealthy interests that patronized him at the RI.

It was for this reason that Davy found himself and his science severely challenged by the resurgence of political radicalism in the late 1810s. As the working-class movement resumed its advance, the weekly journal *The Chemist* enjoyed its brief period of existence from 13 March 1824 to 16 April 1825. Sometimes noted only as the first British periodical to be devoted exclusively to chemistry, it should more accurately be perceived (as Colin Russell has indicated) as the mouthpiece of a newly assertive

---

21 Brande (1819b), p. 10. On the patronage of the institution, see: "The London Institution," *Philosophical Magazine*, 22 (1805), 85–88; and Hays (1974).
22 Torrens (1983), pp. 112–114.   23 See Joyce (1794).
24 See: [John Smith] (1795); Knight (1974).
25 Emblen (1970), esp. pp. 44–45; Desmond (1989), pp. 222–235.

scientific radicalism. The journal was edited by the writer and activist Thomas Hodgskin and published by John Knight and Henry Lacey, publishers of the *Mechanics' Magazine*.[26] In this forum, Davy was unrelentingly denounced as representing "a sort of royal science." His theory of chlorine was once more brought into question, and his inventions of the safety lamp and the anticorrosion treatment for ships' hulls were said to be unoriginal and ineffective. Even the old nitrous oxide incident was dredged up to be thrown against him – he was said to have "fuddled himself pretty often with this gas."[27] But his greatest crime was his elitism and arrogance:

... he has no appearance of labouring for the people. He brings not the science which he pursues down to their level; he stands aloof among dignitaries, nobles, and philosophers, and apparently takes no concern for the improvement of those classes for whom our labours are intended, and to whom we look for support. ... It has certainly long been the fashion for those at the head of science to keep it in a manner inaccessible to the profaning touch of the vulgar, letting them see as much of it as might excite their admiration, without enabling them to estimate its value, or to acquire it by themselves.[28]

From this perspective, Davy's spectacular lectures appeared as pure showmanship, and the fact that they were presented to wealthy aristocratic and middle-class audiences confirmed that they were designed to keep the working population at a respectful distance, rather than to contribute to their education. Davy and his allies, it was said, "would rather have mankind for pupils than fellow-students of the great volume of nature." In *The Chemist*, the reservations about the political alignment of Davy's science, which the Whig Brougham had voiced in previous decades, were given a sharper radical edge. Brougham himself, now encouraging the growth of the Mechanics' Institute movement, endorsed the journal and received recognition in return for his efforts in support of working-class education.[29] Davy was said to have done nothing to aid those who sought to improve themselves while they earned their living in manual trades. His aristocratic values aligned him with the "moral quacks" who always pretended to be acting disinterestedly and denied that material self-interest could be a force for good in society at large. In this respect he was typical of those English scientists "who did not pursue philosophy as a trade, but found it by accident, and then kept to it as an amusement," and he could provide no model for aspiring working men.[30]

By the 1820s, then, Davy was finding himself under attack by radicals

26 On the editorship, see Russell (1983), p. 145.
27 *The Chemist*, 1 (1824), pp. vii, 53–54, 72, 133–135; 2 (1824–25), pp. 46–47, 78–79. The journal was published in weekly parts at 3 pence per issue.
28 *The Chemist*, 1 (1824), vii.
29 Brougham (1825), p. 3; *The Chemist*, 2 (1824–25), 326–327.
30 *The Chemist*, 1 (1824), 190, 252.

on the sidelines of the scientific community. As David Miller has shown, he was also facing opposition within the expert scientific world in his office as President of the Royal Society. To the growing reform group in the society, Davy still seemed to be in the pocket of the RI managers, the wealthy merchants of the city and the "*aristocratical* interest."[31]

There is really no need, however, to look forward as far as the 1820s to detect alternatives to Davy's way of doing chemistry and implicit challenges to him. Although aspects of his style of presentation achieved quite widespread currency, his career offered no model to other aspiring chemists. His talents were so obviously extraordinary and his resources at the RI so clearly unequalled elsewhere that nobody could really hope to emulate him. Other chemists had to explore alternative career paths and accordingly developed quite different modes of chemical practice.

Perhaps closest to Davy in their activities were a small number of gentleman chemists who commanded the resources to pursue independent research. John George Children (1772-1852), a gentleman with a Cambridge education and extensive private means, was a good friend of Davy's for nearly three decades. Children built a private laboratory at Tonbridge, where Davy used to retire for periods of experimental work, interspersed with shooting and fishing in the surrounding countryside. In the years 1811-13 the two men collaborated on a project for manufacturing gunpowder, though Davy later insisted that his commercial interest in the scheme should not be advertised for fear it would damage his reputation.[32] William Hyde Wollaston (1766-1828), another Cambridge man, gave up medical practice to concentrate on chemical research in his own laboratory in London. A process for purifying platinum, which gained him the Copley Medal in 1802, earned him a considerable income for the rest of his life. Wollaston was an ally of Davy's in the controversy about who deserved credit for the invention of the safety lamp, and he subsequently resisted pressure from Babbage and Herschel to stand against his friend for the Presidency of the Royal Society.[33] It is not surprising that *The Chemist* should have named Children and Wollaston among the members of a clique of Davy's closest supporters.[34]

Also connected with this group were the instrument maker William Hasledine Pepys (1775-1856), a longtime supporter of the RI who became its President in 1816, and the Quaker chemical manufacturer William Allen (1770-1843), who lectured there in the early 1800s. Pepys and

31 Miller (1983) (quotation on p. 26). Cf. the rather different perspective of Fullmer (1980). For the wider picture of the Royal Society and reform thereof, see Miller (1981); MacLeod (1983).
32 Fullmer (1964).
33 *Dictionary of Scientific Biography;* Usselman (1978); Miller (1983), pp. 26–28.
34 *The Chemist, 1* (1824), 82.

Allen witnessed some of Davy's crucial work in the RI laboratory, and subsequently attended with him on Children's experiments to develop a more powerful voltaic battery.[35] Pepys, one of the founders of the London Institution, which tried to imitate the RI in attracting genteel audiences, built a battery of 2,000 double plates that was shown there by Davy and Brande in 1819.[36] George John Singer (1786-1817), a private lecturer who operated from Cavendish Square, exhibited his own battery at the Russell Institution from 1810.[37] These men were the closest partners of Davy and propagated his program of big science research; they shared his interest in the use of powerful instruments that required substantial investment of resources and could be deployed in spectacular public displays.

Other chemists, however, including others associated with the RI, followed very different paths. Accum, who was appointed chemical assistant to Davy in March 1801, resigned in September 1803, preferring to develop his own business interests. In the years from 1806 to about 1809 he ran a Philosophical and Chemical Society from his laboratory at 11 Old Compton Street, Soho, a venture that mixed entrepreneurship and scientific sociability in a manner reminiscent of the group formed by Bryan Higgins in the same area in the 1790s.[38] From 1808, Accum was lecturing at the Surrey Institute, where he introduced a middle-class public to new aspects of applied chemistry. His textbooks on mineralogy and analytical methods were successful, and he also probed the market for scientific entertainment with his *Chemical Amusement* of 1817, a collection of simple but striking experiments for readers to repeat at home. Other entrepreneurial activities included the work on gas-lighting projects that gained him an appointment as engineer (and subsequently a director) of the London Gaslight and Coke Company in 1810. Blatantly commercial activities of this kind were anathema to Davy, who privately labelled Accum "a cheat and a Quack."[39]

Davy was even more scathing about Brande, who succeeded him as Professor of Chemistry at the RI in 1813. Brande, the son of an apothecary, was, according to Davy, "a very inferior person [who] followed chemistry . . . [for] as much profit as he could obtain"; he was "mercenary" and incapable of "lofty views."[40] In the inevitable comparisons with his great predecessor, Brande's research was generally judged less striking than Davy's, though Brougham thought his contribution to an-

35 Children (1809, 1815).
36 Ockenden (1937).   37 Inkster (1977), p. 11.
38 Averley (1986), pp. 102–107, 108–113.
39 Cole (1951). For Davy's remarks, see Berman (1978), p. 74 fn. (quoting Fullmer (1967)).
40 Berman (1978), pp. 132–133.

alytical methods was valuable, if somewhat "wanting in originality and invention." More pungently, *The Chemist* portrayed Brande as representative of "a sort of superficial, confident chemistry, which may, perhaps, be styled the *petit maitre* school of this science."[41]

Brande's success as a lecturer was, however, undeniable. He operated from the RI and the London Institution in the 1810s, having already opened up a series of niches in the market for instruction in pharmacy. He first taught pharmaceutical chemistry at the Medical Theatre in Cork Street in 1808, and then moved to the new Medico-Chemical School in Windmill Street. After taking up his RI appointment, he initiated a series of pharmaceutical lectures there, which later earned the plaudits of the medical reformer Thomas Wakley, editor of *The Lancet*. Having been nominated Professor of Chemistry to the Apothecaries' Company in 1812, Brande was well placed to exploit the increased demand for teaching in pharmaceutical chemistry that followed the legal imposition of examination requirements in the 1815 Apothecaries' Act.[42]

Accum and Brande were among the most prominent of dozens of men who were exploiting a burgeoning market for chemical lectures in London.[43] William Nicholson (1753-1815), who had been teaching mathematics and natural philosophy in the capital since the 1770s, was offering weekly lectures on chemistry at his boys' school in Soho Square in the early 1800s.[44] Nicholson, an energetic author and translator, as well as (from 1797) the editor of the *Journal of Natural Philosophy, Chemistry and the Arts*, placed himself at the center of the movement for chemical education, and showed how careers could be made by riding the wave of enthusiasm for the subject. Arthur Aikin also saw the market for chemical lectures in the capital; he began teaching with his brother Charles in 1799 and continued to do so at two sites in the city for several years.[45] Thomas Garnett (1766-1802) was another pioneer in this field. After his brief period at the RI, he resumed his career as a physician and public lecturer in Great Marlborough Street in July 1801. His courses on chemistry, natural philosophy, and medicine were proving quite successful when his life was suddenly cut short by a dose of typhus caught from a patient. D. Gardner, a contributor to the debate on electrolytic decomposition of water, and subsequently a business partner of Accum, was offering courses of lectures at the City Dispensary in 1806. Richard Phillips (1778-1851),

41 [Brougham], review of Brande's Bakerian Lecture (1820), in *Edinburgh Review*, 34 (1820), 431–438 (quotation on p. 431); *The Chemist*, 1 (1824), vi.
42 Berman (1978), pp. 130–134; Obituary notice of Brande, in *Proceedings of the Royal Society of London*, 16 (1867–68), ii-vi. On the effect of the 1815 Act, see: Holloway (1966); Bud and Roberts (1984), pp. 26–27.
43 For surveys of the situation, see: Inkster (1977); Hays (1983).
44 Inkster (1977), p. 5.   45 Torrens (1983), pp. 115–116.

who had served an apprenticeship as an apothecary under William Allen, taught chemistry from his house in Cheapside and from 1817 at the London Hospital.[46]

These private initiatives flourished in conjunction with a dramatic institutional growth of metropolitan science. Accum's group, sometimes known as the London Chemical Society, was a short-lived case of an institution stemming from the entrepreneurial activities of a lecturer. More durable was the London Philosophical Society, formed in 1794 by those who attended the courses of Samuel Varley, a watch maker and scientific instrument maker, at Hatton House. In 1811, this was transmuted into the Philosophical Society of London, which began to offer an extensive program from its premises off Fleet Street.[47] In a similar manner, John Tatum, already in business as a lecturer, became Secretary of the City Philosophical Society on its formation in 1809. He continued to teach a wide range of subjects (including chemistry, mineralogy, and electricity) at the Society's premises in Dorset Street, where the young Michael Faraday attended in 1810.[48]

From the mid 1820s, the growth of the Mechanics' Institutes movement offered further avenues for employment of lecturers, as middle-class cultural interests were spread down the social scale to improve sections of the working class.[49] Chemistry had a prominent place on the timetables of all Mechanics' Institutes; as Brougham explained, it appealed because of its perceived connection with the practical arts and its suitability for teaching by experimental demonstrations.[50] Phillips lectured at the institute founded in Chancery Lane in London in 1824, where his unpretentious manner was well received, according to a report in *The Chemist*. Later the services of John Thomas Cooper (1790-1854), who had already made a name for himself teaching chemistry at the RI and the Russell Institution, were secured.[51]

Outside London also, the rapid increase in the number of Mechanics' Institutes greatly augmented the opportunities for lecturers, whether locally based or itinerant. This was one respect in which the institutes imitated the pattern established by the more genteel Literary and Philosophical Societies, which also continued to grow in number at this time. Brougham reported a substantial interest in chemistry at the institutes in Edinburgh (founded in 1821), Newcastle (1824), Kendal (1824), and Carlisle (1825); and a similar concentration has been noted in the cases

46 Inkster (1977), pp. 6–7, 9, 17; Lythe (1984), pp. 51–53; Cole (1951), p. 134.
47 Inkster (1977), pp. 11–13.   48 Williams (1960), pp. 517–526.
49 On the Mechanics' Institute movement, see: Shapin and Barnes (1977); Royle (1971); Inkster (1975, 1976a).
50 Brougham (1825), p. 11.
51 *The Chemist*, 1 (1824), 122–127; 2 (1824–25), 247–248.

of Derby (1825) and Sheffield (1832).[52] In Derby the institute library quickly acquired the chemical texts of Thomas Thomson, Jane Marcet, Charles Sylvester, and John Murray.[53] In Sheffield, Inkster's work has shown how the focus upon chemistry in the Mechanics' Institute was fostered by the existing Literary and Philosophical Society (founded in 1822) and by the large numbers of itinerant lecturers who regularly plied their trade in the town.[54]

Developments of this kind contributed significantly to the continuing vitality of provincial cultural life in the early nineteenth century, but by this time London had unquestionably become the center of public education in the sciences. As Halévy noted several decades ago, the widespread interest in natural phenomena that had taken root in the English provinces in the late eighteenth century was at last finding a home in the capital.[55] The extraordinary variety of venues for scientific lectures offered multiple opportunities for those able to teach chemistry.

One reason for the concentration of outlets for scientific education in the metropolis was London's increased importance in the market for medical training. Would-be practitioners flocked to the city for the instruction they could obtain at hospitals and private medical schools. Many men made their careers by teaching chemistry in these locations. Furthermore, a group within the profession took a particular interest in the latest advances in the science and developed new applications of it to medical problems. The pursuit of experimental research became a valuable qualification for building a medical career.

We have already noted how Brande, Garnett, Allen, and Phillips owed a measure of their success to an increased demand from medical students for teaching in chemistry and pharmacy. Men knowledgeable in these subjects, even if (like Brande and Allen) they lacked formal medical qualifications, could break into a lucrative market. Lecturers might operate from their own premises, from the private medical schools, or from the major London hospitals. By the beginning of the nineteenth century, Guy's, St. Thomas's, St. Bartholomew's, and the London Hospital were offering courses in chemistry and other medical subjects. William Cullen's pupils had been among the pioneers of this trend, and Edinburgh graduates continued to be prominent among the hospital lecturers. In the 1790s, chemical lecturing at Guy's became well established: William Babington (1756-1833), who was appointed apothecary and then physician to the hospital, took up the teaching of chemistry and materia medica (Figure 12). After 1800 he shared his lectures with Allen and Alexander Marcet

52 Brougham (1825), pp. 17–24; Inkster (1975), pp. 454–455.
53 Inkster (1980), p. 92.
54 Inkster (1975, 1976a, 1976b). See also: Brook (1955); Shapin (1972); Orange (1983).
55 Halévy (1960), p. 571.

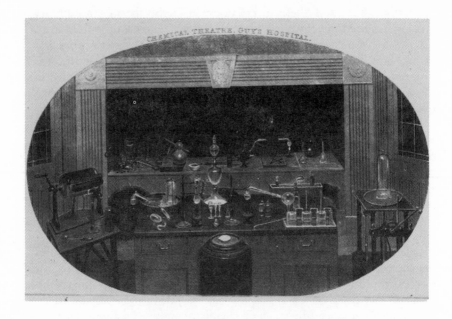

Figure 12. The chemical theater at Guy's Hospital. From Arthur
            Aikin, Syllabus of a Course of Chemical Lectures (Lon-
            don, 1829), frontispiece. [Reproduced by permission of
            The Huntington Library, San Marino, California.]

(1770-1822), an Edinburgh graduate who became physician to the hospital in 1804. John Bostock (1773-1846), another Edinburgh graduate,
continued the tradition into the 1820s. Doctors who taught chemistry
elsewhere in the capital included George Pearson (1751-1828), physician
to St. George's Hospital, who worked from his own laboratory from the
early 1790s to 1805. John Yelloly (1774-1842), also Edinburgh-trained
and physician to the London Hospital from 1807 to 1818, offered a course
on chemistry there.[56]

   Alongside teaching and medical practice, many of these men developed
interests in scientific research. Chemistry offered an obvious field in which
they could hope to acquire the professional kudos that increasingly followed from a reputation in experimental science. Pearson distinguished
himself in the mid-1790s as an early British adherent of Lavoisier's theory and the translator of the French Table of Chemical Nomenclature
(1794). Anthony Carlisle's work with Nicholson on the voltaic decom-

56 S. Lawrence (1985a), pp. 358–361, 367, 370, 504 and fn., 552–553.

position of water was a significant step in his highly successful career in science and medicine. Carlisle went on to become a member of the RI committee on chemical research, Professor of Anatomy at the Royal College of Arts (from 1808), Croonian Lecturer to the Royal Society (in 1804 and 1805), and President of the Royal College of Surgeons (in 1828 and 1837). He was knighted in 1820.[57] At a slightly less exalted level, Bostock and Babington earned recognition for their experimental work on (respectively) physiology and mineralogy.

More immediately than electrochemistry or mineralogy, chemical analysis held out the promise of new applications to medicine. Brande was one of those who led the way in this field, applying the voltaic pile to decompose animal fluids such as blood and urine. Several medical practitioners explored more rudimentary analytical techniques, hoping that they could be applied to bodily fluids to aid diagnosis. Marcet used solution analysis, the blowpipe, and crystallography to try to find tests that would be useful diagnostically. Blood from diabetics and urinary calculi were among the substances to which he applied these methods. The Animal Chemistry Club, a subsidiary group within the Royal Society, was formed in 1808 to encourage this kind of work. The Club brought together chemists (Davy, Brande, Children, and Charles Hatchett) and medical men, including Babington, John Davy, and the surgeon Everard Home. Elsewhere, the same topics were pursued at meetings of the medical societies at Guy's and St. Bartholomew's Hospitals and at the Medico-Chirurgical Society, founded by Marcet and Yelloly in 1805 as a forum for scientific communication among physicians and surgeons.[58]

In some ways the situation of medical men who pursued chemical analysis in the 1800s was comparable with that of the pneumatic practitioners in the 1770s and 1780s. Again, a self-selecting group within the medical profession was exploring a new field of experimental science for its possible therapeutic benefits. In both cases, hardly any were Oxbridge graduates, with Edinburgh training strongly represented. Chemical analysis was, however, less the preserve of a determinedly reformist network than pneumatic therapy had been a couple of decades before. The analysts appear to have been more interested in individual career success than in general social reform. Nor were they firmly rooted in an independent provincial milieu; in fact, the migrations that a number of them made to London provide another index of the increased importance of the capital as the focus of the scientific community. Pearson practised in Doncaster before making the move in 1787. Roget left his post as physician to the

57 Desmond (1989), p. 419.
58 S. Lawrence (1985a), pp. 606–616; idem. (1985b); Coley (1967, 1968).

Manchester Infirmary to come south in 1808. Bostock was established as a physician in Liverpool, but moved to the capital in 1817 to support himself by lecturing while he devoted himself to scientific research.

What these men saw in London was the possibility of using their knowledge of chemistry to make a living that combined teaching and scientific research. It is clear that London offered a large variety of opportunities of this kind, though the wide range of careers being followed makes it inappropriate to speak of a profession of chemistry at the time.[59] Was there nonetheless a chemical community? The question is worth asking in view of recent studies that have emphasized the formation of specialist communities as a correlate of disciplinary consolidation.[60] My emphasis so far has been on the diversity of occupations among chemists, but it is important also to consider the links that bound them together.

Two measures of the degree of consolidation that a disciplinary community has achieved are the publication of journals reporting the results of research and the existence of specialist societies. In this period, the dominance over the national scientific community exerted by the Royal Society and its journal, the *Philosophical Transactions,* came under challenge. Alternative publications and institutions were developed. Although Banks's empire-building approach enabled the Royal Society to subsume many of the important areas of research, and although its fellowship was still sought and gained by many chemists, new structures of support for the chemical community were already beginning to appear. The new journals and societies tended, however, to reflect trends in all of the sciences, rather than confining themselves to a single discipline.

The Royal Society was not a specialist organization itself, of course; it fostered research across the whole range of the sciences. Banks's strategy was to try to subordinate more specialized bodies under its overall auspices. The Linnean and Horticultural Societies fell in line with this scheme, and the Animal Chemistry Club was created as a subordinate institution of this kind. The analytical papers of Brande and other animal chemists were published in the *Philosophical Transactions* during the period of the Club's existence. Other chemical work that appeared there included the papers of Davy and Children on electrochemistry, of Richard Chenevix and Charles Hatchett on mineralogical analysis, and of Allen and Carlisle on chemical aspects of physiology.

It was, however, symptomatic of the degree to which the Banksian "empire" was resented in some quarters that other chemists chose to avoid the *Philosophical Transactions* almost entirely. John Dalton (1766-

59 Compare, for a later period, Russell, Roberts and Coley (1977). For a comparative perspective, see Porter (1978); and, for a review of the general issue, Morrell (1990).
60 See, especially Hufbauer (1982).

1844), the Manchester chemist and teacher of natural philosophy, published his first paper in the *Transactions* only in 1826, having previously appeared mostly in the *Memoirs of the Manchester Literary and Philosophical Society* and in Nicholson's *Journal of Natural Philosophy, Chemistry and the Arts*. Dalton notoriously held himself aloof from the Royal Society, declining to be nominated for a fellowship by Davy in 1810, and apparently accepting in 1822 only when the honor was sprung upon him.[61] Thomas Thomson (1773-1852), though he was made an FRS in 1811, published only two of his approximately 200 papers in the Society's journal, preferring to let them appear in Nicholson's, in his own *Annals of Philosophy*, or in the *Transactions of the Royal Society of Edinburgh*. Richard Phillips (elected FRS only in 1822) published approximately seventy papers, only one of them in the *Philosophical Transactions*, with the rest distributed between Thomson's journal, the *Philosophical Magazine*, and the *Journal of Science and the Arts* published by the RI. William Nicholson himself seems never to have been elected an FRS or had a paper in the society's journal, though about sixty appeared over his name in his own publication.[62]

The appearance of several new periodicals in this period opened up alternative paths to publication for chemists. Nicholson's *Journal* began to appear in April 1797. Its quick success prompted the Scottish journalist and printer Alexander Tilloch (1759-1825) to launch his *Philosophical Magazine* in June of the following year. The two periodicals rivaled each other for a decade, printing original research papers, lifting significant papers from other journals (including the *Philosophical Transactions*), and relaying news of lectures and the meetings of scientific societies. Nicholson, however, lost out in the circulation battle, and was further set back by the appearance of Thomson's *Annals of Philosophy* in a similar format in 1813. He soon announced that his journal was being taken over by Tilloch.[63] Thomson's formula was much the same as the one that had proven successful for the other publishers: original papers, news of societies, and "scientific intelligence." He noted in the first issue that chemistry would preponderate among the contents, because ". . . like all other journals of the present day, our *Annals* must contain a greater proportion of Chemistry, which is making a rapid progress, than of those sciences which are in a great measure stationary."[64] A further competitor entered the lists in 1816, when the RI launched its *Journal,* under Brande's

61 "Dalton" in *Dictionary of Scientific Biography.*
62 "Nicholson" in *Dictionary of Scientific Biography.* Information about publications is extracted from Royal Society (1867–72).
63 *Nicholson's Journal,* 2nd ser., 36 (1813), 387–390.
64 *Annals of Philosophy,* 1 (1813), iv.

. . .

editorship. This contained a slightly higher proportion of original research articles than the other periodicals, with chemistry again strongly
represented, and the usual seasoning of "miscellaneous intelligence."[65]

The first two decades of the nineteenth century saw the flourishing of
general scientific periodicals of this kind. The field of such publications
was thinned out subsequently, as journals covering more specialist concerns arose. The *Annals of Philosophy* (under Phillips's editorship since
1821) folded in 1826, and the RI *Journal* closed in 1831. At the beginning of the century, however, such periodicals had an important role in
focusing and sustaining discussion among contributors at the forefront
of research, particularly perhaps in chemistry. When he gave up running
his journal in 1813, Nicholson recorded that "Chemistry has a second
time, within our own observation, become a new science," and he cited
the discoveries of galvanism, the new elements, and the law of definite
proportions as among the signs of this.[66] It was typical of the period that
this research function went along with an evident desire to serve a relatively wide audience. These journals showed how a commitment to public education was still viewed by many as an integral part of the advancement of science.

As regards the institutional organization of the chemical community,
developments were more limited in scope. A few groups were formed
that designated themselves specialist chemical societies (Accum's London
Chemical Society and the slightly later Lambeth Chemical Society, for
example); but they are best viewed as being in the Enlightenment tradition of polite self-improvement groups.[67] The 1824 Chemical Society
showed the customary emphasis on the values of self-education and
sociability, which it attempted to reproduce at the level of the artisan
class.

More significant as settings for specialist research was a series of bodies that focused the attention of a relatively small number of expert practitioners on particular topics. The Animal Chemistry Club could be said
to have been one of these. It had been preceded by the Askesian Society,
in existence from 1796 to 1807, in which a public education role coexisted with research programs by leading specialists such as Allen, Babington, Pepys, and Phillips on topical subjects like mineralogy and electrochemistry.[68] In the years from 1799 to about 1806, these four men, with
the Aikins and others, were also involved in the British Mineralogical
Society, which had an even more developed research role. Members of
the BMS practised mineralogical analysis and set in train surveys of mineral resources in many parts of the country. Paul Weindling has argued

65 *Journal of Science and the Arts*, 1 (1816), i-iv.
66 *Nicholson's Journal*, 2nd ser., 36 (1813), 388.
67 Averley (1986), pp. 108–120.   68 Inkster (1977).

persuasively that the BMS gave them an opportunity to exploit their specialist skills to make their way into more established scientific institutions.[69] Many of them also played a role in the foundation of the Geological Society in 1807, which rebelled against a position of subordination to the Royal Society and asserted its independence as a specialist research institution.[70] No such foundation was created for chemistry until 1841, when the Chemical Society of London was formed. Although there were continuities between this and the earlier institutions, the 1841 foundation reflected a new degree of professional identity among chemists that had not been present in the earlier period.[71]

Although the short-lived groups of the first two decades of the century were important in consolidating research efforts on particular topics, they showed no evidence that chemists shared a distinct professional identity. Given the diversity of their occupations this is scarcely surprising. The degree of coherence achieved by the chemical community of the 1800s and 1810s owed more to the legacy of the eighteenth-century ideal of public science than to an emerging professionalism.

The identity of the discipline and the reception of Dalton's atomic theory

A chemical philosopher was formerly a sort of wizard, a monster rarely to be seen; and then, in his gown and cap, or enshrined in the cloister of the university. It was his dread lest the vulgar understand him; lest, while he pretend to dazzle, and to be great, he should chance to be useful. These contemptible feelings are now vanished, and chemists are running to an opposite extreme. Not only are treatises of Philosophy and Chemistry met with in every quarter, but beaux, ladies, all are now chemists, or pretend to be so. All are vying with each other in the ardour of experimenting and communicating: monthly, and even weekly journals are teeming with experiments, and with real or supposed discoveries.

Walter Weldon, *Popular Explanation of Chemistry* [1825?][72]

By concentrating on specific topics of experimental research, the chemical community in the early nineteenth century gained a measure of common purpose. Research activity, focused as it was on a variety of quite distinct problems and frequently changing in its aims, could not provide a satisfactory disciplinary identity for chemistry as a whole. The need to consolidate and reorganize the didactic presentation of the subject was a pressing one after the upheaval of Lavoisier's Chemical Revolution and the subsequent dramatic discoveries. Textbooks and lecture courses had to be revised to incorporate the new doctrines and techniques. Those

---

69 Weindling (1983).   70 Weindling (1979); Rudwick (1962–63).
71 Bud and Roberts (1984), pp. 48–51, 100–108.   72 Weldon [1825?], p. 13.

who took the opportunities provided by the burgeoning market for chemical education faced the rhetorical task of furnishing the members of the community with a coherent image of a rapidly changing discipline.

As a textbook writer, Davy was notoriously unsuccessful. His 1802 "Discourse" had encapsulated a rhetorical style by which chemistry could be presented to middle-class audiences as a part of polite culture, but the *Elements of Chemical Philosophy* (1812) entirely failed to answer the needs of the discipline for a reliable textbook. The work was not even completed – only the first volume of part I appeared, and this was so badly received that Davy abandoned the project. He therefore got no further than an exposition of the chemical elements, and the public never had a chance to see how he would have dealt with the simple compounds or the more complex vegetable and animal substances. The damning assessment of the reviewers was that Davy had given far too personal a view of the subject and had indulged his own theoretical and speculative opinions in a way that was inappropriate for a textbook author.

The published volume was arranged by dividing the elements into those that support, and those that undergo, combustion (with the latter class further subdivided into metals and nonmetals). After Lavoisier, combustion could be seen as the crucial operation in chemistry and this way of organizing subject matter became commonplace. More contentious was the way Davy applied Lavoisier's definition of an element as "every body not yet decompounded." Reviewers clearly felt that Davy had arrogated too much authority to himself in determining the elementary status of substances on the basis of his personal command of the instruments of chemical analysis. He told his readers that the most potent analytical instrument was "that constructed by the subscriptions of a few zealous cultivators and patrons of science, in the laboratory of the Royal Institution."[73] It was the power of this instrument that underwrote his most contentious assertion: that chlorine (or oxymuriatic acid) was an element. Other claims were almost as questionable, however. Because sulfur and phosphorus had yielded hydrogen when subjected to voltaic electricity in a fused state, Davy speculated that they might not in fact be elements. And the volume culminated in the "sublime speculation" that all known substances might be different arrangements of the same small number of basic constituents, an opinion that was said to have been endorsed by luminaries like Newton and Boscovich.[74]

Readers could hardly be blamed for feeling that Davy was making too much of his own opinions. Perhaps, as John Bostock suggested in the *Monthly Review,* the great discoverer was too much attached to his own

73 Davy (1812), p. 152.   74 Davy (1812), pp. 480, 489.

theories to make a good textbook author.[75] Most of the work indeed reads as a narrative of Davy's own research, interspersed with a series of arguments against alternative points of view on matters of theory. The doctrine presented was explicitly said to be founded on the author's own experimental work, and the audience at RI lectures were invoked as witnesses:

The greater number of the experiments were made in the laboratory of the Royal Institution; and all that were fitted for demonstration have been exhibited in the Theatre of that useful publick establishment in my annual courses of lectures; and have been received by the members in a manner which I shall always remember with gratitude.[76]

Such references to the institutional basis of Davy's prestige may have increased the antipathy to his text in certain quarters. Thomas Thomson predicted that the book would damage Davy's reputation, and particularly objected to the author's adoption of Count Rumford's theory of the vibratory nature of heat. According to Thomson, Davy had turned his back on forty years' work by Scottish chemists, which supported a material theory of heat, to follow a frivolous fashion.[77] Similarly, Bostock found Davy's idiosyncratic chemical nomenclature objectionable, his doctrine of chlorine fanciful, and his theory of heat completely whimsical.[78] Even the commentator in the conservative *Quarterly Review* thought Davy's statements on chlorine were couched in a "tone somewhat more decisive than the present state of the investigation altogether authorises."[79]

These reactions show how Davy had offended against accepted practices of textbook authorship. The writer of a didactic work was not expected to commit himself so explicitly to contested claims, still less to shape the organization of the book and the nomenclature employed in order to present these claims as uncontentious doctrine. Instead, as Thomson urged in his review, he should appear as the humble compiler of the contributions of others, and in this task the traditional problem of finding a balanced and comprehensive method of exposition would come to the fore. By suggesting revisions of chemical nomenclature to reflect his claimed discoveries, Davy had taken the path of Lavoisier and his French allies, which had been so thoroughly resented by British chemists like Priestley, Keir, and Kirwan. In this respect he seems to have had few followers. Even Brande, whose *Manual of Chemistry* (1819) adopted certain aspects of his presentational rhetoric, declined to use his terminol-

75 [Bostock] (1813), esp. p. 149.   76 Davy (1812), p. ix.
77 [Thomson] (1813a), esp. pp. 373–374.   78 [Bostock] (1813), pp. 152, 155.
79 Anonymous review of Davy's *Elements. Quarterly Review,* 8 (1812), 65–86, esp. p. 75.

ogy; and Davy himself penned a retraction in a draft preface for a projected second edition of his work.[80]

In the early nineteenth century, the prevailing ethic of textbook authorship remained in many respects that of Priestley, in which generosity of credit, comprehensiveness, and balance were the prized qualities. Perhaps because the chemical community was increasingly fragmented in terms of the careers of practitioners and their levels of expertise, there seems to have been a wish to retain the Priestleian conventions by which all participated equally in a public discourse of facts. Nicholson's work provided one thread of continuity between Priestley's era and Davy's. He had declined to offer students a translation of Lavoisier in the late 1780s, because Lavoisier was "almost too artful and too interested an advocate for the beginner to listen to." In his own *First Principles of Chemistry* (1790 and subsequent editions to 1796), Nicholson continued to describe the doctrine of phlogiston alongside Lavoisier's, in order to train the student in patient consideration of alternative interpretations and discourage the hasty adoption of any particular theory.[81]

By the time Thomson produced his *System of Chemistry* in 1802, exposition of the phlogiston theory would have been regarded as eccentric, but Thomson retained aspects of Nicholson's presentational rhetoric, promising adequate recognition of all discoverers of facts, without any theoretical partisanship. He adopted a historical method of exposition, which he justified on Priestley's grounds of didactic efficacy and proper assignment of credit:

By thus blending the history with the science, the facts will be more easily remembered, as well as better understood; and we shall at the same time pay that tribute of respect, to which the illustrious improvers of [chemistry] are justly intitled.[82]

Although Thomson dropped the historical method from his subsequent *Elements of Chemistry* (1810), he and other textbook writers like John Murray and William Henry continued to write in a very different style from Davy's. According to the conventions to which they adhered, authors had to appear not to be advancing only their own interests. The autonomous judgment of the reader was not to be imposed upon by presentation of an unbalanced view, or by the use of a vocabulary of description that predisposed him to adopt a particular theory. In Priestley's work this authorial stance had gone along with an insistence that students should convince themselves of facts by replicating the experiments described. The same position was taken by Samuel Parkes, a devotee of Priestley

---

80 Brande (1819a), esp. pp. xiii, xliv; Davy (1839–40), IV, pp. xi-xv.
81 Fourcroy (1790), I, p. xvi; Nicholson (1790), pp. vii-viii (the discussion of the phlogiston theory remained unchanged in the second edition (1792) and the third (1796)).
82 Thomson (1802), I, p. 13.

and a fellow Unitarian, in his *Rudiments of Chemistry* (1810). Parkes hoped his text would encourage readers to "perform the experiments alone," and he presented chemical doctrine as if it might all be established by the students themselves in this way. The proposal became implausible, however, when it came to Davy's isolation of the alkali metals by a "powerful galvanic battery in a state of intense activity."[83] It is hard to imagine many students being able to reproduce such an experiment.

Parkes's attempt to remain faithful to the Priestleian ethic revealed the quandary of such a commitment as techniques and apparatus were increasing in complexity. The refinement of specialist skills and the growing expense of equipment would require new rhetorical and institutional forms of education. But even as these means were developed, chemical writers reverted continually to the older values of science as a public culture. In this way the self-image of the disciplinary community was consolidated rhetorically in a time of increasing diversity and change. Thus, while forms of specialist training became established, educators also stressed the degree to which techniques should remain accessible to all. And while applications of chemistry to the practical arts were refined, chemists reiterated that improvements in science were to the benefit of the public as a whole.

The rhetoric of Enlightenment public science continued to shape the way chemistry was discussed in relation to the arts. It remained crucial to establishing their social legitimacy that chemists show that the improvements they were arguing for would serve the interests of the public in general. Brande, addressing the London Institution in 1819, noted that discussions of science and the arts would show "how nearly our interests are connected when they often appear widest apart." And William Henry, in his *Elements of Experimental Chemistry*, sketched a picture in which applications of chemistry were driven by "the interests of men" and the ambition of enlightened individuals to become benefactors of mankind, and were opposed only by "the obstacles of ignorance and prejudice."[84]

It was also as important as ever that chemists avoid attracting charges that they were projecting or serving their own individual interests. Accum, advertising his crystallographic models for sale, recorded that "I feel no solicitude as to any imputation of private or commercial views; such attempts are fully consistent with the purest regard to the public welfare." And Nicholson recommended readers of one of his Fourcroy translations to refer also to his journal, with the remark, "By giving this notice, I feel no solicitude of any imputation of personal motives in thus recommending my own work."[85] Although it might be that Accum and

---

83 Parkes (1810), pp. vi, 72 ff.   84 Brande (1819b), p. 23; W. Henry (1826), I, p. xxi.
85 Accum (1813), p. xv; Fourcroy (1804), I, p. xxxii.

Nicholson really were not anxious about such charges, it seems more likely that they were trying to deflect them by giving assurances that their efforts were aimed at the public benefit.

In a similar way, advocates of more specialized technical training in chemistry still made concessions toward a traditional conception of public science. Practical instruction in chemical laboratories became available in a variety of locations in this period. Babington and Allen urged the students at Guy's to attend sessions in the laboratory, to acquire manipulative skill in performing chemical operations, "without which the demonstrations of a Lecture-room will seldom acquire that force which is necessary to fix them in [the] memory."[86] Accum initiated a series of private laboratory classes in parallel with his public lectures, in order to provide the practical training needed by those who wanted to become chemists. He explained that:

... the only effectual means of facilitating the study of that science consist of applying [the student's] hands to the experimental part ... to become familiar with the nature, manipulations, and practical use of the chemical apparatus. This however cannot be done in PUBLIC LECTURES, nor at a distance from the chemical furnaces. He must for that purpose have access to the laboratory of the operative chemist.[87]

Initiatives such as this were clearly important, and by the 1820s were beginning to bear fruit in the recognition of the specialist chemist as a man qualified by practical training in certain techniques. On his appointment as Professor of Chemistry at Glasgow University in 1818, Thomson initiated laboratory training for students, in an attempt to "breed up a set of young practical chemists." Thomson saw the conventional method of teaching – by lectures only, with illustrative experiments – as inadequate for imparting new techniques, particularly those of chemical analysis. What students needed, he proposed, were "the means of acquiring the art of experimental chemistry by regulated practice."[88]

By 1825, the surgeon Walter Weldon saw clear signs that the amateur was being squeezed out of chemistry by increased requirements for resources and expertise:

[Chemistry] requires such an appropriation of time and property; such a variety of expensive and delicate instruments; such an acquisition of manual dexterity; and so much thought and attention for its successful prosecution, as will necessarily confine the *professed pursuit* of it to a few professors, and enthusiastic amateurs, whom fortune and opportunity favour.

But Weldon went on to make the pitch for his *Popular Explanation of Chemistry*, by asserting that "a general knowledge of the outlines of

86 Babington and Allen (1802), p. vii.   87 Accum (1804b), p. ii.
88 [Thomson] (1829–30), p. 276; Morrell (1972), p. 2; and see also Morrell (1969).

chemical science may be readily obtained by any one who will take the trouble to attain it."[89] And in the preceding decades the distinction between specialist training and popular education had been even less clear-cut. Though different levels of achievement were acknowledged to be likely, chemists seemed reluctant to exclude those seeking only general education from the practical training they offered. Henry had envisaged laboratory instruction as suitable for those studying the subject, "either as an art, or as a branch of liberal knowledge." And Babington and Allen noted that "however different the views with which the Gentleman, the Artist, and the Manufacturer may enter upon its study, each will obtain information adapted to the particular line of his pursuit."[90]

Perhaps the only really clear way in which amateurs and specialists were demarcated in this period was in the exclusion of women from specialist training and research. Although Mrs. Fulhame's *Essay on Combustion* (1794) was praised by some writers at the beginning of the century, she seems to have had no female followers in chemical research.[91] It was a common view that women might attend chemical lectures and even perform a few simple experiments at home, but that the laboratory should remain a male preserve. In 1804 a writer remarked that "Your chemists and metaphysicians in petticoats are altogether out of nature – that is, when they make a trade or distinction of such pursuits – but when they take a little general learning as an accomplishment they keep it in very tolerable order."[92] Commenting on his female auditors at the London Institution, Brande noted that "It is not our intention to invite them to assist in our laboratories." Jane Marcet had supported this demarcation when she made it clear in her own text that as a woman she did not claim "the title of a chemist." It was, she acknowledged, inappropriate for a woman to enter "into the minute details of practical chemistry." Her fictional female pupils were accordingly advised to refrain from pedantically using chemical terms in everyday conversation, and not to think of studying pharmacy, which "properly belongs to professional men."[93] As male professional identities became strengthened, women were to be progressively excluded from active participation in chemistry.

For men, the line between public education and specialist training was still an indistinct one in the period 1800-1820. Those who advocated more advanced levels of instruction also felt obliged to emphasize that the tools of the specialist were not exclusive or restricted in availability.

89 Weldon [1825?], p. iii.
90 Henry (1826), I, p. 1; Babington and Allen (1802), p. iii.
91 Parkinson (1807), pp. 30–31. See also Birkbeck's comments, reported in *The Chemist*, 2 (1824–25), 164.
92 Leonard Horner to J.A. Murray, 15 November 1804, quoted in Jones (1871), p. 264.
93 Brande (1819b), p. 36; [Marcet] (1806), I, pp. iii, 2–3. On Marcet, see also: Horrocks (1987).

Notwithstanding his introduction of practical laboratory courses, Accum insisted that much analytical chemistry could be done without a laboratory at all: "All the operations of analysis may be (in a small way) conveniently performed, by the help of a few instruments, in the closet of the amateur of the science."[94] Henry similarly emphasized how much could be accomplished with simple items of apparatus: "I am far from recommending a frivolous regard for show . . . With the aid merely of Florence flasks, of common vials, and of wine glasses, some of the most interesting and useful experiments may be made."[95] Even Davy proposed that one of the signs of current progress in chemistry was that ". . . the apparatus for pursuing original researches is daily improved, the use of it rendered more easy, and the acquisition less expensive."[96]

The popularity of small-scale portable laboratories at this time indicates how assiduously chemists sought to render new techniques widely available to amateur researchers. Henry marketed his "Portable Chemical Chests" as a means of self-instruction for would-be chemists. Three models were available, at 15, 11, and 6½ guineas respectively, and they contained basic vessels, a balance, a blowpipe and spoon for analysis, and a large selection of reagents. Purchasers were advised to set about replicating the processes described in the journals and in the writings of leading chemists.[97] In 1814, Richard Reece advertised a similar range of portable "Chests of Chemistry," available for purchase from the Chemical and Medical Hall in Piccadilly.[98] Accum's *Chemical Amusement* (1818) described a modest selection of apparatus, available at relatively low cost, for "carrying on a general course of Experiments of Study in the small way." Experiments that had a surprising or pleasing result, and that readers performed themselves, were said to be more memorable than "the rapid illustrations inseparable from public and popular courses of lectures."[99] Another such collection, the "Economical Laboratory" devised by Guyton de Morveau, was described by James Parkinson in his *Chemical Pocket-Book*.[100]

These small-scale laboratories were envisaged as a means of popular self-education in chemical techniques; and it is clear that they were also seen as enabling the purchaser to contribute to ongoing research. Henry explained that they could be used "for general and ordinary purposes, . . . and even for the prosecution of new and important inquiries." With simple apparatus of this kind, he remarked, "some of the most interesting

94 Accum (1804a), p. 4.   95 W. Henry (1801), p. 3.   96 Davy (1812), p. 59.
97 W. Henry (1801), pp. iii-vii, 1–3, unnumb. pp. at back of volume.
98 Reece (1814), pp. vii-xi.   99 Accum (1818), pp. v, ix-x.
100 Parkinson (1807), frontispiece. For this and other examples, see: Smeaton (1965–66); Gee (1989).

chemical facts may be exhibited and even ascertained."[101] In the 1820s, *The Chemist* echoed this line with a vehement defense of the validity of the knowledge produced with simple and inexpensive equipment. A correspondent wrote to the journal to assert that "the profoundest of the English chemists discards the fopperies of apparatus, and keeps his laboratory within the compass of a tea-tray; a few glass tubes, a blowpipe, some twenty little phials, and three or four wine glasses, suffice for his experiments." The editor confirmed this claim, and invoked the names of Franklin, Priestley, and Watt to show the great results that could be achieved with modest equipment. What was needed was for the leading chemists to acknowledge the contribution that amateurs could make, for "the humblest intellect may sometimes add a ray of light to the most dazzling genius."[102]

The Chemist's radical defense of an egalitarian economy of discourse in the scientific community was beginning to look rather isolated in the 1820s. But in the previous decades there had been many who held to the Enlightenment values of which Franklin, Priestley, and Watt were symbols. At this time, most chemists still conceived of their community as one in which relatively unskilled individuals, with modest resources, could participate. Notwithstanding Davy's attempts to concentrate instrumental and rhetorical resources in expert hands, advances in technique had not completely displaced the ideal of science as a public culture.

It was in this context that John Dalton's atomic theory first appeared and gained its foothold in chemical discourse. This is not the place to attempt a full account of the process by which Dalton's proposals achieved a degree of acceptance, or the reasons for the continuing opposition that they encountered. Elements of such an account have been assembled by other historians, although a full treatment remains a desideratum.[103] What can be done here is briefly to indicate some respects in which the reception of Dalton's work related to the context I have outlined, in which a developing specialist expertise was offset by a lingering attachment to the Enlightenment ideal of public science.

The story of how Dalton's insights came to be widely known has been told several times, beginning with Thomas Thomson who was a participant in the drama before he wrote its history. In Thomson's account, Dalton's theory followed from the work of certain chemists of the late eighteenth century on the quantitative analysis of salts. The analyses of Torbern Bergman and J.B. Richter could be seen to have pointed toward

101 Henry (1826), I, p. 14.
102 *The Chemist*, 2 (1824–25), pp. 44, 47 and fn.
103 Useful accounts are given by: Greenaway (1966), pp. 148–180; Knight (1967), pp. 16–36; Mauskopf (1969); Whitt (1990).

the law of constant proportions, according to which each substance was associated with a particular quantity and would always enter into chemical combination in a certain proportion of that quantity. Thomson recounted how, on a visit to Manchester in 1804, he had spent time with Dalton, who showed him how the atomic hypothesis could give a simple explanation for the constancy of the proportions in which chemical substances combined. Dalton proposed that the indivisible units (the atoms) of each chemical element had a characteristic weight, and that they formed compounds by uniting in small numbers. For example, water comprised one atom of hydrogen and one of oxygen, nitrous oxide comprised one atom of nitrogen and two of oxygen, and so on. Thomson was particularly struck by the diagrams that Dalton devised to give a simple picture of combinations of small numbers of atoms, diagrams that showed him "at a glance the immense importance of such a theory, when fully developed."[104]

Thomson gave the atomic doctrine its first public airing in the third edition of his *System of Chemistry* (1807). There the theory was linked with the properties of gases, a presentation that reflected its origins in Dalton's work on mixtures of gases and their solubility in water. The following year, Thomson and William Wollaston both published papers in the *Philosophical Transactions* that applied atomic ideas to problems in the analysis of acid salts. Both writers showed how the assumption that each chemical substance had a unit weight and combined in simple multiple proportions could make sense of analytical results. Also in 1808, Dalton himself ventured into print with the first volume of his *New System of Chemical Philosophy*, exhibiting the atomic theory in its full range of implications for chemistry.

Though it found a ready acceptance in certain quarters, Dalton's theory also encountered resistance. Davy's attitude was consistently cool. Already in 1803 he had told the Manchester chemist that speculations about the atomic composition of gases were "rather more ingenious than important"; and despite some softening of his resistance in later years his endorsement was never more than lukewarm.[105] To many chemists, a doctrine that proposed "atoms" that were both chemically distinct and physically indivisible was a speculative excursion beyond the bounds traditionally assigned to chemical theory. As early as 1734, Peter Shaw had written that "Leaving . . . to other Philosophers the sublimer Disquisi-

---

104 Thomson (1830–31), II, pp. 277–308 (quotation on p. 291).
105 Davy's remarks were recalled by Dalton in 1830, reviewing the progress of his work in an address to the Manchester Literary and Philosophical Society. See Thackray (1972), pp. 90–103 (quotation on p. 92). Compare the unsuccessful attempt by Thomson and Wollaston to persuade Davy to support the theory in 1807, recounted in Thomson (1830–31), II, pp. 293–294.

tions of primary Corpuscles, or Atoms, ... genuine chemistry contents itself with grosser Principles, which are evident to the Sense." A robust empiricism of this kind had become in some measure definitive of chemistry, an attitude reinforced by acceptance of Lavoisier's pragmatic definition of an element as the end product of chemical analysis.[106]

Drawing on this tradition, many chemists argued that Dalton's notion of atoms was hypothetical, and that it should be separated from the valid concept of "equivalent weights" and the law of constant proportions. In 1814, Wollaston altered his former stance and withdrew his endorsement of the atomic theory, while publishing a series of measurements of weights of chemical equivalents. Davy praised him for separating "the practical part of the doctrine from the atomic or hypothetical part"; he had already made a similar demarcation in his own *Elements of Chemical Philosophy*. Arguably, however, Davy's differentiation of the law of equivalent weights from the the theory of underlying chemical atoms was dictated less by methodological considerations than by his competing theoretical ideas. Davy's hope that all elements might ultimately be shown to be compounds of hydrogen could not be reconciled with Daltonian chemical atomism.[107]

Thanks to the research that has been done on Dalton's work and its context, particularly by Arnold Thackray, we can now begin to appreciate just how different his form of chemical practice was from Davy's.[108] Dalton's ideas emerged from work on pneumatic chemistry and meteorology; they made the transition to high theory from a distinctly marginal field of experimentation, pursued in his and William Henry's relatively humble Manchester laboratories. His interests in quantification and measurement, and in the phenomena of gases and water vapor, had previously matured in the course of extensive meteorological observations in his native Lake District.[109] As Adam Sedgwick told the British Association for the Advancement of Science (BAAS) meeting in 1833, Dalton had done his work "without any powerful apparatus for making philosophical experiments."[110] Kathleen Farrar has demonstrated that his equipment was not quite as rudimentary as the "few cups, penny ink bottles, rough balances and self-made thermometers" that Manchester legend recorded. Nonetheless, the collection of meteorological instruments, eudiometers, and vessels for pneumatic chemistry that he assembled resembled Priestley's apparatus much more than the well-found facilities of the Royal Institution.[111]

Furthermore, Dalton's atomism was articulated in a didactic setting

106 Shaw and Lavoisier quoted in Whitt (1990), p. 64.
107 Goodman (1969a); Whitt (1990); Davy (1827), part VII, pp. 10–11.
108 Thackray (1972). See also Cardwell (1968).   109 Manley (1968).
110 Sedgwick quoted in W.C. Henry (1854), p. 179.   111 Farrar (1968).

quite different from Davy's. Although Dalton delivered one course of lectures at the RI in 1803–1804, and was pleasantly surprised at how attentive an elite London audience could be, he spent most of his life in Manchester, teaching private pupils in mathematics and natural philosophy and offering several courses of public lectures in the city. The apparatus he used in teaching had none of the spectacular rhetorical potential of the voltaic pile, and he seems to have had no desire to produce dramatic effects with his demonstrations. Indeed, several of those who attended his lectures recorded that the experiments frequently failed altogether to produce the desired result. Davy judged him to be "a very coarse experimenter."[112]

It was in the context of this kind of didactic practice that the atomic theory, with its explanatory diagrams and models, was developed and deployed. Dalton told the Manchester Literary and Philosophical Society in 1830 that "At an early period of my chemical enquiries [I saw] the advantage of a chemical notation . . . most of all in incul[cating] elements of the science to [beginner]s." He had therefore shown his diagrams of various compounds and their constituent atoms in his lectures. This was the way in which his doctrine was articulated; it would, he allowed, "have been very difficult to convey to a large and promiscuous audience without [the use of] suitable diagrams."[113]

Dalton's resolute provincialism was of course another difference between the setting of his work and Davy's. When the Manchester chemist first visited London, he recorded the sardonic impression that it was "worth one's while to see once; but the most disagreeable place on earth for one of a *contemplative turn* to reside in constantly." His view of the metropolis never changed very much. Throughout his life he kept aloof from London institutions; and when the BAAS was founded in 1831 it was Dalton who insisted that the association should remain "ever . . . *Provincials.*" He feared that a metropolitan base would prevent the association from fulfilling its aim of widespread diffusion of the sciences.[114] On the occasion of the award of a Royal Medal to him in 1826, Davy as President of the Royal Society patronizingly suggested that Dalton had been laboring in the obscurity of the country for a quarter of a century, but that the award would "give a lustre to his character . . . [and] make his example more exciting to others."[115] We may assume that Dalton did not see things in that way.

112 W.C. Henry (1854), pp. 48–50, and quotation on p. 217; Thackray (1972), pp. 103–114.
113 Dalton (1830), pp. 95–96 (material in brackets is restoration of missing portions of the text by Thackray).
114 Dalton quoted in W.C. Henry (1854), p. 41. Account of the 1831 BAAS meeting quoted in Thackray (1972), p. 23.
115 Davy (1827), part VII, p. 13.

On the contrary, for him a provincial setting had the advantages of preserving his moral autonomy and simplicity of lifestyle. As Thackray points out, the distance that separated him from the metropolitan scientific establishment was "rather one of class and style" than one of simple geography.[116] Whereas Davy's brother John was privately insulting about Dalton's character ("Mr. Dalton's aspect and manner were repulsive ... there was no gracefulness belonging to him," he recalled),[117] to his supporters, independence, humility, and simplicity were his distinctive virtues. Along with a Priestleian simplicity of experimental style, Dalton assumed aspects of Priestley's public moral persona. William Henry recorded that "In perfect consistency with Mr. Dalton's intellectual qualities, are the moral features of his character; the disinterestedness, the independence, the truthfulness, and the integrity which, through life, have uniformly marked his conduct towards others."[118] Thomson noted that he had supported himself as a teacher and lecturer "if not in affluence, at least in perfect independence" – an independence that he feared to put at risk when he was offered a civil list pension in 1833. Thomson went on:

... there is little doubt that Mr. Dalton, had he so chosen it, might in point of pecuniary circumstances, have exhibited a much more brilliant figure. But he has displayed a much nobler mind by the career which he has chosen – equally regardless of riches as the most celebrated sages of antiquity, and as much respected and beloved by his friends, even in the rich commercial town of Manchester, as if he were one of the greatest and most influential men in the country.[119]

Dalton's rootedness in his provincial context and his clear preference for the sustaining friendship of local institutions rather than the formal recognition of the Royal Society indicate the continuing importance of the provinces in supporting scientific work at this time. His pneumatic experiments with simple apparatus and the deployment of his work in public lectures also show that aspects of Priestley's practice of public science survived in certain locations. It is clear that for those in Dalton's milieu, moral qualities of the kind that Priestley had exemplified continued to underwrite contributions to public experimental discourse. Coincidentally, Dalton (like Priestley) had to rebut a charge of plagiarism by a member of the Higgins family. William Higgins (nephew of Bryan) claimed that the atomic theory had been stolen from his *Comparative View of the Phlogistic and Anti-Phlogistic Theories* (1791); he had to admit, however, that Dalton had applied the doctrine in "a general and popular way," so that it had been associated with his name.[120]

116 Thackray (1972), p. 23.   117 John Davy quoted in W.C. Henry (1854), p. 217.
118 William Henry (author's father) quoted in W.C. Henry (1854), p. 175.
119 Thomson (1830–31), II, p. 286.
120 The dispute between Higgins and Dalton is referred to in: W.C. Henry (1854), pp. 175, 217; Davy (1827), part VII, pp. 9–10; W. Higgins (1814); Nash (1814).

The process by which Dalton's work was generalized and popularized was the result of his own efforts, and those of Thomson, Wollaston, and others. Wollaston's contribution in particular reveals how the reception of the atomic doctrine was also shaped by the development of specialized analytical skills among chemists. In his 1814 paper "A Synoptic Scale of Chemical Equivalents," Wollaston showed how Dalton's ideas could yield a practical routine for calculating the weights of components in the course of analyzing a substance, without explicit commitment to the atomic theory. He demonstrated a slide rule with a pair of adjacent sliding scales: One showed the weights of elements and compounds, the other had a simple numerical scale. Both scales were logarithmic, permitting the multiplication and division of weights to calculate quantities in any particular sample.

Wollaston did his best to make his slide rule as accessible as possible. He disclaimed any intention of advancing atomism as a theory, having (as he put it) "endeavoured to make practical convenience my sole guide." To aid understanding of the scale by chemists without a mathematical education, he explained the concept of logarithms on which it was based. By these means, he sought to make the sliding scale "a specimen of the extreme facility of mechanical approximation," a process that was considerably easier than numerical calculations that were often "more laborious than the accuracy of our data warrants."[121]

The Wollaston scale proved very successful in analytical practice. Although Thomson judged it the most important contribution to spreading the atomic theory, most chemists seem to have used the device without making a corresponding commitment to the existence of atoms, an attitude that Wollaston's presentation had encouraged. Brande, William Prout, Andrew Ure, and Michael Faraday were among the chemists who endorsed the use of the scale, and showed how it could be applied to a variety of analytical problems. Slide rules and scales to be attached to them were sold in large numbers in Britain, on the Continent, and in the United States. As Wollaston had anticipated, the instrument found applications in chemical manufacturing and the pharmaceutical trade. Until the demand arose for greater accuracy in analytical calculations in the 1850s, the scale was an almost universal item of laboratory equipment. In 1826, a fellow of the Royal Society suggested that Wollaston (rather than Dalton) deserved the honor of the Royal Medal for making the doctrine of equivalents into "an instrument of the utmost usefulness and value."[122]

Wollaston had made Dalton's conceptual innovation into a material

121 Wollaston (1814), pp. 7, 15–17, 18.
122 Thomson (1830–31), II, p. 306; Goodman (1969a), pp. 42–53 (quotation on p. 47).

instrument, one that was widely useful for routine analytical calculations. Thomson's view of the matter – that chemists who used Wollaston's slide rule were endorsing the atomic theory – clearly overstates the degree of theoretical commitment needed to find the device useful. On the other hand, Wollaston's view – that he had stripped Dalton's theory of its metaphysical excesses and given it practical application – does not seem entirely fair to Dalton. His work was no more intrinsically metaphysical, no less practical, than Wollaston's. As we have seen, the atomic theory was developed in the context of practical work on problems of meteorology and pneumatics, and it was articulated to give a concrete understanding of such problems to pupils and the audiences at public lectures. Dalton expressed astonishment that Wollaston and Davy could talk about equivalents as if they had no theoretical principles in view, when they were in fact talking about atoms.[123]

Perhaps the best way to regard the translation from Dalton's work to Wollaston's instrument is as a transition from one practical setting to another. Wollaston's achievement was to take the doctrine of equivalents out of the context of provincial pneumatics and didactics, and insert it into the developing enterprise of chemical analysis, as it was being practised in laboratories, factories, and pharmacies. He did this by inventing an instrument that could be used by relatively unskilled practitioners (men who might not even understand the use of logarithms), to yield results of an accuracy in keeping with the standards of the time. Wollaston's device – simple and broadly applicable – was well suited to the period in which it became widely used. It aided the development of chemical analysis and extended its range of applications. It thus helped enhance the expertise of specialist chemists. But, at the same time, it answered to the need for relatively simple equipment and accessible experimental techniques. Hence it took its place in a context of chemical practice that was still shaped by the legacy of Enlightenment ideals of public science.

Mineralogy and the development of chemical analysis

[I]f book-making be a profitable trade, if the title "Author" be distinguished in society, what branch of our knowledge is there that affords so many facilities as mineralogy[?] Philosophers will spring up by its touch, like men from the hands of Deucalion, and stones will be thrown by handfuls in every direction.

Richard Chenevix, *Observations on Mineralogical Systems* (1811)[124]

Probably the most important field in which technical practice advanced rapidly after the Chemical Revolution was inorganic analysis. Davy's iso-

123 Dalton, quoted in Thackray (1972), p. 94.   124 Chenevix (1811), p. 119.

lation of the alkali metals from formerly undecompounded mineral earths was a contribution to this. At least as significant as Davy's voltaic pile, however, were more basic, and more readily reproducible, methods of mineral analysis. "Dry way" techniques of decomposition by fire were supplemented by the use of the blowpipe, and a whole battery of "wet way" tests were applied to substances in solution. Chemical procedures originally developed in mineralogy and the analysis of mineral waters were refined and extended into such new fields as soil analysis.

These techniques became the tools of specialist chemists, imbibed in the course of practical training, and deployed in the building of careers. But even as we trace the enhancement of specialist skills in this field, we continue to hear the echoes of Enlightenment notions of science as an activity in the public realm. Methods of mineral analysis were communicated in ways that enrolled a substantial lay audience in mineralogical pursuits. They were applied by bodies that dedicated themselves to general social improvement – for example, through geology and agricultural chemistry. Instruments were kept inexpensive and simple in order that they could be widely used. The chemical techniques adopted in this period owed their popularity, to a significant degree, to their ready assimilation into traditional public forms of scientific practice.

Eighteenth-century mineralogy had been a subsidiary of Enlightenment natural history. Although the efforts of the great Swedish naturalist Carolus Linnaeus to extend his systematic classification of living forms into the mineral kingdom were judged by many to have been unsuccessful, he was generally taken to have shown the way mineralogy should go. The aim was to differentiate the wild confusion of the mineral realm into distinct species and to situate each in an exhaustive system of classification. This was to be done by amassing collections of mineral samples, each one carefully labeled with name and original location, and displaying them in an ordered arrangement in a cabinet or museum.[125] Collections of this kind enabled practitioners to learn to discriminate species and to assess the mineralogical wealth of particular localities. Public appreciation of the works of the deity was also thereby encouraged. As J.R. Forster (a natural historian at the Warrington Academy) put it in 1768, natural history, "when seriously attended to, will leave the deepest and most lasting impressions of *religion* and *piety* on our hearts."[126]

Although chemical methods had no role in the Linnaean enterprise, they came to exert a greater influence in the course of the eighteenth century. New salts were identified in mineral waters, new metals and semi-metals discriminated. Most important was the analytical identification of dozens of new mineral earths, the work particularly of German

125 See, for example, Meyer (1775).   126 Forster (1768), p. 2.

and Scandinavian chemists of the second half of the century. J.G. Wallerius, professor of chemistry and mineralogy at Uppsala University, and J.H. Pott, an academician and professor of chemistry in Berlin, were among the pioneers in this field in the middle of the century.[127] A.F. Cronstedt, a former pupil of Wallerius and a Swedish mine surveyor, published his *Essay Towards a System of Mineralogy* in 1758, in which he asserted the inadequacy of systems of mineral classification that relied solely on visible external characters. Chemical analysis, Cronstedt proposed, could add other more solid characteristics, which would aid in the construction of a classificatory system. As an instrument for such analysis, he suggested the blowpipe – a narrow tube by which a constant stream of air could be directed at a candle flame to increase its temperature. Already in use among jewelers and glassblowers, the device could be adapted to assay metallic ores by fusing them or to diagnose the contents of mineral samples by heating them in contact with a borax touchstone. Following Cronstedt, the blowpipe became an indispensable item of equipment for the mineralogical chemist.[128]

Torbern Bergman, successor to Wallerius at Uppsala, took the chemical approach to mineral classification further. In his *Outlines of Mineralogy* (1782) he proposed a classification based on chemical analysis, in which the majority component of each mineral determined its genus and the minority component its species. Like Cronstedt, Bergman criticized the Linnaean reliance on external characters, which he said were the result of chance circumstances of formation and hence could not form the basis of a systematic order. In order to exploit the uses of minerals it was necessary to know their chemical composition, for "their component parts . . . being well understood, we know what to expect from them." Cronstedt's methods could, however, be improved upon by "humid" techniques of analysis in solution. By these means, Bergman identified five distinct earths: calcareous, magnesian, argillaceous, siliceous, and barytic. He went on to construct an analytical procedure by which a sample could be routinely tested to determine which of these earths it contained. For him, the use of this kind of test was an example of how instrumentation could valuably supplement the limited power of the senses.[129]

In Britain, Bergman's *Outlines* was translated by Priestley's Birmingham colleague Withering, and Kirwan also worked to transmit Continental mineralogical chemistry. The first edition of Kirwan's *Elements of*

---

127 T.M. Porter (1981); Llana (1985); Laudan (1987), pp. 21–35, 47–69.
128 Cronstedt (1788), pp. v-xxi, 273–318.
129 Bergman (1783b), esp. pp. 6–10. For further remarks on instrumentation, see also: Bergman, "Of the investigation of truth," in Bergman (1784), I, pp. xix-xl. For discussion of Bergman's doctrines, see: T.M. Porter (1981), pp. 560–564; Laudan (1987), pp. 59–63, 78–79.

Mineralogy (1784) complained about the lack of attention that the subject had received in Britain. The native tradition of mineral water analysis had been pursued, the author noted, mainly by "gentlemen in the Medical Line, whose transient attention is soon diverted by their more direct occupations." What was required was serious study of the works of Cronstedt, Bergman, C.W. Scheele, and others, and the recognition that rigorous application of chemical analysis constituted the only basis for a scientific system of mineral classification. Kirwan wrote:

> The only principles . . . that Mineralogy affords are the relations of the bodies it considers with chymical agents. Without referring to these, it can be reckoned at most only a conjectural art. . . . [When] an intire certainty [is] required, such as constitutes the foundation of a *science,* there chemical tests are absolutely requisite, and alone sufficient.[130]

In the 1780s and 1790s, Kirwan had the gratification of seeing Continental methods applied to the analysis of British mineral waters. Bergman's techniques were used on the Buxton Springs by George Pearson in 1784, and on those at Harrogate and Knaresborough by Thomas Garnett in the late 1790s. Kirwan's own *Essay on the Analysis of Mineral Waters* (1799) and Saunders's *Treatise on the Chemical History . . . of . . . Mineral Waters* (1800) provided synoptic summaries of what had been achieved in this field by the end of the century, and further contributions by Marcet, Phillips, Parkes, and Brande followed.[131] Within a few years, a section on methods of mineral water analysis became a standard part of textbooks on chemistry. Already in the second edition of his *Elements* in 1794, Kirwan had reflected that his aspirations for a "scientific" mineralogy based on chemical analysis had largely been achieved. In his *Elements of Experimental Chemistry* (1810), William Henry agreed with him. By applying chemical analysis, Henry wrote, "mineralogy has been advanced from a confused assemblage of its objects, to the dignity of a well methodized and scientific system."[132]

The history of mineralogy in this period cannot, however, be reduced to the unproblematic triumph of a scientific approach. Although the aspiration to make the subject into a science might have been widespread, there was considerable and continuing disagreement as to what form a science of mineralogy should take. This was because the 1780s also saw the emergence of two clear alternatives to the use of chemical analysis as a basis for mineral identification. These were the system for identifying minerals from external characteristics taught by Abraham Gottlob Werner (1750–1817) at the Freiberg mining academy in Saxony, and the

130 Kirwan (1784b), pp. iii, vii-viii, xii.
131 Pearson (1784); Garnett (1790, 1791, 1792); Kirwan (1799); Saunders (1800).
132 W. Henry (1826), I, p. xiv.

French program of crystallography under the leadership of Jean Baptiste Romé de l'Isle (1736–1790) and René Just Haüy (1743–1822).

As originally proposed in his *On the External Characters of Minerals* (1774), Werner's system comprised a method for identifying mineral samples from their sensory qualities. The relevant characteristics included smell, taste, touch (divided into hardness, weight, feel, etc.), and a whole range of visual properties such as color, crystalline form, and transparency. The magnitude of each attribute was to be assessed on a linear scale, so that by judging the position of a sample on as many scales as possible a unique identification could be made. Werner addressed his scheme to an audience of practical miners and gentlemen amateurs and, with this group in view, disdained the use of instruments as too artificial and impractical. The trained use of the senses alone would be sufficient to enable any mineral to be identified, he proposed.[133]

In later works, Werner acknowledged a role for chemical analysis in constructing a comprehensive system of classification for minerals. Such a system should be built upon a knowledge of chemical composition, he agreed. But mineralogists working in the field should be able to identify samples without resort to instrumentation or the chemical laboratory. Werner's attitude to chemical methods in mineralogy was respectful, but his own aims were quite different. Kirwan saw him as complementing the efforts of the chemists. Comparing the chemical analyst M.H. Klaproth with Werner, he wrote, "the former detects the internal principles, the latter depicts the substances to which those principles belong."[134] Other British mineralogists, such as Thomas Thomson, Robert Jameson, and William Phillips, also saw Werner's system as entirely compatible with the use of chemical analysis, although Arthur Aikin and (as we shall see) Richard Chenevix disagreed. The British Wernerians extended their master's important work on the mapping of rocks in the context of local formations, in the long term opening the way to a chronological account of the successive deposition of minerals and the enterprise of historical geology. That story, however, has been told elsewhere and cannot be traced further here.[135]

If the relation of Werner's system to chemical analysis was somewhat problematic, that of crystallography was much more so. Particularly under the leadership of Haüy, the French school made crystalline form, rather than chemical composition, the fundamental feature for determining mineral identity. Haüy asserted that each homogeneous mineral substance possessed an underlying primitive form, from which the visible crystal was built up by regular aggregation of its integrant molecules. The

133 Metzger (1969), pp. 76–79; Laudan (1987), pp. 78–83; and (on the debate surrounding Werner) Melhado (1981), pp. 100–141.

134 Kirwan (1794–96), I, p. vi.   135 Laudan (1987), pp. 87–112.

geometry of the underlying form could be revealed by studying the cleavage planes of the crystal and the angles between them. For Haüy, the geometrical approach, employing precision measurement and mathematical calculation, was the means by which mineralogy could be made properly scientific.[136] To support this claim, he introduced a new instrument – the goniometer. This device (basically a protractor with rotating arms attached) enabled measurements to be made of the angles between the faces of a crystal. The goniometer was recognized as requiring a degree of skill to use it, particularly to align the moving arms with the faces of the crystal and to read the scale accurately. Disputes persisted about the results of measurement and the degree of precision that could legitimately be claimed from using the instrument.[137]

Haüy's claim to have created a new exact science implied a separation of crystallography from the study of noncrystalline materials and the demotion of other sciences such as chemistry to a subordinate status. In Haüy's view, minerals that did not form crystals simply could not be the object of scientific study. A practice that equated science with the measurement of geometrical forms inevitably left a substantial region of formless chaos outside its grasp. Nor could chemistry be admitted to the same level of authority as crystallography, because of the cases where the results of the two approaches came into conflict. For example, some substances such as aragonite and calcite had quite different crystalline forms although they yielded the same chemical analysis. (Both were judged by chemists to be calcium carbonate.) In cases such as this, Haüy proposed that chemical analysis was incomplete and that some overlooked extra component accounted for the different forms.

The marked differences between Haüy's and Werner's systems should be clear, even from this brief exposition. In Werner's scheme, crystalline form was merely one criterion of identification in those cases where it was found; in Haüy's it was the foundation for a scientific approach to mineralogy. The two schools were also at odds over the role of chemical analysis. In crystallography, it was clearly subordinate to skills of measurement and calculation; in Werner's method, it was recognized as important for mineral classification but unnecessary for identification. In Britain in the first two decades of the nineteenth century both systems had articulate advocates. From what has been said already about the status of specialist expertise in this period we might anticipate that arguments over the legitimacy of particular techniques would arise in this dispute. This was indeed the case. And the debate between the competing

136 Metzger (1969), esp. pp. 80–86; Mauskopf (1970, 1976).
137 Goodman (1969b), esp. pp. 152–157.

mineralogical schools also touched revealingly on the role of chemical analysis.

A frontal assault on the Wernerian system from the standpoint of crystallography was launched by Chenevix in his "Observations on Mineralogical Systems," originally published in the *Annales de Chimie* in 1808 and translated into English in 1811. According to Chenevix, the German method took note of far too many mineral characteristics and failed to rank them in a consistent order of priority when making an identification. Werner simply did not recognize the value of crystalline form as a primary feature of mineral identity, nor had he used chemical analysis consistently. Haüy, on the other hand, had made sophisticated use of analysis while grasping its limitations. Chenevix wrote that Haüy, "instead of stopping at the surface, . . . has penetrated into the interior of the mineral, and a new world has presented itself to his view." The French system, he went on, "embraces the whole of mineralogy that can be considered as a science."[138]

Chenevix also dealt with the Wernerian antipathy to the use of instruments and objections to the goniometer. He countered the claim that crystallographic instruments were too difficult to manage and the calculations too complex. Instruments were essential to the progress of science, he asserted, and to give them up would be like giving up civilization. Chenevix suggested sarcastically, "Let us no longer use our telescopes, our microscopes, and our chronometers; they are too difficult to manage." He conceded that Werner's system might be more useful for practical miners, but Haüy's could yield scientific results by which such practitioners could be directed. Thus Werner's method "is addressed to artists, while that of Haüy is intended for philosophers."[139]

Thomas Thomson responded to Chenevix on behalf of British Wernerians in an article in his own *Annals of Philosophy* in 1813. Thomson's view was that chemical analysis fully supported the Wernerian system, whereas it gave only equivocal support to Haüy. Haüy's suggestion that mineral crystals all showed a fundamental primitive form was a mere mathematical hypothesis, quite useless to the practising mineralogist. It selected without good reason just one of the characters that minerals presented to examination, one that was not even applicable to the many minerals that did not occur in crystalline form. Furthermore, Haüy's system boasted a quite spurious degree of precision in its measurements of inter-facial angles. Thomson found more plausible Werner's notion that minerals need not be precisely alike in all respects in order to be classified

138 Chenevix (1811), pp. 20, 29–30.
139 Chenevix (1811), pp. 55, 93–94, 98–99, 110.

as the same species: The assessment of characters on multiple scales allowed for examples to be judged the same if they were sufficiently close to one another on a significant number of scales. Finally, the Wernerian system had the advantage of proven pedagogical success, in Thomson's opinion. The Freiberg school had produced more able mineralogists than Haüy could number among his students, and the fact that Werner did not demand a high degree of mathematical knowledge or a barely attainable level of instrumental skill was clearly relevant to this success. On the issue of instrumentation and its relation to philosophy, Thomson commented:

Now if one man offers to teach us a method of [discriminating minerals] . . . by the assistance of our senses, while another insists upon our calling in the assistance of chemistry and mechanical philosophy, and upon our providing an expensive set of philosophical instruments, I for my part will embrace the first offer, and leave Mr. Chenevix and the philosophers to accept the second.[140]

Thomson's echo of Priestley's objections to expensive apparatus had been anticipated by Thomas Cooper, a disciple of Priestley's who had followed him into exile in the United States in the 1790s. In his *Introductory Lecture* at Carlisle College in Pennsylvania in 1812, Cooper was particularly scathing about the crystallographers, though he also put some distance between his own views and Werner's by insisting that chemical methods be used in mineral identification. "We want at present in this new land, chemists, miners, and practical mineralogists, more than crystallognosts," he announced. Haüy's approach was inappropriate to noncrystalline minerals and required techniques of preparation and measurement that were exceptionally difficult to master. Even skilled lapidaries could not always cut crystals in the manner Haüy demanded, Cooper pointed out: "The degree of manual dexterity required, is so much more than can usually be employed, that the system seems not calculated for common and popular use."[141]

As the comments of Cooper confirm, the dispute between Chenevix and Thomson embraced a wide range of issues. The degree to which the mineral realm could be the subject of a science, the form of instrumentation that was appropriate, the kind of practitioners envisaged, and the pedagogic methods suitable for training them – all were in question. It is clear that participants in the debate saw the issues as connected. Chenevix grasped that Haüy's rhetoric of scientificity could be sustained only by deploying a specific form of instrumentation – the goniometer. This apparatus required skillful handling, so that specialist crystallographers would have to be trained in practical manipulation, as well as in the mathematical techniques required to interpret measurements. Thomson,

140 Thomson (1813b), p. 254.   141 Cooper (1812), pp. 81, 86.

on the other hand, saw Werner's system as having the advantage of greater accessibility to relatively untrained individuals with fewer resources of equipment. At least at this stage of his career (when he was giving up public lecturing in Edinburgh to begin his work as a scientific journalist in London), Thomson appears to have been attracted by a Priestleian view of scientific pedagogy.

The issues in question thus clustered around alternative models of mineralogical practice, which encapsulated distinct visions of the scientific community. Haüy and his allies aimed to create a group of specialist experts, distinguished by their highly trained use of instruments. This vision was opposed by those who wanted to keep open the community of mineralogists to practitioners with only simple apparatus and relatively rudimentary skills. These chemists (including Thomson and Cooper) still saw the science as properly constituted in the public arena. Kirwan's picture of the construction of scientific mineralogy by an unproblematic extension of instrumental practices overlooks the tensions between these ideals. If, however, we highlight the alternative points of view, we can see how the adoption of an instrument such as the goniometer involved a degree of social discipline. Training in the use of novel apparatus resulted in the displacement of other skills – the skills of sensory discrimination that Werner had prized and classical mineralogists had cultivated.

In this debate over mineralogical methods, encoding as it did alternative models of practice and community, the position of chemical analysis is intriguing. As we have seen, both Chenevix and Thomson claimed to use analysis, though both recognized certain limits on its validity. Each side saw the other as resorting to analysis inconsistently, although to the outsider the reservations expressed by both seem much the same. Chenevix took Haüy's line, that analysis could be deceptive when different crystalline forms appeared to have the same composition or when instances of the same form gave different analytical results. Thomson said something very similar: Chemical analysis was a reliable guide to composition, except when samples clearly distinct in external properties (such as diamond and charcoal) gave the same result. In cases like this, Chenevix and Thomson seem to have agreed, chemical investigation must have missed an extra component that caused the difference in physical form.[142]

Although both sides imposed constraints on its interpretation, it is striking that chemical analysis found a common acceptance in principle. This was noted by Berzelius in 1814. Wernerians and Haüyians, he remarked, "both agree in the admission of chemistry to a participation in

---

142 Chenevix (1811), esp. pp. 11 ff., 22–23, 35–37; Thomson (1813b), esp. pp. 245, 249–250.

the foundation of a system of mineralogy, however much they may differ respecting the extent of that participation."[143] Apparently, chemical techniques could be reconciled with both models of mineralogical practice; they had an authority that both sides in the dispute accepted. Perhaps this was because they demonstrated a sufficient degree of expert skill to be undeniable by the crystallographers, and yet seemed attainable enough by unresourced and untrained practitioners to be acceptable also to those like Thomson and Cooper who looked back to a Priestleian ideal of public science. This would accord with what we saw in the last section, where chemical didactics and the reception of Dalton's atomic theory both revealed a community in which specialist expertise was emerging slowly from the traditional forms of Enlightenment public science.

In the years after Chenevix and Thomson came to blows, the dispute they represented wound down rather rapidly. Crystallography became both more accepted and more clearly demarcated from chemical mineralogy, and potential sources of conflict were removed. Its greater acceptance followed in large part from the invention by Wollaston of a new type of goniometer in 1809. This used a reflected beam of light to measure the angles between the faces of a crystal with considerably greater accuracy than Haüy could achieve with his device. Like his slide rule for atomic weight calculations, Wollaston's reflective goniometer was an invention well suited to the general level of chemists' skills and resources. It became widely used, being praised, for example, for its convenience and accuracy in Accum's *Elements of Crystallography* (1813). In 1833, John Herschel said of Wollaston's device, "This simple, cheap, and portable little instrument, has changed the face of mineralogy, and given it all the characters of one of the exact sciences."[144]

As it gained acceptability, crystallography came to be regarded more as an ally than an enemy of chemical mineralogy. Accum apparently saw no conflict between the standpoint represented by his *Manual of Analytical Mineralogy* (1808) and that of his work on crystallography published five years later. In the *Manual* he looked to analytical chemistry to advance mineralogy to the status of a science; in the crystallography volume he proposed the study of crystalline form as a natural complement to analysis.[145] Such relaxed eclecticism became more plausible after the establishment of the doctrines of isomorphism and polymorphism by E. Mitscherlich in 1820–21. These principles explained (respectively) how different chemical substances could have the same crystalline form and

143 Berzelius (1814), p. 6.
144 Goodman (1969b), pp. 154–157 (Herschel quoted on p. 157); Accum (1813), pp. 90–95.
145 Accum [1808?], p. 4; idem., (1813), pp. 355–360. Compare also the view of W. Henry (1826), I, pp. xiv-xv.

how a single substance could have more than one form. With the acceptance of these doctrines, what had previously been contradictions between chemistry and crystallography were removed.[146]

As crystallography and mineralogy developed, therefore, chemical analysis became steadily more entrenched, assuming a secure place in the two increasingly demarcated disciplines. I have suggested that this was a reflection of the ambivalent status of analytical methods – refined enough to command the respect of trained specialists but simple enough to be communicated to wide public audiences. We can confirm this suggestion by examining the primary analytical techniques – the blowpipe and wet way (solution) analysis – in more detail.

Ever since its introduction by Cronstedt, the blowpipe had been presented as a tool for both specialists and amateurs. In the first description of its analytical uses, Gustav von Engestrom had remarked that it could be employed by gentlemen, owners of estates, or scientific mineralogists.[147] Because of its simplicity and portability, it could serve as a substitute furnace for those unable to acquire one or working in locations where one was unavailable. Better than a furnace, the blowpipe rendered the effects of heat directly visible to inspection – they occurred literally under the eye of the operator. Analysis was thus made more immediately subject to the gaze of the practitioner.[148]

The blowpipe was a key component of Cronstedt's portable (or pocket) laboratory (Figure 13). With a candle and stand, a few reagent bottles, a file, and a hammer, it was packed into a handy wooden case for use on mineralogical field trips. It continued to be used in this kind of way for the next half century, as part of the small-scale portable apparatus used to bring chemical analysis directly into the setting of mineralogical field work.[149] As such, it was as much an amateur as a specialist tool. A pocket blowpipe was one of the instruments included in Accum's closet laboratory, described in his *Chemical Amusement* (1818) (Figure 14). It was also said by a correspondent in *The Chemist* (1824–25) to be among the items in the "tea tray" laboratory of the (unnamed) "profoundest of the English chemists."[150]

By the beginning of the nineteenth century, blowpipes could be purchased from such instrument makers as Accum and John Newman of Lisle Street, Soho. Prices ranged from several shillings for the instrument alone to a couple of pounds with accessories. Potential users were guided by written instructions and catalogues of the reactions to be expected

146 Goodman (1969b), pp. 162–165.   147 Engestrom (1772), esp. pp. 310–311.
148 For remarks on this, see: Accum (1820), p. 315; Aikin (1814), p. xl.
149 Smeaton (1965–66).
150 Accum (1818), pp. xi–xii; *The Chemist*, 2 (1824–25), 47 (the chemist referred to here is perhaps Wollaston).

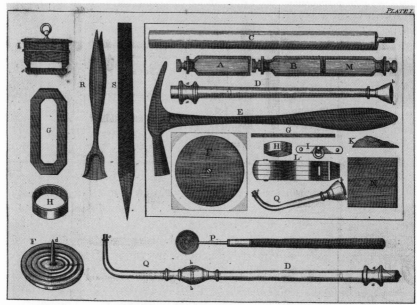

*Figure 13. Cronstedt's "portable laboratory" for examination of
minerals in the field. From A.F. Cronstedt, An Essay
Towards a System of Mineralogy, 2nd edn., ed. J.H. Magellan, 2 vols. (London: Charles Dilley, 1788), vol. I,
plate I. [Reproduced by permission of The Huntington
Library, San Marino, California.]*

with different mineral samples. Arthur Aikin, for example, in his *Manual
of Mineralogy* (1814), described the blowpipe reactions of many of the
minerals he discussed. Readers of such texts were also given advice in the
tricky matter of breathing techniques. To use the blowpipe, the operator
had to blow continuously while taking in short breaths through the nose,
a skill that Berzelius likened to rotating one's right arm and right leg
simultaneously in opposite directions.[151]

Expertise in using the blowpipe and interpreting its results appears to
have been readily acquired by many amateur investigators. Although
Berzelius suggested that some practical training was desirable, books being
in this matter (as in many others) "weak masters," his view does not
seem to have been widely shared.[152] The blowpipe proved an adaptable
and easily replicated technology, and it continued to occupy the area
common to both specialist and amateur realms. Most improvements in
the device were directed at enhancing its use in the field by relatively

151 Berzelius quoted in Oldroyd (1972), p. 222 fn.   152 Berzelius (1822), p. 3.

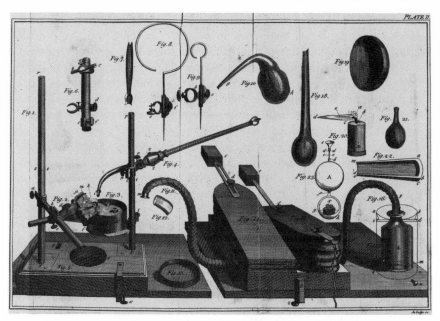

Figure 14. Blowpipe and associated apparatus for mineralogical analysis in the laboratory. From A.F. Cronstedt, An Essay Towards a System of Mineralogy, 2nd edn., ed. J.H. Magellan, 2 vols. (London: Charles Dilley, 1788), vol. I, plate II. [Reproduced by permission of The Huntington Library, San Marino, California.]

unskilled practitioners; some improvements (such as Wollaston's invention of a telescopic version that could be carried in a small pocket) further increased its portability. There was some development of the instrument for more specialist research uses, most notably by the Cambridge professor of mineralogy Edward Daniel Clarke, who used a pipe fed by a mixture of oxygen and hydrogen gases to produce unprecedentedly high flame temperatures.[153] Aikin and William Henry, however, criticized gas-fed blowpipes that were fixed in the laboratory as (in Aikin's words), "more or less cumbrous and imperfect, . . . [detracting] much from the facility with which at all times and in all places the examination of a mineral ought to be conducted."[154]

The commitment to small-scale, readily replicable, techniques also characterized the development of wet-way analytical tests. In analyzing mineral waters, tests with reagents in solution came to be seen as more reliable than exhaustive evaporation. They also had a distinct advantage

153 Oldroyd (1972).   154 Aikin (1814), p. xlii; see also W. Henry (1826), p. 12.

in reproducibility, requiring less in the way of skills and equipment from the analyst. As Henry explained in his *Epitome of Chemistry* (1801), comprehensive quantitative analysis called for a high level of expertise, but qualitative tests could be easily mastered and sufficed for most applications. Reagent tests were readily applied in mineralogical field work and were therefore routinely included in portable laboratories. They were, as Henry wrote, "of so small a bulk, as to add, in the least possible degree to the incumbrances of the traveller."[155]

Accum, like Henry, strove to communicate these techniques to public audiences; his works included didactic texts on analysis such as the *Practical Essay on Analysis of Minerals* (1804) and the *Practical Treatise on ... Chemical Tests* (3rd ed., 1820). In the *Essay* he summed up developments in analytical chemistry, portraying recent trends as directed primarily at improving the accessibility and portability of techniques:

It was once thought that, for the purpose of [analysis] . . . a place or laboratory, regularly fitted up with furnace or other apparatus was absolutely necessary. But this is by no means the case. The great improvements which have been made in analytical chemistry have superseded this necessity; . . . all the apparatus, reagents, and other articles of experiment, necessary for the analysis of minerals, may be comprised in a convenient travelling chest.[156]

In Accum's view, analytical chemistry had been advanced considerably by serving the needs of mineralogy, which was flourishing with widespread lay participation. The study of minerals embraced, he wrote, "a wide circle among the curious and wealthy classes of the community."[157] At the end of the eighteenth century, Kirwan had ascribed the significant growth of the subject, not to the enhanced expertise of specialists, but to "the evident change, I may almost say revolution, that has taken place in the public mind within these last ten years."[158] For at least the first two decades thereafter, participation in mineralogy remained broadly based. Chemical methods played their part in a practice still largely sustained by public interest – a combination of genteel curiosity about natural history and the utilitarian concerns of landowners and mining entrepreneurs. Individual specialists might use their chemical skills to advance themselves in this field, but only by conforming to norms of public instruction and civic improvement.

The activities of the British Mineralogical Society indicate this clearly. Paul Weindling has correctly pointed out how the members of the BMS used specialist techniques of chemical analysis to advance their individual interests; but it is important to note that this had to be done in ways that

155 W. Henry (1801), pp. iv, 131.
156 Accum (1804a), p. 4. See also: Accum [1808?]; idem. (1820).
157 Accum (1813), p. viii.   158 Kirwan (1794–96), I, p. vii.

conformed with a traditional rhetoric of public service.[159] In an announcement in 1804, the members declared that they "were not arrogating to themselves the entire possession of all mineralogical knowledge; but, so far as their abilities as philosophical chemists enabled them, they wished to assist both the miner and the mine-owner." They described themselves, not as specialists or professionals, but as "gentlemen attached to the studies of mineralogy and chemistry," and they explicitly disavowed motives of self-interest: "The consciousness of contributing in their sphere to the public good, and to the improvement of a favourite branch of natural science, is the sole remuneration which the Members of the Society look forward to." The BMS projected itself as dedicated to the disinterested pursuit of the public good because its plans for surveys and mineral analyses depended (as was noted) "essentially on the public concurrence." Individual interests were no doubt served, but in order for this to be possible the society had to constitute itself in accordance with traditional notions of science as a civic enterprise.[160]

By attending to this social dimension of mineralogical practice, we can enhance our understanding of the adoption of chemical analytical methods. To invoke a teleological trend toward a more scientific approach, on the other hand, explains very little. As the controversy between Wernerians and crystallographers shows, the decision to seek a scientific method left many options open. Competing models of scientific practice were possible, encompassing alternative ways of organizing the community of practitioners. Chemical analysis emerged as common ground in this dispute because it could be reconciled with both models of scientific practice. The techniques of the blowpipe and the wet-way reagent tests answered to the need for specialist expertise and for public participation in science. They were developed and refined with constant attention to the circumstances of nonspecialist users and were therefore kept cheap, easily mastered, and mobile. It was for this reason that the instruments of chemical analysis achieved their crucial role in early nineteenth-century mineralogy, an activity poised between the realms of specialist expertise and public science.

Conclusion: Discipline-formation and public science

Has any man, or any society of men, a truth to speak, a piece of spiritual work to do; they can nowise proceed at once and with the mere natural organs, but must first call a public meeting, appoint committees, issue prospectuses, eat a

159 Weindling (1983).
160 Announcements of BMS activities in *Philosophical Magazine, 3* (1799), 318; *6* (1800), 369–372; *12* (1802–1803), 284–287; *19* (1804), 85–90.

public dinner; in a word, construct or borrow machinery, wherewith to speak it and do it. . . . Philosophy, Science, Art, Literature, all depend on machinery. No Newton, by silent meditation, now discovers the system of the world from the falling of an apple; but some quite other than Newton stands in his Museum, his Scientific Institution, and behind whole batteries of retorts, digesters and galvanic piles imperatively "interrogates Nature", – who, however, shows no haste to answer.

Thomas Carlyle, "Signs of the Times" (1829)[161]

Historians have conventionally identified the end of the eighteenth and the beginning of the nineteenth centuries as the time when chemistry acquired the status of a discipline. Lavoisier's achievement is seen as having led to the reorganization of the subject and its establishment on a recognizably scientific basis. In line with the traditional idealist emphasis of the history of science, this is frequently regarded as primarily an achievement of theory – the consequence of correct ideas grasped by a great mind.

Recent studies (many of them cited in the preceding pages) have sought to undermine this perspective, by shifting attention from individuals to communities and from ideas to practice. Disciplinary structures are now perceived as products of the activities of many individuals, situated in defined historical contexts. They are seen as embodied in concrete techniques, specific apparatus, and particular forms of discourse. To use a term of Ludwig Wittgenstein's now gaining increasing currency, scientific disciplines are "forms of life."

This book aims to further this understanding of scientific disciplines. The emphasis on chemistry in Britain is intended to bring into focus the picture of science as a set of practices in a specific social setting. We can now see that the discoveries of Priestley, Lavoisier, Davy, and others became the basis for a new discipline to the extent that they became embodied in instruments, techniques, and modes of discourse, which gained acceptance among a community of practitioners. The formation of a discipline was thus the crystallization of a community around particular forms of practical activity conducted in particular material settings. The social structure of a discipline and its way of pursuing its craft were achieved simultaneously.

One way to understand the changes in chemical practice over the period covered by this book is to think in terms of a contrast between "experience" and "expertise." In the last two chapters we have considered how Davy enforced a definition of his own expertise in relation to the voltaic pile and how a community of electrochemists consolidated around recognition of his discoveries with this instrument. We have also looked at the development of specialist skills in mineral analysis among

161 Carlyle (1971), pp. 65–66.

a wider population of chemists, and the ways in which this complemented evolution of the social and rhetorical dimensions of a specialist community. The quotation from Thomas Carlyle at the beginning of this section casts light on how this process was perceived by a skeptical onlooker. Science now seemed less a matter of individual genius than of material and institutional machinery. Apparatus, research institutions, and bureaucracy appear to Carlyle as the features of contemporary science, an enterprise that tries to elevate mediocrities to the rank of philosophers while claiming continuity with the great Newton.

Carlyle's sardonic observations perhaps point to the gap between expertise, as it was being constructed in early nineteenth-century chemistry, and earlier Enlightenment notions of experience. The distinction is one that brings the issue of the public audience for science to bear on questions of discipline formation.[162] Specialist expertise requires exclusive knowledge and discourse, intensive practical training, elaborate and expensive apparatus. These requirements are in evident conflict with the values of science as a public or civic enterprise, as it had been envisaged by many eighteenth-century writers. From their perspective, experience was prized, with its connotations of accumulating factual information and the individual autonomy necessary for this. "Gentlemanlike facts" were the stuff of Enlightenment experimental philosophy, and the demands of civic gentility set strict curbs on the development of specialization or professionalism.

The contrast between Priestley and Davy encapsulates this distinction. Many aspects of Davy's work were definitive of expertise and specialist skill in his time, and he did not shrink from the labor necessary to enforce recognition of his achievement on recalcitrant individuals. At the same time, Davy stripped public science of the moral and political purpose that Priestley had given it. His presentations at the RI used a rhetoric that was widely imitated elsewhere, and that described experimental science as a genteel, theologically safe, and socially conservative activity. Davy situated his public audience in a very different position from that assigned to it by Priestley. It was to be more passive and participate less; its role was to admire and subscribe in support of scientific work, but not to seek moral enhancement by active involvement in experimentation. It was in this role that public audiences remained very important for the developing discipline of chemistry in the early nineteenth century.

By viewing this situation in the perspective of the preceding Enlightenment, we can grasp the transformation inherent in the formation of a new discipline without the dangers of implicit teleology. After the 1820s, on the other hand, we enter waters more charted by historians. The pro-

162 The connection has been indicated recently by Shapin (1990).

fessionalization of nineteenth-century chemistry has been much discussed. But in these accounts the previous century appears only as a dark hinterland from which a few pioneering individuals emerge to lead to subsequent developments. This book has attempted to do more justice to Enlightenment science in its own terms, to show at least in the case of chemistry that it was a coherent enterprise with its own social rationale. Taking eighteenth-century science more seriously can also teach us something about the emergence of chemistry in a new form in the nineteenth century. It helps us to see this as a truly historical development, rather than as the maturation of an already latent discipline.

One consequence is that we can perceive the strong legacy of Enlightenment notions of science as an activity in the public realm. Continuing insistence on accessibility and simplicity of methods was a sign of this. Notwithstanding the great achievements of Lavoisier and Davy, which relied upon stringent experimental standards and instrumental discipline, many chemists remained committed to democratic diffusion of skills and techniques to all potential practitioners. Thus, although the new techniques of chemistry were indeed taken up and were vital to the reshaping of the subject, the commitment to education and public science also remained important.

This conjuncture of new methods and a continuing commitment to public instruction explains why the first two decades of the nineteenth century were recognized by contemporaries as the heyday of chemistry. A series of startling discoveries was made, the rhetorical unity of the discipline was reconstructed in many textbooks and educational programs, and public audiences were recruited in unprecedented numbers for its support. Although subsequently the attention of the lay public wandered to other subjects, and some gentlemanly specialists turned their attention to such new sciences as geology, chemistry thereafter retained a permanent place in scientific education and in the institutions of the scientific community.

That this was seen at the time as the legacy of a distinctive British experience of the Enlightenment was suggested by Richard Chenevix. Surveying recent French science in the *Edinburgh Review* in 1820, with what seems to have been a personal as well as a chauvinistic animus, Chenevix found much to criticize. In chemistry especially, the French were charged with appropriating the discoveries of others and celebrating them as their own. Lavoisier and his colleagues, in a "spirit of monopoly," were said to have seized the "opportunity of glory" by plagiarizing the work of Black, Priestley, and Cavendish. French perfidy in this respect was traced by Chenevix to excessive centralization of the scientific community and the "eclat which surrounds a small number of individuals." The dazzling splendor with which these few were garbed had obscured the fact, in Chenevix's view, that the French nation as a whole was less well educated in the sciences than the British. In the Paris Academy, "nursed in the hotbeds of despotic vanity," all was "show," but in the country as a whole people remained comparatively ignorant. Britain, on the other hand, had the benefits of the Royal Society, which Chenevix viewed as free and open to all independent gentlemen, and a large number of local voluntary societies. These were groups that "grew, uncontrolled, out of the acquirements of a free and enlightened people." The nature of experimental philosophy in Britain had benefited accordingly:

At least, whatever be his other learned pursuits, the moral and political sciences, the public affairs of his country, fill some portion of the daily thoughts of an Englishman; and, however scientific an assembly of French philosophers may be, an assembly of British philosophers is much more generally enlightened.[163]

For Chenevix, even as late as 1820, the structures of enlightened public science still provided the envisaged framework for the nation's scientific efforts. Though he was not slow to recognize advances in instrumentation and expertise (we have seen that he was a leading exponent of crystallography), he looked with nostalgic favor upon the traditional conception of science as part of a process of civic enlightenment. The ideal of scientific practice as a voluntary activity, pursued by disinterested individuals rather than professional experts, remained alluring. Such an ideal, though undermined by dramatic improvements in expert skills and resources, remained central to the self-image of the British scientific community in this period.

163 [Chenevix] (1820), pp. 395, 402–406, 409–411, 415. For attribution of authorship, see Houghton (1966–87), I, p. 461.

Bibliography

Manuscripts

Glasgow University Library, Department of Special Collections: MSS of William Cullen.
Glasgow University Library, Department of Special Collections: Letters of William Cullen.
Huntington Library, San Marino, California: Letters of Joseph Black.
Huntington Library: "Notes taken at Dr. Higgins's Course of Lectures on Chemistry" (1780, compiled by Devereaux Mytton) (3 vols., call no. HM 46324).
Leeds University Library, Special Collections: William Hey, "Medical and Surgical Cases" (12 vols., MS. 628).
John Rylands University Library of Manchester: "Lectures on Chemistry Delivered 1762 and 1763 by Dr. William Cullen" (MSS CH C.121) (4 vols.).
Royal College of Physicians, London: George Fordyce, "Lectures on Chemistry" (1786) (MSS 146, 147, 148) (3 vols.).
Royal College of Physicians of Edinburgh: MSS of William Cullen.
Royal Institution of Great Britain, London: MSS of Sir Humphry Davy. (Quoted by permission of the Royal Institution.)
Royal Institution of Great Britain, London: MSS of John Davy. (Quoted by permission of the Royal Institution.)
Wellcome Institute for the History of Medicine, London: "Chemical Lectures by William Cullen, M.D." (MS 1918).

Primary source journals

Annales de Chimie (Paris).
Annals of Philosophy, ed. Thomas Thomson, 1813–21; ed. R. Phillips, 1821–26 (London).
The Chemist, 2 vols.: 1 (13 Mar–18 Sep 1824); 2 (25 Sep 1824–16 Apr 1825) (London: Knight and Lacey).
Essays and Observations, Physical and Literary, 3 vols.: 1754, 1755 and 1771 (Edinburgh: G. Hamilton and J. Balfour).
Journal of Science and the Arts, ed. W.T. Brande, 1816–31 (London: The Royal Institution).
The Medical Spectator, 2 vols., 1792–1793 (London: J. Nichols).
Mémoires de Physique et de Chimie de la Société d'Arcueil.
Memoirs and Proceedings of the Manchester Literary and Philosophical Society (Manchester).
Nicholson's Journal (Journal of Natural Philosophy, Chemistry and the Arts, ed. William Nicholson; first series 1797–1802; second series 1802–13) (London).

Observations sur la Physique (Paris).
Philosophical Magazine, ed. Alexander Tilloch, from 1798 (London).
Philosophical Transactions, from 1665 (London: The Royal Society of London).
Transactions of the Royal Society of Edinburgh (Edinburgh).

Newspapers

Anti-Jacobin Review and Magazine
Aris's Birmingham Gazette
Edinburgh Review
European Magazine and London Review
Gentleman's Magazine
Manchester Mercury
Monthly Magazine
Monthly Review (The Monthly Review, Or Literary Journal)
Quarterly Review
York Courant

Books and articles

Accum, Friedrich Christian (1804a). *A Practical Essay on the Analysis of Minerals*. London: for the author.
  (1804b). Chemical demonstrations or private lectures on practical chemistry. In Accum (1804a), pp. i-ix.
  [1808?]. *A Manual of Analytical Mineralogy,* 2nd edn. (of 1804a), 2 vols. London: for the author.
  (1810). *Manual of a Course of Lectures on Experimental Chemistry and Mineralogy*. London: G. Hayden.
  (1813). *Elements of Crystallography After the Method of Haüy*. London: Longman et al.
  (1818). *Chemical Amusement, Comprising A Series of Curious and Instructive Experiments in Chemistry,* 2nd edn. London: Thomas Boys.
  (1820). *A Practical Treatise on the Use and Application of Chemical Tests,* 3rd edn. London: Thomas Boys.
Adair, James Makittrick (1786). *Medical Cautions for the Consideration of Invalids, Those Especially who Resort to Bath*. Bath: R. Cruttwell.
Aikin, Arthur (1814). *A Manual of Mineralogy*. London: Richard and Arthur Taylor.
Albury, William Randall (1972). *The Logic of Condillac and the Structure of French Chemical and Biological Theory, 1780–1801*. Johns Hopkins University Ph.D. dissertation.
Allan, D.G.C. (1973–74). The Society of Arts and government, 1754–1800: Public encouragement of arts, manufactures, and commerce in eighteenth-century England. *Eighteenth-century Studies, 7,* 434–452.
  (1979a). *The Society for the Encouragement of Arts, Manufactures and Commerce: Organisation, Membership and Objectives in the First Three Decades (1755–84)*. University of London Ph.D. dissertation.
  (1979b). *William Shipley, Founder of the Royal Society of Arts: A Biography with Documents,* 2nd edn. London: Scolar Press.
Alston, R.C., Robinson, F.J.G. and Wadham, C. (1983). *Eighteenth-century Subscription Lists: A Check-list*. Newcastle-upon-Tyne: Avero.
Anderson, R.G.W. (1978). *The Playfair Collection and the Teaching of Chemis-*

try at the University of Edinburgh, 1713–1858. Edinburgh: Royal Scottish Museum.

Anderson, R.G.W. and Simpson, A.D.C., eds. (1976). *The Early Years of the Edinburgh Medical School.* Edinburgh: Royal Scottish Museum.

Anderson, R.G.W. and Lawrence, Christopher, eds. (1987). *Science, Medicine and Dissent: Joseph Priestley (1733–1804).* London: Wellcome Trust and The Science Museum.

Anderson, Wilda (1984). *Between the Library and the Laboratory: The Language of Chemistry in Eighteenth-Century France.* Baltimore: Johns Hopkins University Press.

[Anon.] (1794). *The Golden Age: A Poetical Epistle from Erasmus D----n, M.D. to Thomas Beddoes.* London: F. and C. Rivington.

[1796]. *Essays by a Society of Gentlemen at Exeter.* Exeter: Trewman & Son.

(1797). *Critical Examination of the First Part of Lavoisier's Elements of Chemistry.* London: J. Wright.

[1798]. *British Public Characters of 1798.* London: R. Phillips.

(1800). *The Sceptic.* Retford: E. Peart.

(1818). *A Treatise on Soils and Manures, as Founded on Actual Experience ... in which the Theory and Doctrines of Sir Humphry Davy ... are Rendered Familiar to the Experienced Farmer.* London: T. Cadell and W. Davies.

Arden, James (1774). *Analysis of Mr. [John] Arden's Course of Lectures on Natural and Experimental Philosophy,* [1st edn.] N.p.: for the author.

(1782). Ibid., 2nd edn. N.p.: for the author.

Arden, John (1773). *A Short Account of a Course of Natural and Experimental Philosophy.* Birmingham: M. Swinney.

Averley, Gwen (1986). The "social chemists": English chemical societies in the eighteenth and early nineteenth century. *Ambix, 33,* 99–128.

Babington, William and Allen, William (1802). *A Syllabus of A Course of Chemical Lectures read at Guy's Hospital.* London: W. Phillips.

Babington, William, Marcet, Alexander and Allen, William (1811). *A Syllabus of a Course of Lectures read at Guy's Hospital.* London: Royal Free School Press.

Bachelard, Gaston (1980). *La Formation de l'Esprit Scientifique: Contribution à une Psychanalyse de la Connaissance Objective,* 11th edn. Paris: J. Vrin.

Badash, Lawrence (1964). Joseph Priestley's apparatus for pneumatic chemistry. *Journal of the History of Medicine, 19,* 139–155.

Baker, Keith Michael (1981). Enlightenment and revolution in France: Old problems, renewed approaches. *Journal of Modern History, 53,* 281–303.

Banks, John (1775). *An Epitome of a Course of Lectures on Natural and Experimental Philosophy.* Kendal: W. Pennington.

(1789). Ibid. Kendal: W. Pennington.

Bazerman, Charles (1988). *Shaping Written Knowledge: The Genre and Activity of the Experimental Article in Science.* Madison and London: University of Wisconsin Press.

Beddoes, Thomas (1793a). *Observations on the Nature of Demonstrative Evidence . . . and Reflections on Language.* London: J. Johnson.

(1793b). *Observations on the Nature and Cure of Calculus, Sea Scurvy, Consumption, Catarrh and Fever.* London: J. Murray.

[1794a]. *Letters from Dr. Withering of Birmingham, Dr Ewart of Bath, Dr Thornton of London . . .* Bristol: Bulgin and Rosser.

[1794b]. *A Letter to Erasmus Darwin, M.D. on A New Method of Treating Pulmonary Consumption.* Bristol: Bulgin and Rosser.

[1794c]. *A Word in Defence of the Bill of Rights, Against Gagging Bills.* Bristol: N. Biggs.

[1795]. *Where Would be the Harm of a Speedy Peace?* Bristol: N. Biggs.

(1796). *Essay on the Public Merits of Mr. Pitt.* London: J. Johnson.

(1797). *A Lecture Introductury to a Course of Popular Instruction on the Constitution and Management of the Human Body.* Bristol: N. Biggs.

(1799). *Notice of Some Observations made at the Medical Pneumatic Institution.* Bristol: Biggs and Cottle.

(1808). *A Letter to the Right Honourable Sir Joseph Banks . . . On the Causes and Removal of the Prevailing Discontents . . . in Medicine.* London: Richard Phillips.

Beddoes, Thomas and Watt, James (1794–96). *Considerations on the Medicinal Use of Factitious Airs.* (Parts I and II [1794]; part III (1795); *Medical Cases and Speculations, including Parts IV and V of Considerations on . . . Factitious Airs* (1796).) Bristol: Bulgin and Rosser.

(1796). Ibid., parts I and II, 3rd edn. Bristol: Bulgin and Rosser.

Belanger, Terry (1982). Publishers and writers in eighteenth-century England. In Rivers, ed. (1982), pp. 5–25.

Bennet, Abraham (1789). *New Experiments on Electricity.* Derby: John Drewry.

Bensaude-Vincent, Bernadette (1990). A view of the Chemical Revolution through contemporary textbooks: Lavoisier, Fourcroy and Chaptal. *British Journal for the History of Science, 23,* 435–460.

Bergman, Torbern Olaf (1783a). *An Essay on the Usefulness of Chemistry and its Application to the Various Occasions of Life.* London: J. Murray.

(1783b). *Outlines of Mineralogy,* trans. William Withering. Birmingham: Piercy and Jones.

(1784). *Physical and Chemical Essays,* trans. Edmund Cullen, 3 vols. London: J. Murray.

(1785). *A Dissertation on Elective Attractions,* [trans. Thomas Beddoes]. London: J. Murray; reprinted with intro. by A.M. Duncan. London: Frank Cass & Co., 1970.

Berman, Morris (1975). "Hegemony" and the amateur tradition in British science. *Journal of Social History, 8,* 30–50.

(1978). *Social Change and Scientific Organization: The Royal Institution, 1799–1844.* Ithaca, N.Y.: Cornell University Press.

Berthollet, Claude-Louis (1809). Observations sur les proportions des éléments de quelques combinaisons. *Mémoires de Physique et de Chimie de la Société d'Arcueil, 2,* 42–67.

Berzelius, Jöns Jakob (1814). *An Attempt to Establish a Pure Scientific System of Mineralogy,* trans. John Black. London: Robert Baldwin.

(1822). *The Use of the Blowpipe in Chemical Analysis and in the Examination of Minerals,* trans. J.G. Children. London: Baldwin, Cradock and Joy.

Betham-Edwards, M., ed. (1898). *The Autobiography of Arthur Young.* London: Smith, Elder & Co.

Black, Joseph (1756). Experiments on *magnesia alba,* quicklime, and some other alcaline substances. *Essays and Observations, Physical and Literary, 2,* 157–225.

(1803). *Lectures on the Elements of Chemistry, Delivered in the University of Edinburgh . . . now Published from his Manuscripts,* ed. John Robison, 2 vols. Edinburgh: Mundell and Sons.

(1966). *Notes from Doctor Black's Lectures on Chemistry 1767/8,* compiled by Thomas Cochrane, ed. Douglas McKie. Wilmslow, Cheshire: Imperial Chemical Industries.

Blakemore, Steven (1988). *Burke and the Fall of Language: The French Revolution as Linguistic Event*. Hanover, N.H.: University Press of New England.

Bolton, Henry Carrington, ed., (1892). *Scientific Correspondence of Joseph Priestley*. New York: privately printed; reprinted New York: Kraus Reprint Co., 1969.

Borsay, Peter (1977). The English urban renaissance: The development of provincial urban culture about 1680–1760. *Social History, 5*, 581–603.

Bostock, John (1812). Experiment to prove whether water be produced in the combination of muriatic acid gas and ammoniacal gas. *Nicholson's Journal*, 2nd ser., 32, 18–21.

[   ] (1813). Review of Davy's *Elements of Chemical Philosophy*. *Monthly Review*, 2nd ser., 72, 148–158.

(1818). *An Account of the History and Present State of Galvanism*. London: Baldwin, Cradock and Joy.

Boswell, James (1953). *Life of Johnson*. London: Oxford University Press.

Boulton, James T. (1963). *The Language of Politics in the Age of Wilkes and Burke*. London: Routledge and Kegan Paul.

Brande, William Thomas (1819a). *A Manual of Chemistry*. London: John Murray.

(1819b). *An Introductory Discourse, Delivered in the Amphitheatre of the London Institution*. London: Richard and Arthur Taylor.

Brewer, John (1982). Commercialization of politics. In McKendrick, Brewer and Plumb (1982), pp. 197–262.

British Museum (1978). *British Museum Department of Prints and Drawings, Catalogue of Personal and Political Satires*, vol. VI, ed. M. Dorothy George. London: British Museum Publications.

Brock, William H. (1967). The London Chemical Society, 1824. *Ambix, 14*, 133–139.

(1985). *From Protyle to Proton: William Prout and the Nature of Matter, 1785–1985*. Bristol: Adam Hilger.

Brook, M. (1955). Dr. Warwick's chemistry lectures and the science audience in Sheffield 1799–1801. *Annals of Science, 11*, 224–237.

Brooke, John H. (1984). "A sower went forth": Joseph Priestley and the ministry of reform. In Royal Society of Chemistry (1984), pp. 432–460.

Brougham, Henry (1807–1808). Review of Davy's 1806 Bakerian Lecture. *Edinburgh Review, 11*, 390–398.

(1808). Review of Davy's 1807 Bakerian Lecture. *Edinburgh Review, 12*, 394–401.

(1810–11). Review of Davy's paper "Researches on oxymuriatic acid." *Edinburgh Review, 17*, 402–409.

(1811). Review of Davy's paper "On some combinations of oxymuriatic acid." *Edinburgh Review, 18*, 470–480.

(1825). *Practical Observations upon the Education of the People, Addressed to the Working Classes and their Employers*. London: Richard Taylor.

(1872). *The Works of Henry Lord Brougham*, vol. I, *Lives of the Philosophers of the Time of George III*. Edinburgh: Adam and Charles Black.

Brown, John (1795). *The Elements of Medicine of John Brown, M.D.*, ed. Thomas Beddoes, 2 vols. London: J. Johnson.

Brown, Martin (1987). Conspiracy and Dissent: John Robison, Joseph Priestley, and the political crisis of the 1790s. Cambridge University, Department of History, unpublished dissertation.

Brown, Sanborn C. (1979). *Benjamin Thompson, Count Rumford*. Cambridge: MIT Press.

Brownrigg, William (1765). Extract from an essay, entituled, on the uses of a knowledge of mineral exhalations when applied to discover the principles and properties of mineral waters, . . . read before the Royal Society in April 1741. *Philosophical Transactions, 55,* 236–243.

Bud, Robert and Roberts, Gerrylynn K. (1984). *Science versus Practice: Chemistry in Victorian Britain.* Manchester: Manchester University Press.

Burke, Edmund (1796). *A Letter from the Right Honourable Edmund Burke to a Noble Lord.* London: n.p.

(1968). *Reflections on the Revolution in France and on the Proceedings in Certain Societies in London Relative to that Event,* ed. Conor Cruise O'Brien. Harmondsworth: Penguin Books.

Butler, Marilyn (1984). *Burke, Paine, Godwin and the Revolution Controversy (Cambridge English Prose Texts).* Cambridge: Cambridge University Press.

Bynum, W.F. and Porter, Roy, eds. (1985). *William Hunter and the Eighteenth-Century Medical World.* Cambridge: Cambridge University Press.

Camic, Charles (1983). *Experience and Enlightenment: Socialization for Cultural Change in Eighteenth-Century Scotland.* Edinburgh: Edinburgh University Press.

Campbell, R.H. and Skinner, A.S., eds. (1982). *The Origins and Nature of the Scottish Enlightenment.* Edinburgh: John Donald.

Cant, R.G. (1982). The origins of the Enlightenment in Scotland: The universities. In Campbell and Skinner, eds. (1982), pp. 42–64.

Cantor, Geoffrey, 1989. The rhetoric of experiment. In Gooding, Pinch and Schaffer, eds. (1989), pp. 159–180.

Cantor, G.N. and Hodge, M.J.S., eds. (1981). *Conceptions of Ether: Studies in the History of Ether Theories, 1740–1900.* Cambridge: Cambridge University Press.

Cardwell, D.S.L., ed. (1968). *John Dalton and the Progress of Science.* Manchester: Manchester University Press.

Carlyle, Thomas (1971). *Selected Writings,* ed. Alan Shelston. Harmondsworth: Penguin Books.

Carpenter, Kenneth J. (1986). *The History of Scurvy and Vitamin C.* Cambridge: Cambridge University Press.

Carter, Harold B. (1988). *Sir Joseph Banks, 1743–1820.* London: British Museum (Natural History).

Cartwright, F.F. (1952). *The English Pioneers of Anaesthesia (Beddoes, Davy, Hickman).* Bristol: John Wright & Sons.

(1967). The association of Thomas Beddoes, M.D. with James Watt, F.R.S. *Notes and Records of the Royal Society of London, 22,* 131–143.

Cavallo, Tiberius (1781). *A Treatise on the Nature and Properties of Air.* London: for the author.

(1798). *An Essay on the Medicinal Properties of Factitious Airs.* London: for the author.

Cavendish, Henry (1767). Experiments on Rathbone-Place water. *Philosophical Transactions, 57,* 91–108.

(1783). An account of a new eudiometer. *Philosophical Transactions, 73,* 106–135.

(1784). Experiments on air. *Philosophical Transactions, 74,* 119–153.

(1921). *The Scientific Papers of the Honourable Henry Cavendish, F.R.S.,* vol. II, *Chemical and Dynamical,* ed. Edward Thorpe. Cambridge: Cambridge University Press.

Chard, Leslie F. (1975). Joseph Johnson: Father of the book trade. *Bulletin of the New York Public Library, 79,* 51–82.

(1977). Bookseller to publisher: Joseph Johnson and the English book trade, 1760 to 1810. *The Library*, 5th ser., *32*, 138–154.

Chenevix, Richard (1811). *Observations on Mineralogical Systems*. London: J. Johnson.

[ ] (1820). Review of Petit Radel, *Recherches sur les Bibliothèques Anciennes et Modernes*. *Edinburgh Review*, *34* (1820), 383–422.

Children, John George (1809). An account of some experiments performed with a view to ascertain the most advantageous method of constructing a voltaic apparatus. *Philosophical Transactions*, *99*, 32–38.

(1815). An account of some experiments with a large voltaic battery. *Philosophical Transactions*, *105*, 363–374.

Chilton, Donovan and Coley, Noel G. (1980). The laboratories of the Royal Institution in the nineteenth century. *Ambix*, *27*, 173–203.

Chitnis, Anand C. (1976). *The Scottish Enlightenment: A Social History*. London: Croom Helm.

(1986). *The Scottish Enlightenment and Early Victorian English Society*. London: Croom Helm.

Christie, John R.R. (1974). The origins and development of the Scottish scientific community, 1680–1760. *History of Science*, *12*, 122–141.

(1975). The rise and fall of Scottish science. In *The Emergence of Science in Western Europe*, ed. M. Crosland, pp. 111–126. London: Macmillan.

(1979). The Chemical Revolution in Scotland. Paper given at meeting of the British Society for the History of Science, on "Problems and perspectives in the history of chemistry," Leicester, 7 April 1979.

(1981). Ether and the science of chemistry, 1740–1790. In Cantor and Hodge, eds. (1981), pp. 85–110.

(1982). Joseph Black and John Robison. In *Joseph Black 1728–1799: A Commemorative Symposium*, ed. A.D.C. Simpson, pp. 47–52. Edinburgh: Royal Scottish Museum.

Christie, John and Golinski, J.V. (1982). The spreading of the word: New directions in the historiography of chemistry, 1600–1800. *History of Science*, *20*, 235–266.

Clow, Archibald and Clow, Nan L. (1952). *The Chemical Revolution: A Contribution to Social Technology*. London: Batchworth Press.

Cobbett, William (1801). Observations on Priestley's emigration. In Cobbett, *Porcupine's Works*, 12 vols., I, 147–215. London: Cobbett and Morgan.

Cochrane, Archibald, Earl of Dundonald (1795). *A Treatise Shewing the Intimate Connection that Subsists between Agriculture and Chemistry*. London: for the author.

Cole, R.J. (1951). Friedrich Accum (1769–1838): A biographical study. *Annals of Science*, *7*, 128–143.

Coleby, L.J.M. (1952a). John Francis Vigani, first professor of chemistry in the University of Cambridge. *Annals of Science*, *8*, 46–60.

(1952b). John Mickleburgh, professor of chemistry in the University of Cambridge, 1718–56. *Annals of Science*, *8*, 165–174.

(1952c). John Hadley, fourth professor of chemistry in the University of Cambridge. *Annals of Science*, *8*, 293–301.

(1953). Richard Watson, professor of chemistry in the University of Cambridge, 1764–71. *Annals of Science*, *9*, 101–123.

Coley, Noel G. (1967). The Animal Chemistry Club: Assistant society to the Royal Society. *Notes and Records of the Royal Society*, *22*, 173–185.

(1968). Alexander Marcet (1770–1822), physician and animal chemist. *Medical History*, *12*, 394–402.

296   BIBLIOGRAPHY

(1969). John Warltire, 1738/9–1810: Itinerant lecturer and chemist. *West Midlands Studies, 3*, 31–44.
(1982). Physicians and the chemical analysis of mineral waters in 18th-century England. *Medical History, 26*, 123–144.
(1984). The preparation and uses of artificial mineral waters (ca. 1680–1825). *Ambix, 21*, 32–48.
Collins, H.M. (1974). The TEA set: Tacit knowledge and scientific networks. *Science Studies, 4*, 165–186.
(1985). *Changing Order: Replication and Induction in Scientific Practice.* London and Beverly Hills: Sage Publications.
(1987). Certainty and the public understanding of science: Science on television. *Social Studies of Science, 17*, 689–713.
(1988). Public experiments and displays of virtuosity: The core-set revisited. *Social Studies of Science, 18*, 725–748.
Cooper, Thomas (1792). *A Reply to Mr. Burke's Invective Against Mr. Cooper, and Mr. Watt.* London: J. Johnson.
(1812). *The Introductory Lecture of Thomas Cooper, Esq., Professor of Chemistry at Carlisle College, Pennsylvania.* Carlisle, Penn.: Archibald Loudon.
Cooter, Roger (1984). *The Cultural Meaning of Popular Science: Phrenology and the Organisation of Consent in Nineteenth-century Britain.* Cambridge: Cambridge University Press.
Corbin, Alain (1986). *The Foul and the Fragrant: Odor and the French Social Imagination.* Leamington Spa: Berg.
Corry, John (1804). *The Life of Joseph Priestley, LL.D., F.R.S.* Birmingham: Wilks Grafton.
Crane, V.W. (1966). The Club of Honest Whigs: Friends of science and liberty. *William and Mary Quarterly, 23*, 210–233.
Crawford, Adair (1779). *Experiments and Observations on Animal Heat and the Inflammation of Combustible Bodies.* London: Murray and Sewell.
Crellin, J.K. (1974). Chemistry and eighteenth-century British medical education. *Clio Medica, 9*, 9–21.
Cronstedt, Axel Fredric (1788). *An Essay Towards a System of Mineralogy . . . and an Additional Treatise on the Blow-Pipe, by Gustav von Engestrom,* 2nd edn., ed. J.H. Magellan, 2 vols. London: Charles Dilley.
Crook, Ronald E. (1966). *A Bibliography of Joseph Priestley 1733–1804.* London: The Library Association.
Crosland, Maurice P. (1959). The use of diagrams as chemical "equations" in the lecture notes of William Cullen and Joseph Black. *Annals of Science, 15*, 75–90.
(1962). *Historical Studies in the Language of Chemistry.* London: Heinemann.
(1973). Lavoisier's theory of acidity. *Isis, 64*, 306–325.
(1978). *Gay-Lussac: Scientist and Bourgeois.* Cambridge: Cambridge University Press.
(1980). Davy and Gay-Lussac: Competition and contrast. In Forgan, ed. (1980), pp. 95–120.
(1983). Priestley Memorial Lecture: A practical perspective on Joseph Priestley as a natural philosopher. *British Journal for the History of Science, 16*, 223–238.
(1987). The image of science as a threat: Burke versus Priestley and the "Philosophical Revolution". *British Journal for the History of Science, 20*, 277–307.

Cruickshank, William (1800–1801). Some experiments and observations on galvanic electricity. *Nicholson's Journal*, 1st ser., 4, 187–191, 254–264.

(1801a). Some observations on different hydrocarbonates and combinations of carbone with oxygen . . . in reply to some of Dr. Priestley's late objections to the new system of chemistry. *Nicholson's Journal*, 1st ser., 5, 1–9.

(1801b). Some additional observations on hydrocarbonates, and the gaseous oxide of carbon. *Nicholson's Journal*, 1st ser., 5, 201–211.

Cullen, William (1755). Of the cold produced by evaporating fluids, and of some other means of producing cold. *Essays and Observations, Physical and Literary*, 2, 145–175.

(1796). The substance of nine lectures on vegetation and agriculture, delivered to a private audience in the year 1768. In *Additional Appendix to the Outlines of the Fifteenth Chapter of the Proposed General Report from the Board of Agriculture, on the Subject of Manures*, ed. George Pearson. London: W. Bulmer & Co.

Cunningham, Andrew and Jardine, Nicholas, eds. (1990). *Romanticism and the Sciences*. Cambridge: Cambridge University Press.

Daiches, David, Jones, P. and Jones, J., eds. (1986). *A Hotbed of Genius: The Scottish Enlightenment 1730–1790*. Edinburgh: Edinburgh University Press.

Dalton, John (1808–27). *A New System of Chemical Philosophy*, vol. I part i (1808), vol. I part ii (1810), vol. II part i (1827). Manchester: Russell, Bickerstaff and Wilson.

(1830). Chemical observations on certain atomic weights as adopted by different authors. In Thackray (1972), pp. 90–103.

Darnton, Robert (1970). *Mesmerism and the End of the Enlightenment in France*. New York: Schocken Books.

(1971). In search of the Enlightenment: Recent attempts to create a social history of ideas. *Journal of Modern History*, 43, 113–132.

(1979). *The Business of Enlightenment: A Publishing History of the Encyclopédie 1775–1800*. Cambridge, Mass.: Belknap Press of Harvard University Press.

Daumas, Maurice (1955). *Lavoisier: Théoricien et Expérimentateur*. Paris: Presses Universitaires de France.

Daumas, Maurice and Duveen, Denis (1959). Lavoisier's relatively unknown large-scale decomposition and synthesis of water, February 27 and 28, 1785. *Chymia*, 5, 113–129.

Davy, Humphry (1805). On the analysis of soils. *Nicholson's Journal*, 2nd ser., 12, 81–97.

(1812). *Elements of Chemical Philosophy*, part I, vol. I. London: J. Johnson.

(1813). *Elements of Agricultural Chemistry, in a Course of Lectures for the Board of Agriculture*. London: W. Bulmer for Longman, et al.

(1814a). Some experiments and observations on a new substance which becomes a violet coloured gas by heat. *Philosophical Transactions*, 104, 74–93.

(1814b). An account of some new experiments on the fluoric compounds. *Philosophical Transactions*, 104, 62–73.

(1814c). Further experiments and observations on iodine. *Philosophical Transactions*, 104, 487–507.

(1815). Some experiments on a solid compound of iodine and oxygene, and on its chemical agencies. *Philosophical Transactions*, 105, 203–213.

(1827). *Six Discourses Delivered before the Royal Society*. London: John Murray.

(1839–40). *The Collected Works of Sir Humphry Davy*, ed. John Davy, 9 vols. London: Smith, Elder.

Davy, John (1811a). Some remarks on the observations and experiments of Mr. Murray. *Nicholson's Journal*, 2nd ser., *28*, 193–206.

(1811b). An answer to Mr. Murray's observations on the nature of potassium and sodium. *Nicholson's Journal*, 2nd ser., *29*, 35–39.

(1811c). On the nature of oximuriatic gas. *Nicholson's Journal*, 2nd ser., *29*, 39–44.

(1811d). An account of a new gas, with a reply to Mr. Murray's last observations. *Nicholson's Journal*, 2nd ser., *30*, 28–33.

(1812). On the nature of oximuriatic acid. *Nicholson's Journal*, 2nd ser., *31*, 310–316.

(1813). An account of an experiment made in the college laboratory, Edinburgh. *Nicholson's Journal*, 2nd ser., *34*, 68–72.

(1836). *Memoirs of the Life of Sir Humphry Davy*. 2 vols. London: Longman.

ed. (1858). *Fragmentary Remains, Literary and Scientific, of Sir Humphry Davy*. London: John Churchill.

Dawson, Warren R., ed. (1958). *The Banks Letters: A Calendar of the Manuscript Correspondence of Sir Joseph Banks*. London: British Museum (Natural History).

Dear, Peter (1985). *Totius in verba:* Rhetoric and authority in the early Royal Society. *Isis*, *76*, 145–161.

Desmond, Adrian (1989). *The Politics of Evolution: Morphology, Medicine, and Reform in Radical London*. Chicago: University of Chicago Press.

Dickinson, H.T. (1977). *Liberty and Property: Political Ideology in Eighteenth-century Britain*. London: Methuen.

Dickson, Stephen (1796). *An Essay on Chemical Nomenclature*. London and Dublin: J. Johnson, and W. Gilbert.

Dobbin, Leonard (1936). A Cullen chemical manuscript of 1753. *Annals of Science*, *1*, 138–156.

Dobson, Matthew (1779). *A Medical Commentary on Fixed Air*. Chester: J. Monk.

(1787). *Medical Commentary on Fixed Air . . . with an Appendix on the Efficacy of the Solution of Fixed Alkaline Salts Saturated with Fixible Air, in the Stone and Gravel . . . by William Falconer*, 3rd edn. London: T. Cadell.

Donn, Benjamin (1780). *An Enlarged Syllabus of a Course of Lectures in Experimental Philosophy*, 2nd edn. Bristol: for the author.

Donovan, Arthur L. (1975a). British chemistry and the concept of science in the eighteenth century. *Albion*, *7*, 131–144.

(1975b). *Philosophical Chemistry in the Scottish Enlightenment: The Doctrines and Discoveries of William Cullen and Joseph Black*. Edinburgh: Edinburgh University Press.

(1976). Pneumatic chemistry and Newtonian natural philosophy in the eighteenth century: William Cullen and Joseph Black. *Isis*, *67*, 217–228.

(1979). Scottish responses to the new chemistry of Lavoisier. *Studies in Eighteenth-century Culture*, *9*, 237–249.

ed. (1988a). *The Chemical Revolution: Essays in Reinterpretation* (*Osiris*, 2nd series, vol. 4).

(1988b). Introduction. In Donovan, ed. (1988a), pp. 5–12.

Donovan, Michael (1850). Biographical account of the late Richard Kirwan, Esq., President of the Royal Irish Academy. *Proceedings of the Royal Irish Academy*, *4*, lxxxi–cxviii.

Doyle, W.P. (1982). *James Crawford M.D. (1682–1731). Scottish Men of Science* series. Edinburgh.

Duncan, A.M. (1962). Some theoretical aspects of eighteenth-century tables of affinity. *Annals of Science, 18,* 177–194, 217–232.

Durie, Alistair J. (1979). *The Scottish Linen Industry in the Eighteenth Century.* Edinburgh: John Donald.

Duveen, Denis I. and Klickstein, Herbert S. (1954). A letter from Berthollet to Blagden relating to the experiments for a large-scale synthesis of water carried out by Lavoisier and Meusnier in 1785. *Annals of Science, 10,* 58–62.

Eklund, Jon (1976). Of a spirit in the water: Some early ideas on the aerial dimension. *Isis, 67,* 527–550.

Emblen, D.L. (1970). *Peter Mark Roget: The Word and the Man.* London: Longman.

Emerson, Roger L. (1973a). The Enlightenment and social structures. In Fritz and Williams, eds. (1973), pp. 99–124.

(1973b). The social composition of enlightened Scotland: The Select Society of Edinburgh, 1754–1764. *Studies on Voltaire and the Eighteenth Century, 114,* 291–329.

(1979). The Philosophical Society of Edinburgh, 1737–1747. *British Journal for the History of Science, 12,* 154–171.

(1981). The Philosophical Society of Edinburgh, 1748–1768. *British Journal for the History of Science, 14,* 133–176.

(1985). The Philosophical Society of Edinburgh, 1768–1783. *British Journal for the History of Science, 18,* 255–303.

(1986). Natural philosophy and the problem of the Scottish Englightenment. *Studies in Voltaire and the Eighteenth Century, 242,* 243–292.

(1988a). Science and the origins and concerns of the Scottish Enlightenment. *History of Science, 26,* 333–366.

(1988b). The Scottish Enlightenment and the end of the Philosophical Society of Edinburgh. *British Journal for the History of Science, 21,* 33–66.

Engestrom, Gustav von (1772). A treatise on the pocket laboratory, containing an easy method . . . for trying mineral bodies. In A.F. Cronstedt, *An Essay Towards a System of Mineralogy,* ed. Emmanuel Mendes da Costa, pp. 273–318. London: Edward and Charles Dilley.

"Experimentalist, An" (1800–1801). On the chemical effects of the pile of Volta. *Nicholson's Journal,* 1st ser., 4, 472–473, 514–515.

(1801). On the nature of galvanism, electricity, heat, the composition of water, and the theory of chemistry. *Nicholson's Journal,* 1st ser., 5, 231–232.

Falconer, William (1776). *Experiments and Observations.* London: W. Goldsmith.

Farrar, Kathleen R. (1968). Dalton's scientific apparatus. In Cardwell, ed. (1968), pp. 159–186.

Farrar, W.V., Farrar, Kathleen R. and Scott, E.L. (1974). The Henrys of Manchester. Part 2: Thomas Henry's sons: Thomas, Peter and William. *Ambix, 21,* 179–207.

Ferriar, John (1793). An argument against the doctrine of materialism, addressed to Thomas Cooper. *Memoirs of the Manchester Literary and Philosophical Society, 4,* 20–44.

Fitzpatrick, Martin (1984). Priestley in caricature. In Royal Society of Chemistry (1984), pp. 346–369.

(1987). Joseph Priestley and the Millenium. In Anderson and Lawrence, eds. (1987), pp. 29–37.

Fleck, Ludwik (1979). *Genesis and Development of a Scientific Fact*, eds. Thaddeus J. Trenn and Robert K. Merton, trans. Fred Bradley and Trenn, foreword by Thomas S. Kuhn. Chicago: University of Chicago Press.

Fontana, Felice (1779). Account of the airs extracted from different kinds of waters. *Philosophical Transactions*, 69, 432–453.

Foote, George A. (1952). Sir Humphry Davy and his audience at the Royal Institution. *Isis*, 43, 6–12.

Forgan, Sophie, ed. (1980). *Science and the Sons of Genius: Studies on Humphry Davy*. London: Science Reviews.

Forster, John Reinhold (1768). *An Introduction to Mineralogy*. London: J. Johnson.

Foucault, Michel (1970). *The Order of Things: An Archaeology of the Human Sciences*. London: Tavistock Publications.

Fourcroy, Antoine François de (1788). *Elements of Natural History and Chemistry*, trans. William Nicholson, 4 vols. London: G.G.J. and J. Robinson.

(1790). Ibid., 2nd edn., trans. William Nicholson, 3 vols. London: C. Elliot and T. Kay.

(1804). *A General System of Chemical Knowledge*, trans. William Nicholson, 11 vols. London: Cadell and Davies.

Fowler, Richard (1793). *Experiments and Observations Relative to the Influence Lately Discovered by M. Galvani*. Edinburgh: T. Duncan et al.

Fric, René (1959). Contribution à l'étude de l'évolution des idées de Lavoisier sur la nature de l'air et sur la calcination des métaux. *Archives Internationales d'Histoire des Sciences*, 12, 137–168.

Fritz, P. and Williams, D., eds. (1973). *City and Society in the Eighteenth Century*. Toronto: Hakkert.

Fruchtman, Jack (1982). Joseph Priestley on rhetoric and the power of political discourse. *Eighteenth-century Life*, 7 (no. 3) (May 1982), 37–47.

(1983). *The Apocalyptic Politics of Richard Price and Joseph Priestley: A Study in Late Eighteenth-century English Republican Millenarianism*. (*Transactions of the American Philosophical Society*, vol. 73, part 4). Philadelphia: American Philosophical Society.

Fullmer, June Z. (1962). Humphry Davy's adversaries. *Chymia*, 8, 147–164.

(1964). Humphry Davy and the gunpowder manufactory. *Annals of Science*, 20, 165–194.

(1967). Davy's sketches of his contemporaries. *Chymia*, 12, 127–150.

(1969). *Sir Humphry Davy's Published Works*. Cambridge: Harvard University Press.

(1980). Humphry Davy, reformer. In Forgan, ed. (1980), pp. 59–94.

"G.M., Dr." (1800–1801). On the state of galvanism and other scientific pursuits in Germany. *Nicholson's Journal*, 1st ser., 4, 511–513.

Gago, Ramón (1988). The new chemistry in Spain. In Donovan, ed. (1988a), pp. 169–192.

Garfinkle, Norton (1955). Science and religion in England, 1790–1800: The critical response to the work of Erasmus Darwin. *Journal of the History of Ideas*, 16, 376–388.

Garnett, Thomas (1790). *Experiments and Observations on the Horley-Green Spaw*. Bradford, Yorkshire: for the author.

(1791). *Experiments and Observations on the Crescent Water at Harrogate*. Leeds: Thomas Gill.

(1792). *A Treatise of the Mineral Waters at Harrogate*. N.p.: J. Johnson et al.

Garrett, Clarke (1975). *Respectable Folly: Millenarians and the French Revolution in France and England.* Baltimore: Johns Hopkins University Press.

Gascoigne, John (1989). *Cambridge in the Age of the Enlightenment: Science, Religion and Politics from the Restoration to the French Revolution.* Cambridge: Cambridge University Press.

Gay-Lussac, J.L. and Thenard, L.-J. (1809). De la nature et des propriétés de l'acide muriatique et de l'acide muriatique oxigéné. *Mémoires de Physique et de Chimie de la Société d'Arcueil,* 2, 339–358.

    (1810). Observations sur les trois précédens mémoires de M. Davy. *Annales de Chimie,* 75, 290–316.

    (1811). *Recherches Physico-chimiques, faites à l'Occasion de la Grande Batterie Voltaïque donnée . . . à l'École Polytechnique,* 2 vols. Paris: Deterville.

Gee, Brian (1989). Amusement chests and portable laboratories: Practical alternatives to the regular laboratory. In *The Development of the Laboratory: Essays on the Place of Experiment in Industrial Civilization,* ed. Frank A.J.L. James, pp. 37–59. London: Macmillan.

Gelfand, Toby (1985). "Invite the philosopher, as well as the charitable": Hospital teaching as private enterprise in Hunterian London. In Bynum and Porter, eds. (1985), pp. 129–151.

Gerardin, R.L. (1778). Observations sur les eudiomètres. *Observations sur la Physique,* 11, 248–254.

Gibbes, George Smith (1799). *Syllabus of a Course of Chemical Lectures.* Bath: William Meyler.

    (1809). *A Phlogistic Theory Ingrafted upon M. Fourcroy's Philosophy of Chemistry.* Bath: W. Meyler & Son.

Gibbs, F.W. (1951a). Robert Dossie (1717–1777) and the Society of Arts. *Annals of Science,* 7, 149–172.

    (1951b). Peter Shaw and the revival of chemistry. *Annals of Science,* 7, 211–237.

    (1952). William Lewis, M.B., F.R.S. (1708–1781). *Annals of Science,* 8, 122–151.

    (1953). George Wilson (1631–1711). *Endeavour,* 12, 182–185.

    (1960). Itinerant lecturers in natural philosophy. *Ambix,* 7, 111–117.

    (1965). *Joseph Priestley: Adventurer in Science and Champion of Truth.* London: Nelson.

    (1972). Bryan Higgins and his circle. In *Science, Technology and Economic Growth in the Eighteenth Century,* ed. A.E. Musson, pp. 195–207. London: Methuen.

Gilbert, G.N. (1977). Referencing as persuasion. *Social Studies of Science,* 7, 113–122.

Gisborne, Thomas (1798). On the benefits and duties resulting from the institution of societies for the advancement of literature and philosophy. *Memoirs of the Literary and Philosophical Society of Manchester,* 5, 70–88.

Golinski, Jan (1983). Peter Shaw: Chemistry and communication in Augustan England. *Ambix,* 30, 19–29.

    (1984). *Language, Method and Theory in British Chemical Discourse, c. 1660–1770.* University of Leeds Ph.D. dissertation.

    (1987). Robert Boyle: Scepticism and authority in seventeenth-century chemical discourse. In *The Figural and the Literal: Problems of Language in the History of Science and Philosophy, 1630–1800,* eds. Andrew E. Benjamin, Geoffrey N. Cantor and John R.R. Christie, pp. 58–82. Manchester: Manchester University Press.

(1989). A noble spectacle: Phosphorus and the public culture of science in the early Royal Society. *Isis, 80,* 11–39.

(1990). The theory of practice and the practice of theory: Sociological approaches in the history of science. *Isis, 81,* 492–505.

Gooding, David (1985a). "In nature's school": Faraday as an experimentalist. In *Faraday Rediscovered: Essays on the Life and Work of Michael Faraday, 1791–1867,* eds. Gooding and Frank A.J.L. James, pp. 105–135. London: Macmillan.

(1985b). "He who proves, discovers": John Herschel, William Pepys and the Faraday Effect. *Notes and Records of the Royal Society of London, 39,* 229–244.

Gooding, David, Pinch, T. and Schaffer, S., eds. (1989). *The Uses of Experiment: Studies in the Natural Sciences.* Cambridge: Cambridge University Press.

Goodman, D.C. (1969a). Wollaston and the atomic theory of Dalton. *Historical Studies in the Physical Sciences, 1,* 37–59.

(1969b). Problems in crystallography in the early nineteenth century. *Ambix, 16,* 152–166.

Goodwin, Albert (1968). The political genesis of Edmund Burke's *Reflections on the Revolution in France. Bulletin of the John Rylands Library, Manchester, 50,* 336–364.

(1979). *The Friends of Liberty.* London: Hutchinson.

Gough, J.B. (1968). Lavoisier's early career in science: An examination of some new evidence. *British Journal for the History of Science, 4,* 52–57.

(1971). *The Foundations of Modern Chemistry: The Origins and Development of the Concept of the Gaseous State and its Role in the Chemical Revolution of the Eighteenth Century.* Cornell University Ph.D. dissertation.

(1988). Lavoisier and the fulfillment of the Stahlian revolution. In Donovan, ed. (1988a), pp. 15–33.

Graham, James (1789). *A New, Plain, and Rational Treatise on the True Nature and Uses of the Bath Waters.* Bath: R. Cruttwell.

Graham, Jenny (1989). Revolutionary philosopher: The political ideas of Joseph Priestley (1733–1804). *Enlightenment and Dissent,* no. 8, 43–68.

Grant, Alexander (1884). *The Story of the University of Edinburgh During its First Three Hundred Years,* 2 vols. London: Longmans, Green & Co.

Greenaway, Frank (1966). *John Dalton and the Atom.* London: Heinemann.

ed. (1971–75). *The Archives of the Royal Institution of Great Britain in Facsimile: Minutes of the Managers' Meetings 1799–1900,* 7 vols. in 4. Menston, Yorkshire: Scolar Press.

Griffith, W.P. (1983). Priestley in London. *Notes and Records of the Royal Society of London, 38,* 1–16.

Guerlac, Henry (1961). *Lavoisier–the Crucial Year: The Background and Origin of his First Experiments on Combustion in 1772.* Ithaca, N.Y.: Cornell University Press.

(1975). *Antoine Laurent Lavoisier: Chemist and Revolutionary.* New York: Scribners.

(1976). Chemistry as a branch of physics: Laplace's collaborations with Lavoisier. *Historical Studies in the Physical Sciences, 7,* 193–276.

"H.B.K." (1806). Experiments and observations respecting the manner in which the gases are afforded in water by galvanism. *Nicholson's Journal,* 2nd ser., *14,* 50–56.

Habermas, Jürgen (1989). *The Structural Transformation of the Public Sphere:*

An Inquiry into a Category of Bourgeois Society, trans. Thomas Burger and Frederick Lawrence. Oxford: Polity Press.

Hacking, Ian (1983). *Representing and Intervening: Introductory Topics in the Philosophy of Natural Science*. Cambridge: Cambridge University Press.

Haldane, Henry (1800–1801). Experiments and observations made with the newly discovered metallic pile of Sig. Volta. *Nicholson's Journal*, 1st ser., 4, 241–245.

Hales, Stephen (1727). *Vegetable Staticks: Or, an Account of Some Statical Experiments on the Sap in Vegetables*. London: W. and J. Innys and T. Woodward; reprinted, London: Scientific Book Guild, 1961.

Halévy, Elie (1960). *England in 1815* (vol. I of *A History of the English People in the Nineteenth Century*), trans. E.I. Watkin and D.A. Barker. London: Ernest Benn.

Hamlyn, Hilda M. (1947). Eighteenth-century circulating libraries in England. *The Library*, 5th ser., *1*, 197–222.

Hannaway, Owen (1975). *The Chemists and the Word: The Didactic Origins of Chemistry*. Baltimore: Johns Hopkins University Press.

Hans, Nicholas (1951). *New Trends in Education in the Eighteenth Century*. London: Routledge and Kegan Paul.

Harrington, Robert (1781). *A Philosophical Enquiry into the First and General Principles of Animal and Vegetable Life*. London: T. Cadell.

[ ] ("Richard Bewley," pseud.) (1791). *A Treatise on Air, Containing New Experiments and Thoughts on Combustion*. London: T. Evans.

[ ] (1796). *A New System of Fire and Planetary Life*. London: T. Cadell and W. Davies.

(1804). *The Death-Warrant of the French Theory of Chemistry, Signed by Truth, Reason, Common Sense, Honour and Science*. London and Carlisle: Longman, Hurst, Rees and Orme.

(1819). *An Elucidation and Extension of the Harringtonian System of Chemistry*. London: Sherwood, Nealy and Jones.

Harrison, John A. (1957). Blind Henry Moyes, "an excellent lecturer in philosophy." *Annals of Science*, *13*, 109–125.

Hartley, Harold (1972). *Humphry Davy*. Wakefield: EP Publishing.

Haygarth, John (1778). Observations on the population and diseases of Chester, in the year 1774. *Philosophical Transactions*, 68, 131–154.

Hays, J.N. (1974). Science in the City: The London Institution, 1819–40. *British Journal for the History of Science*, 7, 146–162.

(1983). The London lecturing empire. In Inkster and Morrell, eds. (1983), pp. 91–119.

Heilbron, J.L. (1983). *Physics at the Royal Society During Newton's Presidency*. Los Angeles: William Andrews Clark Memorial Library, UCLA.

Heimann, P.M. and McGuire, J.E. (1971). Newtonian forces and Lockean powers: Concepts of matter in eighteenth-century thought. *Historical Studies in the Physical Sciences*, *3*, 233–306.

Henry, Thomas (1781). *An Account of a Method of Preserving Water at Sea from Putrefaction*. Warrington: W. Eyres.

Henry, Thomas and Haygarth, John (1785). On the preservation of sea water from putrefaction by means of quicklime. *Memoirs of the Literary and Philosophical Society of Manchester*, *1*, 41–54.

Henry, William (1800–1801). Experiments on the chemical effects of galvanic electricity. *Nicholson's Journal*, 1st ser., 4, 223–226.

(1801). *An Epitome of Chemistry*, 2nd edn. London: J. Johnson.

(1818). *The Elements of Experimental Chemistry*, 8th edn., 2 vols. London: Baldwin, Cradock and Joy.

(1826). Ibid., 10th edn., 2 vols. London: Baldwin et al.

Henry, William Charles (1854). *Memoirs of the Life and Scientific Researches of John Dalton*. London: Cavendish Society.

Heyd, Michael (1981). The reaction to enthusiasm in the seventeenth century: Towards an integrative approach. *Journal of Modern History, 53*, 258–280.

Hiebert, E.N., Ihde, A.J. and Schofield, R.E. (1974). *Joseph Priestley: Scientist, Theologian and Metaphysician*, eds. L. Kieft and B.R. Willeford. Lewisburg, Penn.: Bucknell University Press.

Higgins, Bryan [c.1775?]. *Syllabus of Doctor Higgins's Course of Philosophical, Pharmaceutical, and Technical Chemistry*. [London?]: n.p.

ed. (1795). *Minutes of the Society for Philosophical Experiments and Conversations*. London: T. Cadell and W. Davies.

Higgins, William (1791). *A Comparative View of Phlogistic and Antiphlogistic Theories*, 2nd edn. London: J. Murray; reprinted in Wheeler and Partington (1960).

(1814). *Experiments and Observations on the Atomic Theory*. Dublin: Graisberry and Campbell; reprinted in Wheeler and Partington (1960).

Holloway, S.W.F. (1966). The Apothecaries' Act, 1815: A reinterpretation. *Medical History, 10*, 107–129, 221–236.

Holmes, Frederic L. (1985). *Lavoisier and the Chemistry of Life: An Exploration of Scientific Creativity*. Madison, Wisconsin: University of Wisconsin Press.

Holmes, Geoffrey (1982). *Augustan England: Professions, State and Society 1680–1730*. London: Allen and Unwin.

Home, Francis (1756a). *The Principles of Agriculture and Vegetation*. Edinburgh: Sands, Donaldson, Murray and Cochran.

(1756b). *Experiments on Bleaching*. Edinburgh: Sands, Donaldson, Murray and Cochran.

Hont, Istvan and Ignatieff, Michael, eds. (1983). *Wealth and Virtue: The Shaping of Political Economy in the Scottish Enlightenment*. Cambridge: Cambridge University Press.

Hoover, Suzanne R. (1978). Coleridge, Humphry Davy, and some early experiments with a counsciousness-altering drug. *Bulletin of Research in the Humanities, 81*, 9–27.

Hopson, Charles Rivington (1781). *An Essay on Fire*. London: J.F. and C. Rivington.

Horrocks, Sally (1987). Audiences for chemistry in Regency Britain: Mrs. Marcet's *Conversations on Chemistry*. Cambridge University, Department of History and Philosophy of Science, unpublished dissertation.

Houghton, Walter E., ed. (1966–87). *The Wellesley Index to Victorian Periodicals*, 4 vols. Toronto: University of Toronto Press.

Howell, W.B. (1930). Doctor George Fordyce and his times. *Annals of Medical History*, new ser., 2, 281–296.

Howell, Wilbur Samuel (1971). *Eighteenth-Century British Logic and Rhetoric*. Princeton: Princeton University Press.

Hufbauer, Karl (1982). *The Formation of the German Chemical Community (1720–1795)*. Berkeley and Los Angeles: University of California Press.

Hughes, Peter (1976). Originality and allusion in the writings of Edmund Burke. *Centrum: Working Papers of the Minnesota Center for Advanced Studies in Language, Style and Literary Theory, 4*, 32–43.

Hulme, Nathaniel (1778). *A Safe and Easy Remedy Proposed for the Relief of the Stone and Gravel, the Scurvy, Gout, &c.*, 2nd edn. London: James Phillips.

Hume, David (1903). *Essays Moral, Political and Literary.* London: Grant Richards.

Hutton, James (1794). *An Investigation of the Principles of Knowledge*, 3 vols. Edinburgh: A. Strahan.

Ingenhousz, John (1776). Easy methods of measuring the diminution of bulk taking place upon the mixture of common air and nitrous air. *Philosophical Transactions*, 66, 257–267.

(1779). *Experiments upon Vegetables, Discovering their Great Power of Purifying the Air in the Sun-shine.* London: P. Elmsley and H. Payne.

(1780). On the degree of salubrity of the common air at sea, compared with that of the sea-shore and that of places far removed from the sea. *Philosophical Transactions*, 70, 354–377.

(1785). Observations sur la construction et l'usage de l'eudiomètre de M. Fontana. *Observations sur la Physique*, 26, 339–359.

(1787). *Expériences sur les Végétaux*, 2 vols., new edn. Paris: Théophile Barrois.

Inkster, Ian (1975). Science and the Mechanics' Institutes, 1820–1850: The case of Sheffield. *Annals of Science*, 32, 451–474.

(1976a). The social context of an educational movement: A revisionist approach to the English Mechanics' Institutes, 1820–1850. *Oxford Review of Education*, 2, 277–307.

(1976b). Culture, institutions and urbanity: The itinerant science lecturer in Sheffield, 1790–1850. In *Essays in the Economic and Social History of South Yorkshire*, eds. S. Pollard and C. Holmes, pp. 218–232. Sheffield: South Yorkshire County Council.

(1977). Science and society in the metropolis: A preliminary examination of the social and institutional context of the Askesian Society of London, 1797–1807. *Annals of science*, 34, 1–32.

(1980). The public lecture as an instrument of science education for adults: The case of Great Britain, c.1750–1850. *Paedogogica Historica: International Journal of the History of Education*, 20, 80–107.

(1983). Introduction: Aspects of the history of science and science culture in Britain, 1780–1850 and beyond. In Inkster and Morrell, eds. (1983), pp. 11–54.

Inkster, Ian and Morrell, Jack, eds. (1983). *Metropolis and Province: Science in British Culture, 1780–1850.* London: Hutchinson.

Jacob, Margaret and Jacob, James, eds. (1984). *The Origins of Anglo-American Radicalism.* London: Allen and Unwin.

Jewson, N. (1974). Medical knowledge and the patronage system in eighteenth-century England. *Sociology*, 8, 369–385.

Jones, Henry Bence (1871). *The Royal Institution: Its Founder and its First Professors.* London: Longmans, Green & Co.

Jones, Peter (1983). The Scottish professoriate and the polite academy. In Hont and Ignatieff, eds. (1983), pp. 89–117.

Jones, William (1834). *The Elements of Chemical Science in Verse, Designed for the Student in Chemistry.* London: John Murray.

Jordanova, Ludmilla (1987). Earth science and environmental medicine: The synthesis of the late Enlightenment. In Jordanova and Porter, eds. (1987), pp. 119–146.

Jordanova, Ludmilla and Porter, Roy, eds. (1979). *Images of the Earth: Essays in the History of the Environmental Sciences* (*BSHS Monographs*, no. 1). Chalfont St. Giles: British Society for the History of Science.

Joyce, Jeremiah (1794). *A Sermon Preached on Sunday, February the 23d, 1794 . . . To which is added an Appendix, Containing an Account of the Author's Arrest for "Treasonable Practices."* London: for the author.

(1807). *Dialogues in Chemistry, Intended for the Instruction and Entertainment of Young People,* 2 vols. London: J. Johnson.

Kames, Lord (Henry Home) (1754). Of the laws of motion. *Essays and Observations, Physical and Literary,* 1, 1–69.

(1776). *The Gentleman Farmer, being an Attempt to Improve Agriculture by Subjecting it to the Test of Rational Principles.* Edinburgh: W. Creech.

Kaufman, Paul (1960). *Borrowings from the Bristol Library 1773–1784: A Unique Record of Reading Vogues.* Charlottesville: Bibliographical Society of the University of Virginia.

(1969). English book clubs and their social import. In Kaufman, *Libraries and their Users: Collected Papers in Library History,* pp. 36–4. London: Library Association.

Keir, James (1779). *A Treatise on the Various Kinds of Permanently Elastic Fluids or Gases,* 2nd edn. London: T. Cadell and P. Elmsley.

(1789). *The First Part of a Dictionary of Chemistry.* Birmingham: Pearson and Rollason.

Kendall, James (1949–52). The first chemical society, the first chemical journal and the Chemical Revolution. *Proceedings of the Royal Society of Edinburgh,* 63A, 346–358, 385–400.

Kent, Alexander, ed. (1950a). *An Eighteenth-Century Lectureship in Chemistry.* Glasgow: Jackson.

(1950b). William Cullen's history of chemistry. In Kent, ed. (1950a), pp. 15–27.

Kipnis, Naum (1987). Luigi Galvani and the debate on animal electricity, 1791–1800. *Annals of Science,* 44, 107–142.

King-Hele, Desmond (1977). *Doctor of Revolution: The Life and Genius of Erasmus Darwin.* London: Faber and Faber.

ed. (1981). *The Letters of Erasmus Darwin.* Cambridge: Cambridge University Press.

Kirwan, Richard (1784a). Remarks on Mr. Cavendish's experiments on air. *Philosophical Transactions,* 74, 154–169.

(1784b). *Elements of Mineralogy.* London: P. Elmsley.

(1789). *An Essay on Phlogiston and the Composition of Acids,* ed. William Nicholson, 2nd edn. London: J. Johnson.

(1794–96). *Elements of Mineralogy,* 2nd edn., 2 vols. London: J. Nichols for P. Elmsley.

(1799). *An Essay on the Analysis of Mineral Waters.* London: J.W. Myers.

(1802). Of chymical and mineralogical nomenclature. *Transactions of the Royal Irish Academy,* 8, 53–76.

(1807). *Logick: Or an Essay on the Elements, Principles and Different Modes of Reasoning,* 2 vols. London: Payne and Mackinlay.

Knight, David M. (1967). *Atoms and Elements.* London: Hutchinson.

(1974). Chemistry in palaeontology: The work of James Parkinson (1755–1824). *Ambix,* 21, 78–85.

(1978). *The Transcendental Part of Chemistry.* Folkestone: Dawson.

(1986). Accomplishment or dogma: Chemistry in the introductory works of Jane Marcet and Samuel Parkes. *Ambix, 33,* 94–98.

Knoefel, Peter K. (1979). Famine and fever in Tuscany: Eighteenth-century Italian concern with the environment. *Physis, 21,* 7–35.

(1984). *Felice Fontana: Life and Works (Studi su Felice Fontana,* no. 2). Trento: Società di Studi Trentini di Scienze Storiche.

Kohler, Robert E. (1972). The origin of Lavoisier's first experiments on combustion. *Isis, 63,* 349–355.

(1975). Lavoisier's rediscovery of the air from mercury calx: A reinterpretation. *Ambix, 22,* 52–57.

Korshin, A.J. (1973–74). Types of eighteenth-century literary patronage. *Eighteenth-century Studies, 7,* 453–473.

Kramnick, Isaac (1986). Eighteenth-century science and radical social theory: The case of Joseph Priestley's scientific liberalism. *Journal of British Studies, 25,* 1–30.

Landriani, Marsilio (1775). Description d'une machine pour mesurer la salubrita de l'air, nommée eudiomètre. *Observations sur la Physique, 6,* 315–316.

Lane, Joan (1984). The medical practitioners of provincial England in 1783. *Medical History, 28,* 353–371.

Langer, Bernard (1971). *Pneumatic Chemistry, 1772–1789: A Resolution of Conflict.* University of Wisconsin Ph.D. dissertation.

Latour, Bruno (1983). Give me a laboratory and I will raise the world. In *Science Observed: Perspectives on the Social Study of Science,* eds. Karin D. Knorr-Cetina and Michael Mulkay, pp. 141–170. London: Sage Publications.

(1987). *Science in Action: How to Follow Scientists and Engineers through Society.* Milton Keynes: Open University Press.

(1990). Essay review: Postmodern? No, simply amodern! Steps towards an anthropology of science. *Studies in History and Philosophy of Science, 21,* 145–171.

Latour, Bruno and Woolgar, Steve (1979). *Laboratory Life: The Social Construction of Scientific Facts.* London and Beverly Hills: Sage.

Laudan, Rachel (1987). *From Mineralogy to Geology: The Foundations of a Science, 1650–1830.* Chicago: University of Chicago Press.

Lavoisier, Antoine Laurent (1789). *Traité Élémentaire de Chimie, Présenté dans un Ordre Nouveau, et d'Après les Découvertes Modernes,* 2 vols. Paris: Cuchet.

(1790). *Elements of Chemistry, in a New Systematic Order, Containing all the Modern Discoveries,* trans. Robert Kerr. Edinburgh: William Creech.

Lavoisier, Antoine and Laplace, Pierre Simon de (1982). *Memoir on Heat, Read to the Royal Academy of Sciences, 28 June 1783,* trans. Henry Guerlac. New York: Neale Watson Academic Publications.

Law, J. and Williams, R.J. (1982). Putting facts together: A study of scientific persuasion. *Social Studies of Science, 12,* 535–558.

Lawrence, Christopher John (1984). *Medicine as Culture: Edinburgh and the Scottish Enlightenment.* London University Ph.D. dissertation.

(1990). The power and the glory: Humphry Davy and Romanticism. In Cunningham and Jardine, eds. (1990), pp. 213–227.

Lawrence, Susan C. (1985a). *Science and Medicine at the London Hospitals: The Development of Teaching and Research 1750–1815.* University of Toronto Ph.D. dissertation.

(1985b). "Desirous of improvements in medicine": Pupils and practitioners in the medical societies at Guy's and St. Batholomew's hospitals, 1795–1815. *Bulletin of the History of Medicine, 59,* 89–104.

LeGrand, Homer E. (1972). Lavoisier's oxygen theory of acidity. *Annals of science, 29,* 1–18.

——— (1974). Ideas on the composition of muriatic acid and their relevance to the oxygen theory of acidity. *Annals of Science, 31,* 213–225.

——— (1975). The "conversion" of C.-L. Berthollet to Lavoisier's chemistry. *Ambix, 22,* 58–70.

Lehmann, William C. (1971). *Henry Home, Lord Kames and the Scottish Enlightenment: A Study in National Character and in the History of Ideas.* The Hague: Martinus Nijhoff.

Lennard, Reginald (1931). The watering places. In *Englishmen at Rest and Play: Some Phases of English Leisure 1558–1714,* ed. R. Lennard, pp. 3–78. Oxford: Clarendon Press.

Levere, Trevor H. (1971). *Affinity and Matter: Elements of Chemical Philosophy 1800–1865.* Oxford: Clarendon Press.

——— (1977). Dr. Thomas Beddoes and the establishment of his Pneumatic Institution: A tale of three presidents. *Notes and Records of the Royal Society of London, 32,* 41–49.

——— (1980). Humphry Davy, "the sons of genius", and the idea of glory. In Forgan, ed. (1980), pp. 33–58.

——— (1981a). Dr. Thomas Beddoes at Oxford: Radical politics in 1788–1793 and the fate of the Regius chair in chemistry. *Ambix, 28,* 61–69.

——— (1981b). *Poetry Realized in Nature: Samuel Taylor Coleridge and Early Nineteenth-Century Science.* Cambridge: Cambridge University Press.

——— (1984). Dr. Thomas Beddoes (1750–1808): Science and medicine in politics and society. *British Journal for the History of Science, 17,* 187–204.

——— (1990). Lavoisier: Language, instruments, and the Chemical Revolution. In *Nature, Experiment, and the Sciences,* eds. T.H. Levere and W.R. Shea, pp. 207–223. Netherlands: Kluwer Academic Publishers.

Levi, Primo (1985). *The Periodic Table,* trans. Raymond Rosenthal. London: Michael Joseph.

Linder, Bertel and Smeaton, W.A. (1968). Schwediauer, Bentham and Beddoes: Translators of Bergman and Scheele. *Annals of Science, 24,* 259–273.

Lindsay, Jack, ed. (1970). *Autobiography of Joseph Priestley.* Bath: Adams and Dart.

Llana, James W. (1985). A contribution of natural history to the Chemical Revolution in France. *Ambix, 32,* 71–91.

Lödwig, T.H. and Smeaton, W.A. (1974). The ice calorimeter of Lavoisier and Laplace and some of its critics. *Annals of Science, 31,* 1–18.

Lucas, Charles (1756). *An Essay on Waters.* London: A. Millar.

Lundgren, Anders (1988). The new chemistry in Sweden: The debate that wasn't. In Donovan, ed. (1988a), pp. 146–168.

——— (1990). The changing role of numbers in 18th-century chemistry. In *The Quantifying Spirit in the Eighteenth Century,* eds. Tore Frangsmyr, J.L. Heilbron and Robin E. Rider, pp. 245–266. Berkeley and Los Angeles: University of California Press.

Lythe, S.G.E. (1984). *Thomas Garnett (1766–1802): Highland Tourist, Scientist and Professor, Medical Doctor.* Glasgow: Polpress.

Macbride, David (1764). *Experimental Essays.* London: A. Millar.

MacKie, J.D. (1950). Glasgow University in the eighteenth century. In Kent, ed. (1950a), pp. 28–40.

MacLeod, Roy M. (1983). Whigs and savants: Reflections on the reform movement in the Royal Society, 1830–1848. In Inkster and Morrell, eds. (1983), pp. 55–90.

Magellan, Jean-Hyacinthe (1777). *Description of a Glass Apparatus for Making Mineral Waters*. London: for W. Parker.

(1783). Ibid., 3rd edn. London: for the author.

Malton, Thomas (1777). *An Essay Concerning the Publication of Works on Science and Literature by Subscription*. London: n.p.

(1778). *A Compleat Treatise on Perspective*. London: for the author.

Manley, Gordon (1968). Dalton's accomplishment in meteorology. In Cardwell, ed. (1968), pp. 140–158.

[Marcet, Jane] (1806). *Conversations on Chemistry, in which the Elements of that Science are Familiarly Explained and Illustrated by Experiments*, 2 vols. London: Longman, Hurst, Rees, Orme and Brown.

(1817). Ibid., 5th edn, 2 vols. London: Longman et al.

Martine, George (1787). *Essays and Observations on the Construction and Graduation of Thermometers*, 4th edn. Edinburgh: Alexander Donaldson.

Martineau, Harriet (1849–50). *The History of England During the Thirty Years' Peace, 1816–1846*, 2 vols. London: Charles Knight.

Mauskopf, Seymour H. (1969). Thomson before Dalton: Thomas Thomson's considerations of the issue of combining weight proportions prior to his acceptance of Dalton's chemical atomic theory. *Annals of Science*, 25, 229–242.

(1970). Minerals, molecules and species. *Archives Internationales d'Histoire des Sciences*, 23, 185–206.

(1976). Crystals and compounds: Molecular structure and composition in nineteenth-century French science. *Transactions of the American Philosophical Society*, new ser., 66 (part iii), 1–82.

McClellan, James E. (1985). *Science Reorganized: Scientific Societies in the Eighteenth Century*. New York: Columbia University Press.

McElroy, D.D. (1969). *Scotland's Age of Improvement: A Survey of Eighteenth-Century Literary Clubs and Societies*. Pullman, Washington: Washington State University Press.

McEvoy, John G. (1975). *Joseph Priestley: Philosopher, Scientist and Divine*. University of Pittsburgh Ph.D. dissertation.

(1978–79). Joseph Priestley, "aerial philosopher": Metaphysics and methodology in Priestley's thought 1772–1781. *Ambix*, 25 (1978), 1–55, 93–116, 153–175; 26 (1979), 16–38.

(1979). Electricity, knowledge and the nature of progress in Priestley's thought. *British Journal for the History of Science*, 12, 1–30.

(1983). Enlightenment and Dissent in science: Joseph Priestley and the limits of theoretical reasoning. *Enlightenment and Dissent*, no. 2, 47–67.

(1987). Causes and laws, powers and principles: The metaphysical foundations of Priestley's concept of phlogiston. In Anderson and Lawrence, eds. (1987), pp. 55–71.

(1988a). The Enlightenment and the Chemical Revolution. In *Metaphysics and Philosophy of Science in the Seventeenth and Eighteenth Centuries: Essays in Honour of Gerd Buchdahl*, ed. R.S. Woolhouse, pp. 307–325. Dordrecht: Kluwer Academic Publishers.

(1988b). Continuity and discontinuity in the Chemical Revolution. In Donovan, ed. (1988a), pp. 195–213.

McEvoy, J.G. and McGuire, J.E. (1975). God and nature: Priestley's way of rational dissent. *Historical Studies in the Physical Sciences*, 6, 325–404.

McGuire, J.E. (1972). Boyle's conception of nature. *Journal of the History of Ideas, 33*, 523–542.

McKendrick, Neil (1973). The role of science in the Industrial Revolution: A study of Josiah Wedgwood as a scientist and industrial chemist. In *Changing Perspectives in the History of Science: Essays in Honour of Joseph Needham*, eds. Mikulas Teich and Robert M. Young, pp. 274–319. London: Heinemann.

McKendrick, Neil, Brewer, John and Plumb, J.H. (1982). *The Birth of a Consumer Society: The Commercialization of Eighteenth-century England*. London: Hutchinson.

McKie, Douglas (1951). Mr. Warltire, a good chymist. *Endeavour, 10*, 46–49.

   (1952). *Antoine Lavoisier: Scientist, Economist, Social Reformer*. London: Constable.

   (1956–57). Priestley's laboratory and library and other of his effects. *Notes and Records of the Royal Society of London, 12*, 114–136.

   (1961). Joseph Priestley and the Copley Medal. *Ambix, 9*, 1–22.

McKie, Douglas and Heathcote, Niels H. de V. (1935). *The Discovery of Specific and Latent Heats*. London: Edward Arnold; reprinted New York: Arno Press, 1975.

McKie, Douglas and Kennedy, David (1960). On some letters of Joseph Black and others. *Annals of Science, 16*, 129–170.

McLachlan, H. (1931). *English Education Under the Test Acts*. Manchester: Manchester University Press.

   (1943). *Warrington Academy: Its History and Influence*. Manchester: Chetham Society.

McNeil, Maureen (1987). *Under the Banner of Science: Erasmus Darwin and His Age*. Manchester: Manchester University Press.

Melhado, Evan M. (1981). *Jacob Berzelius: The Emergence of his Chemical System*, Lychnos-Bibliotek, no. 34. Stockholm: Almqvist and Wiksell International.

   (1983). Oxygen, phlogiston and caloric: The case of Guyton. *Historical Studies in the Physical Sciences, 13*, 311–334.

   (1985). Chemistry, physics, and the Chemical Revolution. *Isis, 76*, 195–211.

Metzger, Hélène (1969). *La Genèse de la Science des Cristaux*, 2nd edn. Paris: Albert Blanchard.

Meyer, D.L. (1775). *A Method of Making Useful Mineralogical Collections*. London: Lockyer Davis.

Meyer, Gerald Dennis (1955). *The Scientific Lady in England 1650–1760: An Account of her Rise with Emphasis on the Major Roles of the Telescope and Microscope*. Berkeley and Los Angeles: University of California Press.

Miles, Wyndham D. (1961). "Sir Humphrey Davie, the prince of agricultural chemists." *Chymia, 7*, 126–134.

Millburn, John R. (1976). *Benjamin Martin: Author, Instrument-maker and "Country Showman."* Leyden: Noordhoff.

   (1983). The London evening courses of Benjamin Martin and James Ferguson, eighteenth-century lecturers on experimental philosophy. *Annals of Science, 40*, 437–455.

   (1985). James Ferguson's lecture tour of the English Midlands in 1771. *Annals of Science, 42*, 397–415.

Miller, David P. (1981). *The Royal Society of London 1800–1835: A Study of the Cultural Politics of Scientific Organization*. University of Pennsylvania Ph.D. dissertation.

(1983). Between hostile camps: Sir Humphry Davy's presidency of the Royal Society of London, 1820–1827. *British Journal for the History of Science,* 16, 1–47.

(1989). "Into the valley of darkness": Reflections on the Royal Society in the eighteenth century. *History of Science,* 27, 155–166.

Milner, Isaac [c. 1784]. *A Plan of a Course of Experimental Lectures, Introductory to the Study of Chemistry and Other Branches of Natural Philosophy.* Cambridge: J. Archdeacon.

Money, John (1977). *Experience and Identity: Birmingham and the West Midlands, 1760–1800.* Manchester: Manchester University Press.

(1988–89). Joseph Priestley in cultural context: Philosophic spectacle, popular belief and popular politics in eighteenth-century Birmingham. *Enlightenment and Dissent,* no. 7 (1988), 57–81; no. 8 (1989), 69–89.

Morrell, J.B. (1969). Thomas Thomson: Professor of chemistry and university reformer. *British Journal for the History of Science,* 4, 245–265.

(1971a). The University of Edinburgh in the late eighteenth century: Its scientific eminence and academic structure. *Isis,* 62, 158–171.

(1971b). Professors Robison and Playfair, and the *theophobia gallica:* Natural philosophy, religion and politics in Edinburgh, 1789–1815. *Notes and Records of the Royal Society,* 26, 43–63.

(1972). The chemist breeders: The research schools of Liebig and Thomas Thomson. *Ambix,* 19, 1–46.

(1976). The Edinburgh Town Council and its University, 1717–1766. In Anderson and Simpson, eds. (1976), pp. 46–65.

(1990). Professionalisation. In Olby, Cantor, Christie and Hodge, eds., (1990), pp. 980–989.

Morris, R.J. (1969). Lavoisier on fire and air: The memoir of July 1772. *Isis,* 60, 374–380.

(1972). Lavoisier and the caloric theory. *British Journal for the History of Science,* 6, 1–38.

Mossner, Ernest Campbell (1970). *The Life of David Hume.* Oxford: Clarendon Press.

Moyes, Henry (n.d.) *Heads of a Course of Lectures on the Philosophy of Chemistry and Natural History.* n.p.

(1781). *Heads of a Course of Lectures Upon the Most Important Subjects of Philosophical Chemistry.* [York: n.p.]

Mullett, Charles F. (1946). Public baths and health in England, 16th–18th century. *Supplements to the Bulletin of the History of Medicine,* no. 5, 1–85.

Murdoch, Alexander (1980). *The People Above: Politics and Administration in mid-Eighteenth-Century Scotland.* Edinburgh: John Donald.

Murray, John (1801). *Elements of Chemistry,* 2 vols. Edinburgh: T. Maccliesh.

(1806). *A System of Chemistry,* 4 vols. Edinburgh: Longman, Hurst, Rees and Orme.

(1811a). Observations and experiments on the nature of oxymuriatic acid. *Nicholson's Journal,* 2nd ser., 28, 132–152.

(1811b). Observations and experiments on the alkaline metalloids. *Nicholson's Journal,* 2nd ser., 28, 241–249.

(1811c). On the nature of oxymuriatic acid. *Nicholson's Journal,* 2nd ser., 28, 294–310.

(1811d). Further observations and experiments on oxymuriatic acid. *Nicholson's Journal,* 2nd ser., 29, 187–202.

(1811e). On the nature of oximuriatic gas. *Nicholson's Journal,* 2nd ser., 30, 226–235.

(1812a). Experiments on muriatic acid gas. *Nicholson's Journal*, 2nd ser., *31*, 123–133.

(1812b). Experiments on the existence of water in muriate of ammonia. *Nicholson's Journal*, 2nd ser., *32*, 185–197.

(1813). On the existence of combined water in muriatic acid gas. *Nicholson's Journal*, 2nd ser., *34*, 264–276.

(1814). *Elements of Chemistry*, 3rd edn., 3 vols. Edinburgh: John Anderson.

(1818). Experiments on muriatic acid gas, with observations on its chemical constitution. *Transactions of the Royal Society of Edinburgh, 8*, 287–328.

(1819). *System of Chemistry*, 4th edn., 4 vols. Edinburgh: Francis Pillans.

Musson, A.E. and Robinson, Eric (1969a). *Science and Technology in the Industrial Revolution*. Manchester: Manchester University Press.

(1969b). James Watt and early experiments in alkali manufacture. In Musson and Robinson (1969a), pp. 352–371.

(1969c). Science and industry in the late eighteenth century. In Musson and Robinson (1969a), pp. 87–189.

Myers, Greg (1985). Texts as knowledge claims: The social construction of two biology articles. *Social Studies of Science, 15*, 593–630.

Nangle, Benjamin Christie (1934–55). *The Monthly Review: Index of Contributors and Articles*, 2 vols. Oxford: Clarendon Press.

Nash, John (1814). The discovery of the atomic theory claimed for Mr. Higgins. *Philosophical Magazine, 43*, 54–57.

Neve, Michael (1983). Science in a commercial city: Bristol 1820–60. In Inkster and Morrell, eds. (1983), pp. 179–204.

Nicholson, Francis (1923–24). The Literary and Philosophical Society 1781–1851. *Memoirs and Proceedings of the Manchester Literary and Philosophical Society, 68*, 97–148.

Nicholson, William (1790). *The First Principles of Chemistry*. London: G. and J. Robinson.

(1795). *A Dictionary of Chemistry, Exhibiting the Theory and Practice of that Art*, 2 vols. in 1. London: G.G. and J. Robinson.

(1792). *Ibid.*, 2nd edn. London: Robinson.

(1796). *First Principles of Chemistry* 3rd edn. London: Robinson.

(1800–1801). Account of the new electrical or galvanic apparatus of Sig. Alex. Volta, and experiments performed with the same. *Nicholson's Journal*, 1st ser., *4*, 179–187.

Nooth, John Mervin (1775). The description of an apparatus for impregnating water with fixed air. *Philosophical Transactions, 65*, 59–66.

Ockenden, L.C. (1937). The great batteries of the London Institution. *Annals of Science, 2*, 183–184.

Olby, R.C., Cantor, G.N., Christie, J.R.R. and Hodge, M.J.S., eds. (1990). *Companion to the History of Modern Science*. London: Routledge.

Oldroyd, D.R. (1972). Edward Daniel Clarke, 1769–1822, and his role in the history of the blow-pipe. *Annals of Science, 29*, 213–235.

Orange, Derek (1983). Rational dissent and provincial science: William Turner and the Newcastle Literary and Philosophical Society. In Inkster and Morrell, eds. (1983), pp. 205–230.

Outram, Dorinda (1980). Politics and vocation: French science, 1793–1830. *British Journal for the History of Science, 13*, 27–43.

Parascandola, John and Ihde, Aaron J. (1969). History of the pneumatic trough. *Isis, 60*, 351–361.

Parkes, Samuel (1810). *The Rudiments of Chemistry, Illustrated by Experiments*. London: for the author.

(1816). *The Chemical Catechism with Notes, Illustrations and Experiments*, 7th edn. London: Baldwin, Cradock and Joy.

(1818). Ibid. 8th edn. London: Baldwin, Cradock and Joy.

Parkinson, James (1807). *The Chemical Pocket Book or Memoranda Chemica*, 4th edn. London: H.D. Symonds.

Partington, James R. (1961–70). *A History of Chemistry*, 4 vols. London: Macmillan.

Paulsen, Ronald (1983). *Representations of Revolution (1789–1820)*. New Haven, Conn.: Yale University Press.

Peach, W. Bernard and Thomas, D.O., eds. (1983). *The Correspondence of Richard Price*, vol. I (July 1748–March 1778). Durham, N.C. and Cardiff: Duke University Press and University of Wales Press.

Pearson, George (1784). *Observations and Experiments for Investigating the Chemical History of the Tepid Springs of Buxton*, 2 vols. London: n.p.

(1794). *A Translation of the Table of Chemical Nomenclature*. London: J. Johnson.

(1797–98). Experiments and observations made with the view of ascertaining the nature of the gas produced by passing electric discharges through water. *Nicholson's Journal*, 1st ser., 1, 241–248, 299–305, 349–355.

Peart, Edward (1795). *The Anti-Phlogistic Doctrine of M. Lavoisier Critically Examined and Demonstratively Confuted*. London: W. Miller and Mrs Murray.

Percival, Thomas (1772–73). *Essays Medical and Experimental*, 2 vols. London: J. Johnson.

(1774–76). Observations on the state of population in Manchester, and other adjacent places. *Philosophical Transactions*, 64 (1774), 54–66; 65 (1775), 322–335; 66 (1776), 160–167.

(1807). *The Works, Literary, Moral and Medical of Thomas Percival M.D.*, 4 vols., new edn. Bath: Richard Cruttwell.

Perkin, Harold (1969). *The Origins of Modern English Society, 1780–1880*. London: Routledge and Kegan Paul.

Perrin, Carlton E. (1969). Prelude to Lavoisier's theory of calcination – some observations on *mercurius calcinatus per se*. *Ambix*, 16, 140–151.

(1982). A reluctant catalyst: Joseph Black and the Edinburgh reception of Lavoisier's chemistry. *Ambix*, 29, 141–176.

(1986). Lavoisier's thoughts on calcination and combustion, 1772–1773. *Isis*, 77, 647–666.

(1988). Research traditions, Lavoisier and the Chemical Revolution. In Donovan, ed. (1988a), pp. 53–81.

(1989). Document, text and myth: Lavoisier's crucial year revisited. *British Journal for the History of Science*, 22, 3–25.

Phillipson, Nicholas (1973). Towards a definition of the Scottish Enlightenment. In *City and Society in the Eighteenth Century*, eds. P. Fritz and D. Williams, pp. 125–147. Toronto: Hakkert.

(1975). Culture and society in the eighteenth-century province: The case of Edinburgh and the Scottish Enlightenment. In *The University in Society*, ed. Lawrence Stone, 2 vols., II, pp. 407–488. Princeton: Princeton University Press.

(1981). The Scottish Enlightenment. In Porter and Teich, eds. (1981), pp. 19–40.

Pinch, Trevor (1986). *Confronting Nature: The Sociology of Solar Neutrino Detection*. Dordrecht: Reidel.

Pinch, Trevor and Collins, H.M. (1984). Private science and public knowledge:

The Committee for the Scientific Investigation of the Claims of the Paranormal and its use of the literature. *Social Studies of Science*, 14, 521–546.

Playfair, John (1805). Biographical account of the late James Hutton. *Transactions of the Royal Society of Edinburgh*, 5 (part iii), 39–99.

Playfair, Lyon (1858). *A Century of Chemistry in the University of Edinburgh*. Edinburgh: Murray and Gibb.

Pocock, J.G.A. (1975). *The Machiavellian Moment*. Princeton: Princeton University Press.

(1980). Post-Puritan England and the problem of the Enlightenment. In *Culture and Politics from Puritanism to the Enlightenment*, ed. Perez Zagorin, pp. 91–111. Berkeley and Los Angeles: University of California Press.

(1985). *Virtue, Commerce and History*. Cambridge: Cambridge University Press.

Porter, Roy (1978). Gentlemen and geology: The emergence of a scientific career, 1660–1920. *The Historical Journal*, 21, 809–836.

(1980). Science, provincial culture and public opinion in Enlightenment England. *British Journal for Eighteenth-century Studies*, 3, 20–46.

(1981). The Enlightenment in England. In Porter and Teich, eds. (1981), pp. 1–18.

(1982a). *English Society in the Eighteenth Century*. Harmondsworth: Penguin.

(1982b). Was there a medical Enlightenment in eighteenth-century England? *British Journal for Eighteenth-century Studies*, 5, 49–63.

(1985a). Laymen, doctors and medical knowledge in the eighteenth century: The evidence of the *Gentleman's Magazine*. In *Patients and Practitioners: Lay Perceptions of Medicine in Pre-industrial Society*, ed. R. Porter, pp. 283–314. Cambridge: Cambridge University Press.

(1985b). William Hunter: A surgeon and a gentleman. In Bynum and Porter, eds. (1985), pp. 7–34.

(forthcoming). *Doctor of Society: Thomas Beddoes, Medicine and Reform*. London: Routledge.

Porter, Roy and Teich, Mikulas, eds. (1981). *The Enlightenment in National Context*. Cambridge: Cambridge University Press.

Porter, Theodore M. (1981). The promotion of mining and the advancement of science: The chemical revolution of mineralogy. *Annals of Science*, 38, 543–570.

Poynter, F.N.L., ed. (1966). *The Evolution of Medical Education in Britain*. London: Pitman Medical.

Priestley, Joseph (1767). *The History and Present State of Electricity, With Original Experiments*. London: J. Dodsley et. al.

(1769). *A Familiar Introduction to the Study of Electricity*, 2nd edn. London: J. Dodsley, et al.

(1772a). *The History and Present State of Discoveries Relating to Vision, Light and Colours*, 2 vols. London: J. Johnson.

(1772b). *Directions for Impregnating Water with Fixed Air*. London: J. Johnson.

(1772c). Observations on different kinds of air. *Philosophical Transactions*, 62, 147–264.

(1775). *Philosophical Empiricism: Containing Remarks on a Charge of Plagiarism Respecting Dr. H--s*. London: J. Johnson.

(1775–77). *Experiments and Observations on Different Kinds of Air*, 2nd edn, 3 vols. London: J. Johnson.

(1777). *A Course of Lectures on Oratory and General Criticism*. London: J. Johnson.

(1778). *Miscellaneous Observations Relating to Education*. Bath: R. Cruttwell.

(1779–86). *Experiments and Observations Relating to Various Branches of Natural Philosophy*, 3 vols. London: J. Johnson.

(1783). Experiments relating to phlogiston, and the seeming conversion of water into air. *Philosophical Transactions, 73,* 398–434.

(1790). *Experiments and Observations on Different Kinds of Air, and Other Branches of Natural Philosophy*, 3 vols. Birmingham: Thomas Pearson.

(1791a). Further experiments relating to the decomposition of dephlogisticated and inflammable air. *Philosophical Transactions, 81,* 213–222.

(1791b). *Letters to the Right Honourable Edmund Burke, Occasioned by his Reflections on the Revolution in France*. Birmingham: Thomas Pearson.

(1794). *Heads of a Course of Experimental Philosophy, Particularly Including Chemistry, Delivered at the New College in Hackney*. London: J. Johnson.

(1795): *An Answer to Mr. Paine's Age of Reason*. London: J. Johnson.

(1796). *Considerations on the Doctrine of Phlogiston, and the Decomposition of Water*. Philadelphia: Thomas Dobson; reprinted, ed. William Foster, Princeton: Princeton University Press, 1929.

(1800). *The Doctrine of Phlogiston Established and that of the Composition of Water Refuted*. Northumberland, Penn.: for the author.

(1802). Observations and experiments relating to the pile of Volta. *Nicholson's Journal*, 2nd ser., *1,* 198–204.

(1803). *The Doctrine of Phlogiston Established*, 2nd edn. Northumberland, Penn.: for the author.

(1806). *Memoirs of Dr. Joseph Priestley to the Year 1795, Written by Himself.* London: J. Johnson.

[1817–31]. *The Theological and Miscellaneous Works of Joseph Priestley, LL.D. F.R.S.*, ed. John Towill Rutt, 25 vols. in 26. Hackney: George Smallfield.

Prout, William (1834). *Chemistry, Meteorology and the Function of Digestion Considered with Reference to Natural Theology (Bridgewater Treatise no. 8)*, 2nd edn. London: William Pickering.

Ramsay, William (1918). *Life and Letters of Joseph Black*. London: Constable.

Randolph, George (1745). *An Enquiry into the Medicinal Virtues of Bristol-water*. Oxford: James Fletcher.

Rappaport, Rhoda (1967). Lavoisier's geologic activities, 1763–1792. *Isis, 58,* 375–384.

Reece, Richard (1814). *The Chemical Guide, or Complete Companion to the Portable Chest of Chemistry*. London: Longman et. al.

Reid, William Hamilton (1800). *The Rise and Dissolution of the Infidel Societies in this Metropolis*. London: J. Hatchard.

Rice-Oxley, L., ed. (1924). *Poetry of the Anti-Jacobin*. Oxford: Basil Blackwell.

Riley, James C. (1987). *The Eighteenth-century Campaign to Avoid Disease*. London: Macmillan.

Risse, Guenther B. (1974). "Doctor William Cullen, Physician, Edinburgh": A consultation practice in the eighteenth century. *Bulletin of the History of Medicine, 48,* 338–351.

Rivers, Isabel, ed. (1982). *Books and their Readers in Eighteenth-century England*. Leicester: Leicester University Press.

Roberts, Lissa (1991a). A word and the world: The significance of naming the calorimeter. *Isis, 82,* 199–222.

(1991b). Setting the table: The disciplinary development of eighteenth-century chemistry as read through the changing structure of its tables. In *The Liter-*

ary Structure of Scientific Argument: Historical Studies, ed. Peter Dear, pp. 99–132. Philadelphia: University of Pennsylvania Press.

Robinson, Bryan (1743). *Dissertation on the Aether of Sir Isaac Newton*. Dublin: George Ewing and William Smith.

Robinson, Eric (1963). Benjamin Donn (1729–1798), teacher of mathematics and navigation. *Annals of Science, 19*, 27–36.

Robinson, Eric and McKie, Douglas, eds. (1970). *Partners in Science: Letters of James Watt and Joseph Black*. London: Constable.

Robinson, F.J.G. and Wallis, P.J. (1975). *Book Subscription Lists: A Revised Guide*. Newcastle: Harold Hill and Son, for the Book Subscription Lists Project.

Robison, John (1797). *Proofs of a Conspiracy against all the Governments and Religions of Europe, Carried on in Secret Meetings of Free Masons, Illuminati and Reading Societies*. Edinburgh: William Creech et al.

Roche, Daniel (1978). *Le Siècle des Lumières en Province: Académies et Académiciens Provinciaux, 1680–1789*, 2 vols. Paris and The Hague: Mouton.

(1980). Talent, reason and sacrifice: The physician during the Enlightenment. In *Medicine and Society in France: Selections from the Annales E.S.C.*, eds. Robert Forster and Orest Ranum, pp. 66–88. Baltimore: Johns Hopkins University Press.

Ross, Ian S. (1972). *Lord Kames and the Scotland of his Day*. Oxford: Clarendon Press.

Rouse, Joseph (1987). *Knowledge and Power: Toward a Political Philosophy of Science*. Ithaca, N.Y.: Cornell University Press.

Rousseau, George S. (1967). Matt Bramble and the sulphur controversy in the XVIIIth century: Medical background of Humphry Clinker. *Journal of the History of Ideas, 28*, 577–589.

(1982). Science books and their readers in the eighteenth century. In Rivers, ed. (1982), pp. 197–255.

Rousseau, George S. and Porter, Roy, eds. (1980). *The Ferment of Knowledge: Studies in the Historiography of Eighteenth-Century Science*. Cambridge: Cambridge University Press.

Rowbottom, Margaret E. (1968). The teaching of experimental philosophy in England, 1700–1730. In *Actes du XIe Congrès International d'Histoire des Sciences, Varsovie-Cracovie 1965*, 6 vols., IV, pp. 46–57. Wroclaw, Poland.

Royal Society (1867–72). *Catalogue of Scientific Papers (1800–1863)*, compiled by the Royal Society, 6 vols. London: Eyre and Spottiswood.

(1985). *The Public Understanding of Science: Report of a Royal Society ad hoc Group endorsed by the Council of the Royal Society*. London: The Royal Society.

Royal Society of Chemistry (1984). *Oxygen and the Conversion of Future Feedstocks: Proceedings of the Third BOC Priestley Conference*. London: Royal Society of Chemistry.

Royal Society of Edinburgh [1739]. Proposals for the regulation of a society for improving arts and sciences, and particularly natural knowledge. Reprinted in *Transactions of the Royal Society of Edinburgh. General Index to the First Thirty-Four Volumes (1783–1888)*, pp. 22–26. Edinburgh: Neill & Co., 1890.

Royle, Edward (1971). Mechanics' Institutes and the working classes, 1840–1860. *The Historical Journal, 14*, 305–321.

Rudwick, Martin J.S. (1962–63). The foundation of the Geological Society of

London: Its scheme for co-operative research and its struggle for independence. *British Journal for the History of Science, 1*, 325–355.

Rupp, Theophilus Lewis (1798). Remarks on Dr. Priestley's experiments and observations relating to the analysis of atmospherical air. *Memoirs of the Manchester Literary and Philosophical Society, 5*, 123–162.

Russell, Colin A. (1959–63). The electrochemical theory of Sir Humphry Davy. *Annals of Science, 15* (1959), 1–13, 15–25; *19* (1963), 255–271.

——— (1983). *Science and Social Change, 1700–1900*. London: Macmillan.

Russell, Colin A., Roberts, G.K. and Coley, N.G. (1977). *Chemists by Profession: The Origins and Rise of the Royal Institute of Chemistry*. Milton Keynes: Open University Press.

Saunders, William (1766). *A Syllabus of Lectures on Chemistry and Pharmacy*. [London: n.p.]

——— (1800). *A Treatise on the Chemical History and Medical Powers of Some of the Most Celebrated Mineral Waters*. London: William Phillips.

Schaffer, Simon (1980). Natural philosophy. In Rousseau and Porter, eds. (1980), pp. 55–91.

——— (1983). Natural philosophy and public spectacle in the eighteenth century. *History of Science, 21*, 1–43.

——— (1984). Priestley's questions: An historiographical survey. *History of Science, 22*, 151–183.

——— (1987). Priestley and the politics of spirit. In Anderson and Lawrence, eds. (1987), pp. 39–53.

——— (1990). Measuring virtue: Eudiometry, enlightenment and pneumatic medicine. In *The Medical Enlightenment of the Eighteenth Century*, eds. Andrew Cunningham and Roger French, pp. 281–318. Cambridge: Cambridge University Press.

Scheele, Carl Wilhelm, et al. (1897). *The Early History of Chlorine*, Alembic Club reprints, no. 13. Edinburgh: William Clay.

Schiebinger, Londa (1989). *The Mind has no Sex? Women in the Origins of Modern Science*. Cambridge: Harvard University Press.

Schofield, Robert E. (1959). The Society of Arts and the Lunar Society of Birmingham. *Journal of the Royal Society of Arts, 107*, 512–514, 668–671.

——— (1961). Boscovich and Priestley's theory of matter. In *Roger Joseph Boscovich*, ed. Lancelot L. Whyte, pp. 168–172. London: Allen and Unwin.

——— (1963). *The Lunar Society of Birmingham: A Social History of Provincial Science and Industry in Eighteenth-century England*. Oxford: Clarendon Press.

——— ed. (1966). *A Scientific Autobiography of Joseph Priestley (1733–1804)*. Cambridge: M.I.T. Press.

——— (1967). Joseph Priestley, natural philosopher. *Ambix, 14*, 1–15.

——— (1970). *Mechanism and Materialism: British Natural Philosophy in an Age of Reason*. Princeton: Princeton University Press.

——— (1974). Joseph Priestley and the physicalist tradition in British chemistry. In Hiebert, et al., (1974), pp. 92–117.

——— (1983). Joseph Priestley: Theology, physics and metaphysics. *Enlightenment and Dissent*, no. 2, 69–81.

Scott, E.L. (1970). The "Macbridean doctrine" of air: An eighteenth-century explanation of some biochemical processes, including photosynthesis. *Ambix, 17*, 43–57.

Secord, James A. (1985). Newton in the nursery: Tom Telescope and the philosophy of tops and balls, 1761–1838. *History of Science, 23*, 127–151.

Senebier, Jean (1783). *Recherches sur l'Influence de la Lumière Solaire pour Métamorphoser l'Air Fixe en Air Pur par la Végétation*. Geneva: B. Chirol.

Sennett, Richard (1977). *The Fall of Public Man*. Cambridge: Cambridge University Press.

Shapin, Steven (1972). The Pottery Philosophical Society, 1819–1835: An examination of the cultural uses of provincial science. *Science Studies*, 2, 311–336.

(1982). History of science and its sociological reconstructions. *History of Science*, 20, 157–211.

(1984). Pump and circumstance: Robert Boyle's literary technology. *Social Studies of Science*, 14, 481–520.

(1988). The house of experiment in seventeenth-century England. *Isis*, 79, 373–404.

(1990). Science and the public. In Olby, Cantor, Christie and Hodge, eds. (1990), pp. 990–1007.

Shapin, Steven and Barnes, Barry (1977). Science, nature and control: Interpreting Mechanics' Institutes. *Social Studies of Science*, 7, 31–74.

Shapin, Steven and Schaffer, Simon (1985). *Leviathan and the Air-Pump: Hobbes, Boyle and the Experimental Life*. Princeton: Princeton University Press.

Shaw, John Stuart (1983). *The Management of Scottish Society, 1707–1764*. Edinburgh: John Donald.

Shaw, Peter [1734a]. *Chemical Lectures, Publickly Read at London in the Years 1731 and 1732, and Since at Scarborough in 1733*. London: Schuckburgh and Osborn.

(1734b). *An Enquiry into the Contents, Virtues, and Uses of the Scarborough Spa-waters*. London: for the author.

(1735). *A Dissertation on the Contents, Virtues and Uses of Cold and Hot Mineral Springs*. London: Ward and Chandler.

Shelley, Mary (1968). Frankenstein, or the modern Prometheus. In *Three Gothic Novels*, ed. Peter Fairclough, pp. 258–497. Harmondsworth: Penguin Books.

Shepherd, C.M. (1982). Newtonianism in Scottish universities in the seventeenth century. In Campbell and Skinner, eds. (1982), pp. 65–85.

Sher, Richard (1985). *Church and University in the Scottish Enlightenment: The Moderate Literati of Edinburgh*. Edinburgh: Edinburgh University Press.

Shinn, Terry and Whitley, Richard, eds. (1985). *Expository Science: Forms and Functions of Popularisation*, Sociology of the Sciences Yearbook, no. 9. Dordrecht: Reidel.

Siegfried, Robert (1972). Lavoisier's view of the gaseous state and its early application to pneumatic chemistry. *Isis*, 63, 59–78.

Siegfried, Robert and Dobbs, B.J.T. (1968). Composition, a neglected aspect of the Chemical Revolution. *Annals of Science*, 24, 275–293.

[Simmons, Samuel Fort, ed.] (1783). *The Medical Register for the Year 1783*. London: J. Johnson.

Singer, Dorothea Wade (1948–50). Sir John Pringle and his circle. *Annals of Science*, 6, 127–180, 229–261.

Smeaton, W.A. (1965–66). The portable laboratories of Guyton de Morveau, Cronstedt and Göttling. *Ambix*, 13, 84–91.

Smith, Adam (1980). The principles which lead and direct philosophical enquiries, illustrated by the history of astronomy. In *Essays on Philosophical Subjects*, eds. W.P.D. Wightman and J.C. Bryce, pp. 31–105. Oxford: Clarendon Press.

Smith, Barbara and Moilliet, J.L. (1967). James Keir of the Lunar Society. *Notes and Records of the Royal Society of London*, 22, 144–154.

[Smith, John] (1795). *Assassination of the King! . . . to which is added a Letter from Mr. Parkinson Containing the Particulars of his Examination before the Privy-Council.* London: J. Smith.

Smith, Olivia (1984). *The Politics of Language 1791–1819.* Oxford: Clarendon Press.

Smith, Robert (1778). *The Elementary Parts of Dr. Smith's Compleat System of Opticks.* Cambridge: J. Archdeacon.

Smith, W.D.A. (1982). *Under the Influence: A History of Nitrous Oxide and Oxygen Anaesthesia.* London: Macmillan.

Snelders, H.A.M. (1988). The new chemistry in the Netherlands. In Donovan, ed. (1988a), pp. 121–145.

Speck, W.A. (1982). Politicians, peers and publications by subscription 1700–1750. In Rivers, ed. (1982), pp. 47–68.

Stansfield, Dorothy A. (1984). *Thomas Beddoes M.D. 1760–1808: Chemist, Physician, Democrat.* Dordrecht: D. Reidel.

Stansfield, Dorothy A. and Stansfield, Ronald G. (1986). Dr. Thomas Beddoes and James Watt: Preparatory work 1794–96 for the Bristol Pneumatic Institute. *Medical History, 30,* 276–302.

Stewart, John (1754). Some remarks on the laws of motion and the inertia of matter. *Essays and Observations, Physical and Literary, 1,* 70–140.

Stewart, Larry (1986a). Public lectures and private patronage in Newtonian England. *Isis, 77,* 47–58.

(1986b). The selling of Newton: Science and technology in early eighteenth-century England. *Journal of British Studies, 25,* 178–192.

(forthcoming). *Rhetoric and the Rise of Public Science, 1660–1750: Technology, Projectors and Natural Philosophy in Newtonian Britain.*

Sudduth, William M. (1980). The voltaic pile and electrochemical theory in 1800. *Ambix, 27,* 26–35.

Sutherland, L.S. and Mitchell, L.G. eds. (1986). *The History of the University of Oxford,* vol. V: *The Eighteenth Century.* Oxford: Clarendon Press.

Sutton, Geoffrey (1981). The politics of science in early Napoleonic France: The case of the Voltaic pile. *Historical Studies in the Physical Sciences, 11,* 329–366.

Sylvester, Charles (1804). Observations and experiments to elucidate the operation of the galvanic power. *Nicholson's Journal,* 2nd ser., 9, 179–182.

(1806). Repetition of the experiment in which acids and alkali are produced in pure water by galvanism. *Nicholson's Journal,* 2nd ser., 15, 50–52.

(1809). *An Elementary Treatise on Chemistry.* Liverpool: E. and W. Smith.

Thackray, Arnold (1970). *Atoms and Powers: An Essay on Newtonian Matter-Theory and the Development of Chemistry.* Cambridge: Harvard University Press.

(1972). *John Dalton: Critical Assessments of His Life and Science.* Cambridge: Harvard University Press.

(1974). Natural knowledge in cultural context: The Manchester model. *American Historical Review, 79,* 672–709.

Thomas, D.O. (1987). Progress, liberty and utility: The political philosophy of Joseph Priestley. In Anderson and Lawrence, eds. (1987), pp. 73–80.

Thompson, E.P. (1978). Eighteeenth-century English society: Class struggle without class? *Social History, 3,* 133–165.

Thomson, John (1832–59). *An Account of the Life, Lectures and Writings of William Cullen M.D.*, 2 vols., vol. II ed. by David Craigie. Edinburgh: William Blackwood & Sons.

Thomson, Thomas (1802). *A System of Chemistry*, 4 vols. Edinburgh: Bell and Bradfute.

(1813a). Review of Davy's *Elements of Chemical Philosophy*. *Annals of Philosophy*, *1* (1813), 371–377.

(1813b). Some observations in answer to Mr. Chenevix's attack upon Werner's mineralogical method. *Annals of Philosophy, 1*, 241–258.

(1817). *A System of Chemistry*, 5th edn., 4 vols. London: Baldwin, Cradock and Joy.

(1829–30). History and present state of chemical science. *Edinburgh Review, 50*, 256–276.

(1830–31). *The History of Chemistry*, 2 vols. London: Henry Colburn and Richard Bentley.

Torrens, Hugh S. (1979). Geological communication in the Bath area in the last half of the eighteenth century. In Jordanova and Porter, eds. (1979), pp. 215–247.

(1983). Arthur Aikin's mineralogical survey of Shropshire 1796–1816, and the contemporary audience for geological publications. *British Journal for the History of Science, 16*, 111–153.

Treneer, Anne (1963). *The Mercurial Chemist: A Life of Humphry Davy.* London: Methuen.

Turner, A.J. (1977). *Science and Music in Eighteenth-century Bath.* Bath: University of Bath.

Turner, E.S. (1967). *Taking the Cure.* London: Michael Joseph.

Turner, G.L'E. (1986). The physical sciences. In Sutherland and Mitchell, eds., (1986), pp. 659–681.

[University of Edinburgh] (1867). *List of Graduates in Medicine in the University of Edinburgh from MDCCV to MDCCCLXVI.* Edinburgh: Neill & Co.

Ure, Andrew (1821). *A Dictionary of Chemistry on the Basis of Mr Nicholson's.* London: Thomas & George Underwood et al.

Usselman, Melvyn C. (1978). The Wollaston/Chenevix controversy over the elemental nature of palladium: A curious episode in the history of chemistry. *Annals of Science, 35*, 551–579.

Viseltear, Arthur J. (1968). Joanna Stephens and the eighteenth century lithontriptics: A misplaced chapter in the history of therapeutics. *Bulletin for the History of Medicine, 42*, 199–220.

Volta, Alessandro (1800). On the electricity excited by the mere contact of conducting substances of different kinds. *Philosophical Transactions, 90*, 403–431.

Walker, Adam (1766). *Analysis of a Course of Lectures on Natural and Experimental Philosophy*, [1st edn.] Kendal: for the author.

(1771). *Syllabus of a Course of Lectures on Natural and Experimental Philosophy*, [1st edn.] Liverpool: W. Nevett.

(1777). *A Philosophical Estimate of the Causes, Effects and Cure of Unwholesome Air in Large Cities.* [London]: for the author.

[1780?]. *Analysis of a Course of Lectures on Natural and Experimental Philosophy*, 4th edn. [London]: for the author.

[c. 1785?]. Ibid., 8th edn. [London]: C. Buckton.

[1796?]. *Syllabus of a Course of Lectures on Natural and Experimental Philosophy.* N.p.

(1807). *Analysis of a Course of Lectures in Natural and Experimental Philosophy*, 14th edn. London: J. Barfield.

Walker, Ezekiel (1823). *Philosophical Essays Selected from the Originals Printed in the Philosophical Journals*. Lynn, Norfolk: John Wade.

Wall, Martin (1783). *Dissertations on Select Subjects in Chemistry and Medicine*. Oxford: D. Prince and J. Cooke.

Watson, Richard (1783). *Chemical Essays*, 3rd edn. Dublin: R. Moncrieffe.

Watson, Robert Spence (1897). *The History of the Literary and Philosophical Society of Newcastle-upon-Tyne (1793–1896)*. London: Walter Scott.

Webster, Charles (1986). The medical faculty and the physic garden. In Sutherland and Mitchell, eds. (1986), pp. 683–723.

Webster, John (1811). *Elements of Chemistry*. London: for the author.

Weindling, Paul (1979). Geological controversy and its historiography: The prehistory of the Geological Society of London. In Jordanova and Porter, eds. (1979), pp. 248–271.

(1983). The British Mineralogical Society: A case-study in science and social improvement. In Inkster and Morrell, eds. (1983), pp. 120–150.

Weld, Charles Richard (1848). *A History of the Royal Society*, 2 vols. London: John Parker.

Weldon, Walter [1825?]. *A Popular Explanation of the Elements and General Laws of Chemistry*. London: J. Barfield.

Wetzels, Walter D. (1990). Johann Wilhelm Ritter: Romantic physics in Germany. In Cunningham and Jardine, eds. (1990), pp. 199–212.

Wheeler, T.S. and Partington, J.R. (1960). *The Life and Work of William Higgins, Chemist (1763–1825)*. New York: Pergamon Press.

White, William (1778). Experiments upon air, and the effects of different kinds of effluvia upon it. *Philosophical Transactions*, 68, 194–220.

(1782). Observations on the bills of mortality at York. *Philosophical Transactions*, 72, 35–43.

Whitley, Richard (1983). From the sociology of scientific communities to the study of scientists' negotiations and beyond. *Social Science Information*, 22, 681–720.

(1985). Knowledge producers and knowledge acquirers: Popularisation as a relation between scientific fields and their publics. In *Expository Science: Forms and Functions of Popularisation (Sociology of the Sciences Yearbook*, no. 9), eds. Terry Shinn and Richard Whitley, pp. 3–28. Dordrecht: Reidel.

Whitt, L.A. (1990). Atoms or affinities? The ambivalent reception of Daltonian theory. *Studies in History and Philosophy of Science*, 21, 57–89.

Wightman, William P.D. (1955). William Cullen and the teaching of chemistry, I. *Annals of Science*, 11, 154–165.

(1956). William Cullen and the teaching of chemistry, II. *Annals of Science*, 12, 195–205.

Williams, L. Pearce (1960). Michael Faraday's education in science. *Isis*, 51, 515–530.

Williams, W.J. and Stoddart, D.M. (1978). *Bath: Some Encounters with Science*. Bath: Kingsmead Press.

Wilson, Arthur M. (1983). The Enlightenment came first to England. In *England's Rise to Greatness, 1660–1763*, ed. Stephen B. Baxter, pp. 1–28. Berkeley and Los Angeles: University of California Press.

Wilson, George (1709). *A Compleat Course of Chymistry*, 3rd edn. London: John Bayley.

Wollaston, William Hyde (1814). A synoptic scale of chemical equivalents. *Philosophical Transactions, 104*, 1–22.

Wood, J. Russell (1948–50). A biographical note on William Brownrigg, MD, FRS, (1711–1800). *Annals of science, 6*, 186–196.

(1948–51). The scientific work of William Brownrigg, MD, FRS, (1711–1800). *Annals of science, 6* (1948–50), 436–447; 7 (1951), 77–94, 199–206.

Woodhouselee, Lord (Alexander Frazer Tytler) (1807). *Memoirs of the Life and Writings of the Honourable Henry Home of Kames*, 3 vols. Edinburgh: William Creech.

Wright, Thomas and Evans, R.H. (1851). *Historical and Descriptive Account of the Caricatures of James Gillray*. London: Henry Bohn.

Wynne, Brian (1982). *Rationality and Ritual: The Windscale Inquiry and Nuclear Decisions in Britain*, B.S.H.S. Monographs, no. 3. Chalfont St. Giles, Bucks.: British Society for the History of Science.

Yolton, John W. (1984). *Thinking Matter: Materialism in Eighteenth-Century Britain*. Oxford: Basil Blackwell.

Ziemacki, Richard L. (1974). *Humphry Davy and the Conflict of Traditions in Early Nineteenth-century British Chemistry*. Cambridge University Ph.D. dissertation.

Ziman, John (1968). *Public Knowledge: An Essay Concerning the Social Dimension of Science*. Cambridge: Cambridge University Press.

Index

Académie des Sciences (Paris), 138, 143, 287

Accum, Friedrich Christian, 126, 141, 241, 245, 246, 247, 248, 254, 259–260, 262

works of
Chemical Amusement, 246, 262, 279
Elements of Crystallography, 278
Manual of Analytical Mineralogy, 278
Practical Essay on Analysis of Minerals, 282
Practical Treatise on . . . Chemical Tests, 282

acids, 26, 208, 209, 211. see also individual acids

Adelphi Theatre (London), 175

affinity tables, use by Cullen, 21, 25

aggregates and aggregation, Cullen's theories on, 23–24

agricultural chemistry, Davy's lectures on, 198, 199

agriculture
chemistry applied to, 28–29, 31, 32, 33, 195, 199
Cullen's and Black's contributions to, 12
Cullen's interest in, 31–33
Cullen's lectures on, 17
Lord Kames' interest in, 31, 32, 35

Aikin, Arthur, 243, 247, 250, 254, 273, 281
Manual of Mineralogy, 280

Aikin, Charles, 247, 254

Aikin, John, 75, 254
as lecturer at Warrington Academy, 54, 95

air and airs, 91–128
analysis of, 117–128
dephlogisticated, 73, 78, 86, 88, 97, 98, 99, 135, 136, 143
fixed, 93, 97, 107, 109, 110, 135
Black's thesis on, 43
inflammable, 97, 134, 135, 136, 144, 213
Lavoisier's theories of, 131
medical uses of, 105, 109, 110, 111

nitrous, 97
Priestley's discoveries on, 8, 49, 50, 64–65, 77, 97, 106
public lectures on, 96, 103
therapeutic uses of, 157–166

Albrighton (England), 157

Albury, W. R., 147

alchemy, Enlightenment chemistry compared to, 179

Alcock, Nathan, as lecturer at Oxford, 53

Alderson, John, 161

Alemani, 209

alkali earths, Davy's work on, 204, 213

alkali manufacture
applied chemistry in, 29, 34
Black's and Watt's work on, 40
Cullen's work on, 35, 37

alkali metals, 213, 214
Davy's isolation of, 239, 259, 270

alkalis, 26, 27, 208, 209, 211

Allen, John, 207, 241

Allen, William, 239, 245–246, 247, 249, 252, 254, 260, 261

Alumni Cantabrigienses, 209

Amicus [pen name], 174, 175

ammonia, 227, 228, 230–231

analysis, chemical. see chemical analysis

Anderson, Wilda, 147

Anderson's Institution (Glasgow), 191

Animal Chemistry Club (London), 251, 252, 254

animal motion, Galvani's discoveries and, 216

Annales de Chimie, 275

Annals of Philosophy, 253, 254, 275

Anti-Jacobin Review and Magazine, 172, 173, 174, 179

antiphlogistic theory, 134, 136, 146, 173, 213

antiseptics, 106, 109

Apothecaries' Act of 1815, 247

Apothecaries' Company, 247

apparatus
Black's, 47
in laboratories, 262

apparatus (*cont.*)
  Lavoisier's, 129
  Priestley's, 83–84, 93
  in public lectures, 99
  simplicity of, 117
aqua mephitica alkalina, 115
aragonite, 274
Arden, John, as public lecturer, 96, 97, 98, 99, 100, 101
Argand lamp, 199
Argyll, Duke of. *see* Campbell, Archibald (Duke of Argyll)
Aris's Birmingham Gazette, 97, 98, 99, 100, 101, 102, 103
aristocrats
  as members of Royal Society of London, 55
  as patrons of science, 6–7, 12, 14–15, 54–55, 56, 60, 63–64, 66, 70, 184, 190, 193, 207, 237
  as readers of Priestley's books, 75, 76
  role in 18th century English cultural life, 57
Aristotle, four elements of, 20
arts
  Adam Smith's views on, 30
  chemistry applied to, 28, 29, 48, 60, 103, 195–196, 198
  Cullen's views on, 31, 48
Ashmolean Museum, 53
Askesian Society, 254
atomic theory, Dalton's, 255–269

Babbage, Charles, 245
Babington, William, 208, 239, 241, 249, 251, 254, 260, 261
Bachelard, Gaston, 3
Bacon, Francis, 85, 104
  inductive method of, 25, 29
    Cullen's application to chemistry, 19–20
  views on secrecy in science, 4
Bakerian Lecture(s)
  Brande's, 247
  Davy's, 189, 203–204, 209, 211–212, 214, 215, 217
balance of nature, Priestley's view of, 78
balances
  chemical
    Dalton's, 265
    Lavoisier's, 138–139
    in portable laboratories, 262
  for soil analyses, 199
Banks, John, as public lecturer, 97, 98, 101
Banks, Sir Joseph, 63, 192, 205, 217–218
  as President of Royal Society of London, 55, 69, 124, 158, 162–163, 190, 252

Barfoot, Michael, 36
Barruel, Abbé, 185
Bath (England), 53, 59, 67, 97, 111, 213
  mineral waters of, 62
Bath Philosophical Society, 100
battery, voltaic pile as, 205, 215–216, 218, 219, 246, 259
Baumé, Antoine, 54, 132
Beddoes, Anna Edgeworth [Thomas's wife], 162, 168
Beddoes, Thomas, 51, 116, 141, 190
  acceptance of Lavoisier's chemical theories by, 153–154, 155
  as Black's student, 154
  collaboration with Watt, 157
  critics of, 173–175, 176, 179, 184, 186, 193
  as Davy's mentor, 234
  discovery [with Davy] of and experiments on nitrous oxide as an intoxicant, 9, 152, 156, 166–175
  as lecturer at Oxford, 54, 154
  ostracism by Royal Society, 158, 162, 163
  Pneumatic Institution of, 157–166, 188
  uses of nitrous oxide by, 153–187, 237, 243
  views on importance of chemistry, 8–9, 196, 242
  voltaic pile and, 203
  works of
    Considerations on the Medicinal Use of Factitious Airs, 157, 160, 162, 164, 166, 174
    Essay on the Public Merits of Mr. Pitt, 159
    Letter to Erasmus Darwin, 157, 165
    Notice of Some Observations made at the Medical Pneumatic Institution, 167
    Observations . . . Nature and Cure of Calculus, Sea Scurvy, Consumption, Catarrh, and Fever, 157, 158
    Observations on the Nature of Demonstrative Evidence, 155, 170
    writing style of, 163–164, 166, 216–218
Bedford, Duke of, 165, 184
Bell, Benjamin, 162
Bell, John, 162
Bennet, Abraham, *New Experiments on Electricity*, 76
Bentham, Jeremy, 181
Bentley, T., 96
Bergman, Tobern, 263–264, 272
  Outlines of Mineralogy, 271
Berlin, 271

Berman, Morris, 57, 190
Berthollet, Claude-Louis, 132, 133, 222
Berzelius, Jöns Jakob, 224, 277–278, 280
Beverley (England), 96, 97
Bewley, Richard [Robert Harrington], 151–152
Bewley, William, 72–73, 76, 115–116
Bewley's julep, 115–116, 157
Birkbeck, George, 242, 261
Birmingham (England), 59, 97, 98, 99, 100, 101, 103, 110, 271
  Priestley in, 65
Black, Alexander [brother of Joseph], 39, 40
Black, George, Jr., 46
Black, James [brother of Joseph], 39, 40
Black, Joseph, 34, 54, 58, 71, 107, 108, 116, 141, 162, 170
  as chemistry professor, 15, 38, 39, 49
  Cullen as professor and mentor of, 12, 27, 38
  French chemistry and, 286–287
  Henry Home (Lord Kames) and, 39
  James Watt and, 39–40
  lack of publications by, 13, 41, 43, 44
  lectures by, 13, 37, 41, 45–47
  as pioneer of Scottish chemistry, 12–13
  portrait of, 42
  reaction to Lavoisier's chemistry, 48, 134
  studies on heat, 37, 45, 47
  views on importance of chemistry, 7
  as a writer, 41
black-boxing, of instruments, 206, 214, 216
Blagden, Sir Charles, 135, 149, 201
Blair, Hugh, 13
bleaching
  chemistry applied to, 28–29, 31, 33, 34, 195
  Cullen's work on, 33–34, 37
  Home's chemical theory of, 33
blood, chemical analysis of, 251
blowpipe
  description and use of, 271, 279
  in portable laboratories, 279
  use in chemical analysis, 251, 262, 263, 270, 283
Board for Improving Agriculture (Scotland), 32
Board of Agriculture, 191, 198
Board of Trustees for the Encouragement of Fisheries, Arts and Manufactures (Edinburgh), 33, 34
Boaz, Sieur Herman, 103
Boerhaave, Hermann, 20, 25, 60
  Cullen's attack on, 18
book clubs, 75

books, role in Enlightenment, 14
Boscovich, Roger, 256
Bostock, John, 112, 209, 228, 234, 250, 251, 252, 256–257
Boswell, James, Life of Johnson, 93
botany, classification and, 20
Boulton, Matthew, 40, 41, 43, 57, 67, 69, 76, 100, 116, 119, 161, 164
  role in the Lunar Society, 70
Boulton, Robinson, 170
Boyle, Robert, 7, 11, 25, 107, 197
  Priestley compared to, 87
  views on dissemination of scientific knowledge, 5, 19
Brande, William Thomas, 220, 221, 241, 242–243, 246–247, 249, 251, 252–253, 259, 261, 268, 272
  Bakerian Lecture of, 247
  Manual of Chemistry, 257–258
breathing machines
  use in nitrous oxide experiments, 167–170
  of Watt and Boulton, 164–165
Breslaw, 122
Bristol (England), 97, 113, 162, 164, 167, 170, 174, 187, 188, 216
  Davy's departure from, 9
  Enlightenment in, 73
  library at, borrowing of Priestley's books from, 73–74
  Pneumatic Institution in, 157–166
British Association for the Advancement of Science (BAAS), 265, 266
British Mineralogical Society, 254–255, 282–283
Brocklesby, Dr. Richard, 88, 89
Brougham, Henry, 190, 193, 212, 217, 224, 234, 244, 246–247, 248
  as Black's student, 45–46
Brown, John, 159
Brownrigg, William, 107–108, 109, 113
Brunonianism, pneumatic medicine and, 159, 168
Burke, Edmund, 150, 193
  as critic of chemists and chemistry, 9, 176–179, 181, 184, 185–186
  Letter to a Noble Lord, 176, 177, 178, 184
  Reflections on the Revolution in France, 176
Buxton Springs (England), 272

calcite, 274
Calne (Wiltshire, England), 64, 67, 97, 99, 100
caloric theory, Lavoisier's, 154
calorimeter, Lavoisier's, 140–142

calorimetry, Lavoisier's work on, 132, 133, 137
Cambridge University, 61, 75, 111, 160, 209, 245, 281
  chemistry lectures at, 53
  medical education at, 52
Campbell, Archibald (Duke of Argyll), 37
  as patron of Cullen, 16, 34, 35–36
cancer, pneumatic therapy of, 161
Canning, George, 172
Canton, John, 68, 69, 75
carbon dioxide, use in pneumatic medicine, 159
carbonic acid, 227
carbon monoxide, 136
Carlisle (England), 151, 173, 213
  Mechanics' Institute at, 248
Carlisle, Anthony, 205, 206, 207, 250–251, 252
Carlisle College (Pennsylvania), Cooper's Introductory Lecture at, 276
Carlyle, Thomas, 194, 284, 285
cartoons
  Priestley as subject of, 179–183
  ridicule of science lectures by, 201, 202
Cashel, Archbishop of (Ireland), 174
Cavallo, Tiberius, 121, 123, 127
  Complete Treatise on Electricity, 74
  dispute with Magellan, 122–124
  Treatise on the Nature and Properties of Air, 74, 122
Cavendish, Henry, 12, 54, 62, 75, 108, 109, 117, 133, 139, 151
  on air, 135, 136
  disagreement with Lavoisier's nomenclature, 149
  eudiometer of, 124, 125, 126
  experiments of, 135, 136
  French chemistry and, 286
chemical analysis, 236–287
  Davy's views on, 199
  equivalents in, 269
  mineralogy role in development of, 269–283
  techniques of, 10
  voltaic pile use in, 9, 208, 214, 222–223
  Wollaston's slide rule and scale use in, 268
Chemical and Medical Hall (Piccadilly, London), 241, 262
chemical community
  education, analysis, and, 236–287
  in London, 251–252
  structure of, 8, 10
chemical education, for women, 241
chemical laboratories, for instruction, 260
chemical manufacturing, Wollaston's slide rule and scale use in, 258

chemical reagents
  development of, 281–282
  in portable laboratories, 279, 282
Chemical Revolution
  controversy surrounding, 8
  development of, 129–152
  in England, 129–152
  inorganic analysis in, 269–270
  Lavoisier's, 8, 48, 255
chemical societies, 254
Chemical Society (Edinburgh University), 154
  formation by Black, 39
Chemical Society of London, 255
  The Chemist, 238, 243, 244, 247, 248, 263, 279
Chemin, 139
chemistry. see also agriculture; alkali manufacture; bleaching; medicine
  as an academic discipline, 13–25, 255–267, 283–287
  at Dissenting Academies, 54
  at English universities, 52, 53–54
  applications to other sciences, 17–18
  applied, 28, 29, 34, 39, 191, 195–196, 198–199, 237
  Burke's criticism of, 176–177
  as early adjunct of academic medicine, 15, 60, 61
  historiography of, 7, 12
  history of, in England, 7–8, 51
  Lavoisier's theory of, 173
  poetry based on, 240
  professionalization of, 286
  at the Royal Institution (London), 191–192
  in the Scottish Enlightenment, 11–49
  specialist expertise in, 10, 12, 15, 36, 199, 233, 237–255, 261–262, 277, 285
  Davy as model for, 10
  as a study for gentlemen, 11–49
Chenevix, Richard, 252, 269, 273, 277
  "Observations on Mineralogical Systems," 269, 275
  Thomson's dispute with, 275–278
  views of French chemistry, 286
Cheshire (England), Priestley in, 63
Chester (England), 59, 111
Chesterfield, Lord, Letters to his Son, 74
Chester Infirmary, 111
Children, John George, 214, 245, 246, 251, 252
China, 116
chlorine, 209
  Davy's discovery of, 9, 189, 218–235, 244, 256
Christie, John, 13, 23, 133, 134

Christie, John [bleaching entrepreneur], 34
City Dispensary (London), 207, 247
City Philosophical Society, 248
Clarke, Edward Daniel, 281
classification
  Cullen's work on, 20, 24–25, 26
  in early chemistry, 20
clay, Black's work on, 39
clergymen
  in intellectual clubs and societies, 14, 56, 57
  as subscribers to Priestley's books, 75
Clerk, David, 27, 28
"Club of Honest Whigs," 68
clubs
  intellectual, 14, 69, 75
  role in dissemination of chemical knowledge, 13, 14
  of scientists, critics of, 184, 185
clysters, use in pneumatic medicine, 110
Cobbett, William, 77, 88
  "Observations on Priestley's Emigration," 77
Cochrane, Archibald (Lord Dundonald), 40, 41
  Treatise on Agriculture and Chemistry, 198
Cochrane, Thomas, 45
Coleridge, Samuel Taylor, 167, 192–193
Collins, H. M., 229, 231, 234
commercial processes, secret nature of, 4
Comus, 122
Condillac, Étienne Bonnot de, 147
conservatism, 8–9, 172
  French Revolution and, 8
consumption. see tuberculosis
Cook, Captain James, 107, 112
Cooper, John Thomas, 248
Cooper, Thomas, 163, 185–186, 276, 277, 278
Copley Medal, 55, 69, 119, 245
  award to Priestley, 107, 108
Corpus Christi College (Oxford), 53, 174
Cort, Henry, 40, 41
Cottle, Joseph, 173
Crawford, Adair, 47, 134
Crawford, James, as predecessor of Cullen, 16
Crisp, Nicholas, 56
Cronstedt, A. F.
  blowpipe of, 279
  Essay Towards a System of Mineralogy, 271, 272
  portable laboratory of, 279, 280
Croonian Lecturer, Carlisle as, 251
Crosland, Maurice P., 67, 149, 176, 177, 232
Cruickshank, William, 136, 207

Crump, George, 157
crystallography, 259
  use in chemical analysis, 251, 279
Cullen, Robert [William Cullen's brother], 31
Cullen, Robert [William Cullen's son], 23
Cullen, William, 13, 43, 49, 50, 56, 71, 120, 207
  ether theory of, 22–24
  former students of, 38, 58, 112, 249
  lectures by, 13, 16, 18, 19, 20, 21, 24–25, 29, 31, 38, 44
  medical education of, 16
  as pioneer of Scottish chemistry, 11–13, 17, 22, 37, 48, 52
  professional medical career of, 15–16
  as professor at Edinburgh University, 11, 15, 16–17, 38–39
  as professor at Glasgow University, 15, 16, 17, 18, 26, 32, 37
  views on importance of chemistry, 7, 28–29
  work on bleaching, 33–34, 37
Cumberland (England), 107
Cuthbertson, John, 209

Daer, Lord, 165
Dalton, John, 195, 233, 234
  atomic theory of, 255–269
  on gases, 264
  as lecturer at Manchester Academy, 54
  lectures of, 266
  provincialism of, 266–267
  Royal Society of London and, 252–253
  works of
    New System of Chemical Philosophy, 264
    New System of Chemistry, 222
Darnton, Robert, 156
Darwin, Erasmus, 57, 58, 67, 76, 82, 157, 159, 160, 162, 164, 172, 174
Daumas, Maurice, 138
Davy, Humphry, 51, 88, 126, 161, 188–235, 251, 252
  biographies of, 190
  career of, 187, 188, 190–203
  as chemical pioneer, 188, 284, 286
  chemical theory of, 190
  controversy with Lavoisier, 9, 256
  controversy with Murray, 225–232
  critics of, 173, 174, 175
  discoveries of, literature reports on, 239
  discovery [with Beddoes] of and experiments on nitrous oxide as an intoxicant, 9, 152, 156, 166–171, 174, 175
  on electrochemistry, 252, 284

Davy, Humphry, (cont.)
  experiments of, 10, 201, 210, 227, 228–229, 235, 257
  on gases, 172
  on heat, 257
  instruments used by, 10, 46, 187, 263
  laboratories of, 189, 219–221, 238, 245
  lectures of, 9, 10, 187, 188, 190–204, 209, 218, 238, 239, 244, 285
  Napoleonic medal awarded to, 215, 219
  patrons of, 237, 245
  portrait of, 192
  as President of the Royal Society of London, 245, 266
  reaction to Dalton's atomic theory, 264, 265, 266, 269
  religious views of, 197
  view of public science, 285
  voltaic pile of, 189, 203–218, 233, 237, 238, 239, 270, 284
  works of, 257–258
    "An account of some experiments on galvanic electricity made . . . in the Royal Institution," 202
    "Discourse Introductory to a Course of Lectures on Chemistry," 195, 197–198, 199–200, 239, 240, 256
    Elements of Agricultural Chemistry, 199
    Elements of Chemical Philosophy, 203, 228, 256, 257, 265
    "On a combination of oxymuriatic gas and oxygen gas," 23, 221
    Researches, chemical and philosophical, chiefly concerning nitrous oxide, 167, 169
    "Researches on the oxymuriatic acid, its nature and combinations," 223
Davy, John [brother of Humphry], 197, 219, 221, 222, 225–226, 227, 228–230, 231, 251, 267
  "An Account of an Experiment made in the College Laboratory, Edinburgh," 228
  Memoirs of the Life of Sir Humphry Davy, 1, 176
DeLuc, Jean André, 41
Dent, W., 181, 182, 186
Derby (England), Mechanics' Institute at, 249
Derby Philosophical Society, 57, 82
Desaguliers, Jean Theophilus, scientific lectures and displays by, 6
Deskford, Lord, 34
Devon and Exeter Hospital, 161

diabetes, chemical testing for, 251
Dickson, Stephen, 150
didactics
  chemical, Cullen's use of, 19, 22, 26–27
  of chemical textbooks, 257, 258
  of Dalton's atomism, 265–266
  Scottish tradition of, 48
  use by specialist chemists, 239, 241
diffusion model, of science popularization, 94–95
disease
  environmental causes of, 105, 106, 107
  eudiometry and, 112, 118–119
Dissenters, support by Priestley of, 63, 65, 68
Dissenting Academies
  chemistry as a discipline at, 54
  Priestley as chemical lecturer at, 54, 63, 65, 95, 243
D.N.B. [Dictionary of National Biography], 96, 111, 160, 161, 174, 207
Dobson, Matthew, 110, 111, 116
  as Cullen's student, 38, 58, 112
  Medical Commentary on Fixed Air, 111
doctors
  in intellectual clubs and societies, 14, 56, 57, 58
  pneumatic therapy use by, 160–163
  as subscribers to Pneumatic Institution, 162
  as subscribers to Priestley's books, 75
  use of chemical analysis by, 251
Doncaster (England), 97, 251
Donn, Benjamin, 97, 98, 101, 103
Donovan, Arthur, 13, 17, 23, 34, 43, 133, 138
Dossie, Robert, 56
Downman, Dr. Hugh, 186
"Dr. Phlogiston," Priestley's depiction as, 180, 181
dry way chemical tests, 270
Dublin Medico-Philosophical Society, 106
Duncan, Andrew, 162
Dundonald, Lord. see Cochrane, Archibald (Lord Dundonald)
dyeing, chemistry applied to, 195

earths
  alkali. see alkali earths
  Bergman's classification of, 271
Edgeworth, Richard Lovell, 162, 167, 168
Edinburgh, 207, 277
  Cullen's and Black's civic activities in, 12
  Mechanics' Institute at, 248
  stimulating intellectual environment at, 13

Edinburgh Philosophical Society, 32, 33, 40

Edinburgh Review, 199, 204, 212, 217

Edinburgh University, 25, 34, 40, 54, 58, 111, 112, 154, 160, 161, 162, 175, 228, 231, 249, 250, 251
  Black as professor of chemistry at, 12, 13, 15, 39, 40, 41
  Cullen as professor of chemistry at, 11, 13, 15, 26, 34, 37
  Cullen's students at, 38, 39
  distinguished early faculty members at, 13
  Hope as professor of chemistry at, 42–43

Edinburgh University Medical School, founding of, 16

education
  chemical community's role in, 236–287
  chemistry's role in, 241

Educational Society of Bristol, 75

electricity, 7
  animal, 205
  galvanic, 208
  Priestley's work and books on, 63, 68, 69, 71, 72, 92
  relation to chemistry, 48

electrochemistry, 254
  Davy's work on, 187, 189, 210, 216, 237, 252
  experimental work on, 251

electrodes, use in electrolysis of water, 205, 210

electrolysis of water, 205, 206–207, 211–212, 213–214, 216
  contaminant effects on, 210, 211

elements. *see also* individual chemical elements
  isolation of, 9, 222, 225

Elements of Agricultural Chemistry, 198

Eller, J. T., 131

Ellis, Charles, 172

Ellis, 226

Encyclopédie, publishing of, 76

Enfield, William, 96

Engestrom, Gustav von, 279

England
  Enlightenment in. *see* English Enlightenement
  history of chemistry in, 7, 284
  provincial
    Priestley's chemical program in, 8, 49, 87
    societies in, 56–58, 69
    resistance to Lavoisier's theories in, 130–137

English Enlightenment, 6–10, 75, 155, 162
  fate of science in, 176–187, 235
  Priestley's role in, 50–90
  in the provinces, 6, 66, 75, 218, 234
  Scottish Enlightenment compared to, 51, 52, 58, 59
  social function of science in, 196–197, 242, 287
  uses of chemistry in, 52–63, 238, 286

Enlightenment
  end of science in, 176–187
  in England. *see* English Enlightenment
  in Europe, 14, 69–70
  in France, 8
  historiography of, 6–7
  role in the extension of scientific knowledge to public realm, 6, 259
  in Scotland. *see* Scottish Enlightenment
  as a social movement, 14

Epsom spa, 62

equivalents, chemical, 265, 269

Essays by a Society of Gentlemen at Exeter, 161

ether
  Cullen's ideas on, 22, 23
  Newton's ideas on, 22

euchlorine, 230

eudiometer(s), 93
  Cavendish's, 124, 125
  Dalton's, 265
  Fontana's, 118–123
  Landriani's, 118, 119, 122
  manufacture of, 93, 126, 127–128
  problems of standardization of, 93, 117, 125
  Volta's, 126, 135

eudiometry, 117, 135
  application to medicine, 93, 118–119, 120, 132
  Cavendish's work on, 124
  failure of, 93, 121, 125, 126, 127
  Ingenhousz's work on, 120–121
  Magellan's work on, 122–124
  origins of, 117
  Priestley's work on, 117–120

European Enlightenment, 14, 69–70

European Magazine and London Review, 96, 97, 99

evaerometer, Fontana's, 118

Ewart, Dr. William, 116

Exeter (England), 186

experimenters' regress, 229

experiments, 2, 3
  in Accum's *Chemical Amusement,* 246
  in lectures, 45–46, 201–203, 219–220
  publication of results of, 3
  replication of, 9

experiments (*cont.*)
  by students, 19
  witnessing of, 4, 224
experimentum crucis, on ammonia and
  muriatic acid, 227, 228, 231

Falconer, William, 111, 112, 115, 116,
  120
  as Cullen's student, 38
Faraday, Michael, 189, 216, 231, 248, 268
Farrar, Kathleen, 265
Ferguson, Adam, as colleague of Cullen,
  13
Ferguson, James, *Introduction to Electricity,* 74
Ferguson, John, early scientific road tours
  of, 6–7
Ferriar, John, 58, 161, 163, 172
Firmian, Count, 119
Fitzpatrick, Martin, 179
fixed air, Black's thesis on, 43
Fleck, Ludwik, 1, 235
fluorine compounds, Davy's studies on,
  232
Fontana, Felice, eudiometer and eudiomet-
  rics of, 118, 119, 120, 121, 122, 125,
  126, 127, 128
Fontanist method, 121, 124, 127
Fordyce, George
  chemical and medical career of, 60–61
  as Cullen's student, 38, 39, 60
  views on Cullen's importance to chemistry, 18
Forster, J. R., 270
Fortin, Nicholas, 138, 139
Fothergill, John, as Cullen's student, 39
Foucault, Michel, 22
Fourcroy, Antoine François de, 132, 133,
  144, 241, 259
  *Elements of Natural History and Chem-
  istry,* 143
Fox, Charles James, 181
France, 12, 22, 41, 70, 178, 207, 215, 219,
  243
  awareness of Priestley's work in, 94
  chemistry in, 154, 231, 234, 286–287
  crystallography in, 273–274, 275
  Enlightenment in, 8
  provincial academies and societies in,
  70, 76
Franklin, Benjamin, 68, 75, 124, 263
Freiberg (Germany), 272, 276
Freind, John, as lecturer at Oxford, 53
French Revolution, 55, 155, 172, 173,
  176, 179, 185, 193, 198
  conservative reaction to, 8
  Priestley's support of, 65

Frere, Hookham, 172
Fric, René, 131
Fulhame, Mrs., *Essay on Combustion,*
  261
Fullmer, June Z., 190
Fyfe, Andrew, 62

gagging bills, of Pitt government, 159
Galileo, 107
Galton, Mary Anne, 70
Galton, Samuel, 67, 70
Galvani, Luigi, 204, 205, 216
galvanism, 200, 209
  Davy's work on, 203–218
  experiments based on, in public lectures,
  202–203
  Lavoisier's theory of, 173
Gardner, D., 207, 247
Garnett, Thomas, 161, 191, 206, 247,
  249, 272
gases. *see also* air and airs, nitrous oxide
  atomic composition of, 264
  Dalton's work on, 264
  Priestley's work on, 8, 49, 50, 54
  therapeutic use of, 237
gasogene, 115
Gay-Lussac, Joseph Louis, 214, 215
  Davy's rivalry with, 214, 219, 222, 224,
  233
  iodine discovery by [with Thenard], 232
  Recherches Physico-Chimiques, 222,
  224
General Hospital (Birmingham), 161
Geneva (Switzerland), 126
genius
  Davy as public example of, 188–235
  Davy's notions of, 195
gentleman farmers, Lord Kames's views of,
  32–33, 198
gentlemanly science
  chemistry's role in, 11–49
  in the public realm, 25–37
Gentleman's Magazine, 174, 175
gentlemen, chemistry as a study of, 11–49
gentlemen chemists, 245
Geoffroy, E. F., affinity table of, 21–22
Geological Society, 255
Germany
  acceptance of Lavoisier's theories in,
  130, 133
  chemical analysis in, 270–271, 272–273
  chemical work in, 12, 154, 207, 208
Gibbes, George Smith, 213, 214
Giddy, Davies, 154
Gillray, James, 186, 201
Gisborne, Rev. Thomas, 186

Glasgow, 191
  Cullen's and Black's civic activities in, 12
Glasgow University, 25, 27, 47, 260
  Black as lecturer at, 39, 47
  Cullen as medical student at, 16
  Cullen as professor at, 15, 16, 17, 18, 26, 28, 32, 37
glass making, chemistry applied to, 195
Godwin, William, 185
goniometer
  Haüy's, 274, 276
  Wollaston's, 278
Goodchild, John, 37
Gooding, David, 189
Goodwyn, Edmund, 157
Gordon, John, 199
Gough, J. B., 131, 132, 138
Griscom, John, 233
Guerlac, Henry, 131
Gulstonian Lecturer, Saunders honored as, 61
gunpowder sermon, Priestley's, 177
Guy's Hospital (London), 61, 160, 161, 175, 239, 241, 249, 251
  chemical theater at, 250
Guyton de Morveau, Louis Bernard, 132, 133, 177
  "Economical Laboratory" of, 262

Hackney Academy, Priestley as lecturer at, 54, 65, 95, 243
Hadley, John, as lecturer in chemistry at Cambridge, 53
Haldane, Colonel Henry, 207
Hales, Stephen, 62, 108
  Vegetable Staticks, 107
Hall, Sir James, 162
Halévy, Elie, 249
Hamilton (Scotland), Cullen as a doctor in, 16
Harrington, Robert, 173, 175, 179, 191, 213, 214, 217
  The Death-Warrant of the French Theory of Chemistry, 153, 191
  A Treatise on Air, 151–152
Harrison, John, 139
Harrogate (England), 161
  spa at, 62, 272
Hartley, David, 172
Harveian Orator, Saunders honored as, 61
Hassenfratz, J. H., 141, 177
Hatchett, Charles, 226, 251, 252
Hauksbee, Francis, scientific lectures by, 6, 60
Hawkesworth, John, Voyages, 74

Haüy, René Just, 273, 274, 275
  crystallographic system of, 273–278
  goniometer of, 274, 276
  Werner's crystallography compared to, 273, 274
Haygarth, John, 111, 115
  as Cullen's student, 38, 111, 112
heat, 7. see also caloric
  Black's work on, 37, 41, 45, 47, 54
  Crawford's work on, 47
  Cullen's investigations on, 22, 24
  instrumentation for, 47
  Irvine's work on, 47
  latent, 41, 47
  specific, 41
  theories of, 48, 53, 131, 257
heat transfer, Black's work on, 37, 41–42, 44, 45, 48
Henry, Thomas, 54, 111, 115, 162, 164
Henry, William, 207, 233, 234, 258, 261, 262, 265, 266, 267, 281
  Elements of Experimental Chemistry, 233, 240, 259, 272
  Epitome of Chemistry, 282
  "Portable Chemical Chests" of, 262
Herschel, John, 245, 278
Hessian crucibles, 199
Hey, William, 75, 110, 111, 113
Higgins, Bryan, 134, 144, 149, 246
  Priestley's dispute with, 52, 88–90, 91
Higgins, William [nephew of Bryan], 134, 144, 145, 146
  Comparative View of the Phlogistic and anti-Phlogistic Theories, 267
history of science, 2, 4, 5
Hobbes, Thomas, views on dissemination of scientific knowledge, 5
Hodgskin, Thomas, 244
Holland, 52
Holmes, Frederic L., 138
Home, Everard, 251
Home, Francis, 34
  as academic chemist, 15
  Experiments on Bleaching, 33
  Principles of Agriculture and Vegetation, 33
Home, Henry (Lord Kames)
  Black's relationship with, 39–40
  as friend and patron of Cullen, 16, 17, 23, 26, 27, 31, 32, 34, 35, 36
  The Gentlemen Farmer, 32, 35, 39
  interest in agriculture, 31–33
Hope, John, 40, 41
Hope, Thomas Charles, 126, 175, 228, 229, 231
  as successor to Black, 42–43
Hopson, Charles, 134

Horner, Leonard, 261
Horticultural Society (London), 252
hospitals. *see also* individual hospitals
    in London, chemistry courses at, 249,
    250
Howard, H., 192
Hufbauer, Karl, 133
Hull Infirmary, 161
Hulme, Nathaniel, 111, 112
    A Safe and Easy Remedy . . . for the Re-
        lief of the Stone and Gravel, the
        Scurvy, Gout, etc., 115
Hume, David, 24, 27, 71, 76
    Essays, 11
    as friend of Cullen, 13, 23
    History of England, 74
Hunter, William, 23, 31, 44, 61
Hutton, Charles, 96
Hutton, James, 40, 41, 134
    Investigation of Principles of Knowl-
        edge, 11
hydrochloric acid, 209
hydrogen, 134, 142, 153, 213, 281
    production by electrolysis of water, 205,
        208, 211–214, 216

ice calorimeter, 133, 141
induction, Cullen's application to chemis-
    try, 19–20
Industrial Revolution, 7
Ingenhousz, Jan
    eudiometric work of, 120–122, 124,
        126, 127
    Experiments upon Vegetables, 120–121
instrument makers, 47, 106, 116, 127,
    164, 248
instruments. *see also* calorimeter; eudiome-
    ters; thermometers; voltaic pile
    black-boxing of, 206, 214, 216
    Cullen's and Black's development of, 49
    Davy's development of, 203–218
    development of, 93, 118
    Lavoisier's, 137–144
    in mineralogy, 274, 276, 277
    for pneumatic medicine, 164
    social construction of, 9–10, 93, 189
    for study of heat, 47
    use in science lectures, 206
    Wollaston's slide rule and scale, 268–
        269
iodine, 232
    Davy's studies on compounds of, 232
    French discovery of, 190, 232
Irvine, William
    as academic chemist, 15
    as Black's student, 39, 47
    theory of heat capacities of, 47

isomorphism, of crystals, 278–279
Italy, 118, 119, 122, 209
    eudiometric research in, 118, 119

Jameson, Robert, 273
Johnson, Joseph, as Priestley's publisher,
    71, 75
Johnson, Samuel, 93, 94
Jones, 122
Jones, William, *Chemical Science in Verse*,
    240
*Journal of Natural Philosophy, Chemistry
    and the Arts. see* Nicholson's *Journal*
journals, 252–253. *see also* individual
    journal titles
Joyce, Jeremiah, 243
    Dialogues on Chemistry, 239

Kames, Lord. *see* Home, Henry (Lord
    Kames)
Katterfelto, 122
Kaufman, Paul, 73
Keir, James, 57, 58, 70, 76, 123, 127, 129,
    136, 149, 162, 257
    The First Part of a Dictionary of Chem-
        istry, 146, 147, 150
Kendal (England), 97
    Kendal Academy, 96, 97
    Mechanics' Institute at, 248
Kerr, Robert, 149
kidney stones
    impregnated water as therapy for, 110,
        115
    Marcet's chemical studies on, 251
    pneumatic therapy for, 157
King, J., 201
Kinglake, Dr., 167
Kingston-upon-Thames (England), 60
Kipnis, Naum, 205
Kirwan, Richard, 124, 129, 134, 135, 141,
    145, 146, 149, 150, 151, 257, 273
    on chemical analysis, 282
    works of
        An Essay on Phlogiston and the Com-
            position of Acids, 135, 137, 144
        Elements of Mineralogy, 271–272
        *Essay on the Analysis of Mineral
            Waters*, 272
Klaproth, Martin Heinrich, 273
Knaresborough (England), 272
Knight, John, 244

laboratory(ies). *see also* apparatus; experi-
    ments; instruments
    Children's, 245
    Cullen's student instruction in, 38–39,
        49

portable, 60, 262–263, 279
privacy in, 2–3
private, 60
at the Royal Institution, 219–221
Lacey, Henry, 244
Lambeth Chemical Society, 254
LaMettrie, Julien Offray de, *L'Homme Machine*, 172
The Lancet, 247
Landriani, Marsilio, eudiometer of, 118, 119, 122, 128
Langer, Bernard, 131
language. *see* nomenclature; rhetoric
Laplace, Pierre Simon de, 133, 138, 139, 140, 141, 142
Latour, Bruno, 5, 92, 95
  Science in Action, 91
Lauderdale, Lord, 207
laughing gas. *see* nitrous oxide
Laurence, Richard, 174
Lavoisier, Antoine Laurent, 250, 258
  acidity theory of, 224
  Beddoes's views of, 171
  Black's reaction to, 46
  British resistance to theories of, 130–137, 139–140, 145–152, 187, 257
  as chemical pioneer, 284, 286
  Chemical Revolution of, 8, 48, 129–152, 237, 255
  criticism of, 173, 286
  Davy's disagreement with, 9, 256
  experiments of, 131, 132, 135, 141, 145, 153–154
  precision, 142
  instruments used by, 46, 126, 137, 138–144, 155, 171
  Priestley's disagreement with, 78, 86, 87, 128, 129, 136, 207
  studies on heat, 140–141
  studies on water composition, 133, 134, 135, 136, 143, 153–154
  style of, 46
  works of
    Mémoire sur la Chaleur, 140
    Méthode de Nomenclature Chimique, 133, 149
    "Rapport sur les nouveaux caractères chimiques," 143
    "Réflexions sur le phlogistique," 133
    Traité Élémentaire de Chimie, 133, 137, 139, 145, 147
law of constant proportions, 263–264
lawyers
  in intellectual clubs and societies, 14, 56
  as subscribers to Priestley's books, 75

lay people
  access to science by, 3, 4, 104
  view of science of, 2, 4
lectures
  Black's, 13, 37, 41–42, 44–47
  Brande's, 247
  chemical, in London, 247–248
  Cullen's, 13, 16, 17, 18, 19, 20, 21, 24–25, 29, 31, 38, 44
  Dalton's, 266
  Davy's, 190–203, 219
  experiment use in, 219–220
  Faraday's, 189
  by former Cullen students, 38
  instrument use in, 206
  by Priestley, 50–51
  public, 92, 94–104
Leeds (England), Priestley in, 63, 75
Leeds Infirmary, 110, 111
Leeds Literary and Philosophical Society, 110–111
Lemery, Nicholas, 25
Leslie, P. D., 134
Levere, Trevor, 155, 158
Levi, Primo, *The Periodic Table*, 153, 203
Lewis, William, 60
Leyden Jar, 73
Leydon, 112
Libavius, Andreas, 19
liberal education, chemistry's role in, 241
Lichfield, Earl of, as patron of Oxford, 53
Linnaeus, Carolus, classification system of, 20, 270, 271
Linnean Society (London), 252
Literary and Philosophical Societies, 248
lithontriptics, 115
Liverpool (England), 97, 110, 111, 252
Liverpool Literary and Philosophical Society, 228
Locke, John, 172
London, 75, 99, 188, 207, 219
  chemical community in, 251–252
  specialist careers in, 237, 238–255
  chemical lecturers in, 6, 38, 60–61, 247–250
  Dalton's dislike of, 266
  eudiometric tests on, 120
London Chemical Society, 242, 248, 254
London Gaslight and Coke Company, Accum as director of, 246
London Hospital, 248, 249, 250
London Institution, 241, 242, 245, 246, 259, 261
London Review, 96, 97, 99
"Loves of the Triangles," 172
Lucas, Charles, 108

Lunar Society of Birmingham, 76, 123,
    146, 158, 161, 162
  membership of, 58
  Priestley's association with, 57, 63, 64,
    65, 66–70, 146
Lundgren, Anders, 139
Lynn (England), 213

Macbride, David, 109, 110, 111, 113
  on cause of disease, 106–107
  Experimental Essays, 74, 106, 107
MacGowan, 40
Mackenzie, Sir George, 228
Macquer, Pierre-Joseph, 21, 22, 146
Magellan, John Hyacinthe, 41, 42, 43, 44,
    115, 122, 123, 127
  Description of a Glass-Apparatus, 122
  dispute with Cavallo, 122–124
magnesia alba, Black's thesis on, 43
Malton, Thomas, Compleat Treatise on
    Perspective, 75
Manchester (England), 75, 96, 97, 111,
    112, 207, 222, 233, 253, 264, 265,
    266
Manchester Academy, chemistry teaching
    at, 54
Manchester Circulating Library, 75
Manchester Infirmary, 110, 111, 161, 252
Manchester Literary and Philosophical So-
    ciety, 57, 58, 111, 112, 161, 186, 264,
    266
Manchester Mercury, 102
manufacturing, Cullen's and Black's con-
    tributions to, 12
Marcet, Alexander, 239, 249, 251, 272
Marcet, Jane
  Conversations on Chemistry, 194, 218,
    232, 239, 241–242, 249
  views on women in chemistry, 261
Martin, Benjamin
  philosophical shows of, 59
  scientific road tours of, 6–7
Martin, David, 42
Martine, George, Essays . . . on the Con-
    struction and Graduation of Ther-
    mometers, 47
Martineau, Harriet, 195
masonic societies, in provincial France, 70
materialism, English, Davy and, 172
materia medica, Cullen's lectures on, 44
mathematics, Newton's work in, 11
matter theory, 19, 23
May, John, 159
McEvoy, John, 65, 130–131, 142
mechanical philosophy, 7
  chemistry and, 18–19
Mechanics' Institutes, 242, 243, 248, 249

Mechanics' Magazine, 244
mechanism, Priestley's view of, 65
medical chemistry. see also pharmacy;
    pneumatic medicine
  culmination of Enlightenment, 153–187
medical education
  chemistry's role in, 16–17
  Cullen's, 16
  at English universities, 52
  at London hospitals and medical
    schools, 60, 249
  at Scottish universities, 15
Medical Society (Edinburgh), Cullen as
    founder-member of, 39
The Medical Spectator, 116, 151
medicine. see also doctors
  academic, chemistry as early adjunct of,
    13, 15, 60, 61
  chemistry and, 48
  Cullen's and Black's contributions to, 12
  pneumatic. see pneumatic medicine
  specialist professors in, 15
Medico-Chemical School (London), 247
Medico-Chirurgical Society, 251
Melhado, Evan, 131
Memoirs of the Manchester Literary and
    Philosophical Society, 253
mercury, Black's studies on expansion of,
    47
Mesmerism, 171, 204
  Lavoisier's theory of, 173
metallurgy, chemical basis of, 12, 29, 195
meteorology, 265
Mégnié, Pierre, 138
mice, Priestley's use for air experiments, 83
Mickleburgh, John, as lecturer in chemistry
    at Cambridge, 53
Miller, David, 245
Mill Hill Chapel (Leeds), Priestley as min-
    ister at, 63
Milton, Lord, 36–37
mineral earths, 270
  Cronstedt's classification of, 271
mineralogy, 246, 251, 252, 254
  blowpipe use in, 279–281
  chemical basis of, 12
  classification and, 20
  role in development of chemical analysis,
    269–283
  voltaic pile applied to, 238
  Wernerian, 273–276
minerals, analysis of, 237
mineral waters, 107–108, 161
  analysis of, 61, 62, 109, 270, 272, 281
  artificial, 77, 115, 116
  therapeutics based on, 62, 105, 113, 115
Mitchill, Samuel Latham, 166

Mitscherlich, E., 278–279
Money, John, 65, 96, 104
Monro, Alexander, II, 162
Monthly Review, 72, 151, 154, 171, 172, 228, 233, 256–257
Morris, R. J., 132
Morrison, Sir Jeremiah, 174
mortality statistics, 112
Morton, Earl of, 71
Moyes, Henry, 98, 101, 102
muriatic acid, 26, 209, 223, 224, 225, 227, 228, 230–231, 233. *see also* hydrochloric acid
Murray, J. A., 261
Murray, John [private lecturer in Edinburgh, d. 1820], 258
  controversy with Davy brothers, 222, 225–233
  Elements of Chemistry, 230, 249
  System of Chemistry, 230

Nantwich (England), Priestley's school at, 96
Napoleonic medal, given to Davy, 215, 219
natural history, Cullen's linkage of chemistry with, 20
natural philosophy, 4, 7, 17, 66, 80, 101
  Enlightenment public life and, 6
  growth of interest in, 59
  Priestley's work on, 50, 51, 79–80
natural sciences, specialist professors in, 15
natural theology, Davy's link of chemistry to, 197, 239–240
Naudin, 138
Newcastle (England), 96
  Mechanics' Institute at, 248
Newcastle Literary and Philosophical Society, 57–58
Newman, John, 279
New Meeting House (Birmingham), Priestley as minister at, 65
newspapers, role in Enlightenment, 14
Newton, Sir Isaac, 7, 11, 14, 21, 22, 256
  Opticks, 22
  as President of Royal Society of London, 55
Nicholson, Mr., 96
Nicholson, William, 129, 136, 143, 144, 145, 149, 150, 169, 205, 206, 207, 250–251, 252, 253
  A Dictionary of Chemistry, Exhibiting the Theory and Practice of that Art, 129, 150, 241
  First Principles of Chemistry, 258
  Journal of Natural Philosophy, Chemis-

try and Arts. see Nicholson's *Journal*
  role as chemical educator, 247
Nicholson's *Journal*, 204, 207, 208, 210, 222, 247, 253, 254, 259–260
niter, 31
nitric acid, 209
nitrogen, 210
nitrous acid, 26, 136
nitrous air, 109
nitrous-air eudiometer, 93, 121–122, 127, 134
nitrous air test, 85, 86, 88, 93, 117, 127
nitrous oxide, 152, 195, 201
  as anesthetic, 175
  composition of, 264
  physiological effects of, 9, 167–170, 175, 201
  respiration of, 152, 156, 166–175, 201, 204, 237
  role in culmination of Enlightenment medical chemistry, 153–187
  skepticism of claims for, 9, 172–173, 187, 216
nomenclature, chemical, 133, 148–149, 150, 152, 219, 257
Nooth, John Mervin, water-impregnation apparatus of, 113–115
Northumberland, Duke of, 71, 112
nosology, Cullen's work on, 20
Notcutt, William Russell, 167
Nottingham Infirmary, 165

oils, 29
Oldenburg, Henry, as Secretary of the Royal Society of London, 5
optics, Priestley's work and books on, 63, 71
Oxford University, 61, 75, 111, 154, 157, 174
  chemistry lectures at, 53–54
  medical education at, 52
oxygen, 132, 142, 153, 171, 174, 213, 227, 281
  Priestley's discovery of, 78
  production by electrolysis of water, 205, 208, 211–214, 216
  respiration of, 78
  use in pneumatic medicine, 159
oxymuriatic acid, 189, 209, 223, 224, 225, 227, 230, 256. *see also* chlorine

Pacchiani, 209
Paine, Thomas, 181
Paris, 78, 134, 138, 219
Paris Academy. *see* Académie des Sciences
Parker, William, 115, 127

Parkes, Samuel, 272
  Chemical Catechism, 232, 240
  Rudiments of Chemistry, 239, 258–259
Parkinson, James, 243
  Chemical Pocket Book, 126, 239, 262
Parr, Bartholomew, 161
patronage, aristocratic
  of English scientists, societies, and uni-
    versities, 53, 54–55, 56, 60, 63–64,
    66, 207
  of Scottish scientists, 15, 16, 35–36
Pearson, George, 115, 149, 250, 251, 272
Pearson, Richard, 160–161
  A Short Account of the Nature and
    Properties of Different Kinds of
    Airs, 161
Peart, Edward
  Anti-Phlogistic Doctrine Examined,
    150–151
  On the Composition and Properties of
    Water, 171
Peel, William, 209
Pennsylvania, Priestley's retirement in, 65,
  276
Pepys, William Hasledine, 126, 214, 221,
  246, 254
  as President of the Royal Institution, 245
Percival, Thomas, 110, 111, 115, 160
  as Cullen's student, 38, 58, 112
  Essays Medical and Experimental, 111
periodicals, role in Enlightenment, 14
Perrin, Carlton E., 133, 138
pharmaceutical chemistry, 247
  Wollaston's slide rule and scale use in,
    258
pharmacy, 261
  chemical basis of, 12, 28
phenomeno-technics
  chemists' use of, 7
  definition of, 3–4
  of gases, 51, 92
Phillips, Richard, 247–248, 249, 253, 254,
  272
Phillips, William, 273
Philosophical and Chemical Society, 246
philosophical chemistry, Cullen as advocate of, 29, 31, 36, 37
philosophical farmers, Davy's views of,
  198
Philosophical Magazine, 201, 204, 207,
  209, 213
Philosophical Society of Edinburgh
  Cullen's papers submitted to, 26, 27–28
  Essays and Observations, 27–28
Philosophical Society of London, 248
Philosophical Transactions. see under
  Royal Society of London

philosophy, influence on Cullen, 23
philosophy of science, 2
phlogiston, 120, 139, 181
  as principle of combustion, 23, 134
phlogiston theory, 118, 131, 134, 136–
    136, 146, 151, 154, 213, 258
  overthrow of, 131, 133, 135, 140
phosgene, 230
  Davy's discovery of, 227
phosphorus, 239, 256
photogen, 213
physicians. see doctors
physiology, 251
Pitt, William, 70, 159, 185
plagiarism, Black's fear of, 43, 44
platinum-purification process, 245
Playfair, John, 228, 236
  "Biographical Account of Hutton," 236
Plummer, Andrew, as predecessor of Cul-
    len, 16, 27
pneumatic chemistry, 48, 53, 97, 98, 108,
    128, 152
  Dalton's atomic theory and, 265
  Lavoisier's work on, 132
  phlogiston theory in, 134
  Priestley's views on, 9, 98
Pneumatic Institution (Bristol), 157–166,
    167, 172, 173, 175, 187, 188, 201
pneumatic medicine, 9, 128, 133, 159,
    160, 251
  apparatus for, 112–114, 115, 116
  birth of, 105–117
  criticism of, 172–173
  public health reform and, 105, 106
  therapeutic techniques in, 109–110,
    111, 157, 161
pneumatics, 7
pneumatic trough, 108
poetry, chemistry themes in, 173, 240
polymorphism, of crystals, 278–279
porcelain manufacture, chemistry applied
    to, 195
Port, J. H., 271
portable laboratories, 262–263, 282
  blowpipe as part of, 279
  Cronstedt's, 279, 280
  Shaw's, 60
Porter, Roy, 6
Portugal, 41
potash, 222
potassium, 214
  Davy's discovery of, 9, 189, 212, 222,
    234
Price, Richard, 68, 70, 179, 181
Priestley, Joseph, 12, 58, 61, 62, 112, 114,
    133, 151, 152, 155, 161, 172, 211,
    263, 277

on airs [gases], 8, 49, 69, 73, 74, 77–78,
80, 83, 86, 87–88, 97, 98, 131,
135, 143–144, 237
apparatus of, 83–84, 265, 276
bibliography of works, 63
biography of, 63
Boyle compared to, 87
career of, 51–52, 63–76
cartoons depicting, 179–183
chemistry of, in public education, 93–
105
critics of, 173, 176, 177, 178, 181, 184,
185–186, 187, 193
as Davy's mentor, 234
disagreement with Lavoisier, 129, 130,
136–137, 142, 143–144, 145, 148,
149, 151–152, 154, 187, 207,
257
discoveries of
discussion in provincial societies, 57,
58
dissemination by itinerant lecturers,
59, 93–105
Earl of Shelburne as patron of, 63–65,
67, 70, 94, 95
on electricity, 68, 69, 72, 81, 207
English Enlightenment and, 50–90, 156,
158, 235
as entrepreneur, 71–72
eudiometry and, 93
experiments of, 77–90, 91, 94, 143–
144, 145
French chemistry and, 286
at Hackney Academy, 54, 65, 95
laboratory of, 65, 67, 99
lectures by, 50–51, 79, 82, 83, 95, 237
letters of, 63
in London, 65, 67, 99
as member of Lunar Society of Birmingham, 57, 63, 64, 65, 66–70, 146
as a minister, 63, 65, 71
moral code of, 72, 78, 85, 86, 87, 90,
91, 114, 117, 187, 267
on optics, 63, 71
patrons and patronage of, 63–65, 67,
70, 94, 95, 112, 184
personality of, 85, 87, 89
as pioneer in chemistry, 50, 51
pneumatic chemistry of, 105
portrait of, 64
private financial support of, 67, 68, 112
public lecturers and, 93–105
public readership of, 73–76
role in popularization of chemistry, 8, 9,
49, 194
Society of Arts and, 56, 63
subscribers to works by, 72, 74–75
in the United States, 65, 67, 88, 129,
207, 276
view of science as public culture, 66, 67,
78, 117, 196, 197, 216–217, 235,
242, 267, 278, 284, 285
at Warrington Academy, 54, 63, 78, 79,
82, 95
on water impregnation, 77, 109
water-impregnation machine of, 112–
115
works of, 63, 64–65, 66, 71, 72, 83–84,
91
   An Answer to Mr. Paine's Age of Reason, 77
   Chart of Biography, 71
   Directions for Impregnating Water
with Fixed Air, 77
   Experiments and Observations on
Different Kinds of Air, 72, 74,
78, 80, 81, 84, 98, 99, 109–110,
113–114, 117, 119
   Experiments and Observations Relating to Various Branches of Natural Philosophy, 80, 81, 116
   Familiar Introduction to the Study of
Electricity, 71, 72
   The History and Present State of Dis-
coveries Relating to Vision, Light
and Colours, 72, 74, 76, 78, 79,
96
   History and Present State of Electric-
ity, 66, 68, 72, 74, 78, 79, 85
   Lectures on Oratory, 83
   Letters to . . . Edmund Burke, 185
   Memoirs, 68, 74
   Miscellaneous Observations Relating
to Education, 95
   New Chart of History, 71
   Philosophical Empiricism, 50, 88, 94
as writer, 71, 73–75, 77–90, 91, 216–
217, 258
writing style of, 51, 52, 77, 168
Priestley, William [Joseph's son], 181
Pringle, Sir John, 106, 113, 119, 120, 124
   Copley Medal address of, 108, 109, 119,
197
   Observations on the Nature and Cure of
Hospital and Jayl Fevers, 106
   as President of Royal Society of London,
55, 69
printed materials, role in Enlightenment,
14
professional groups
   members of, role in the Enlightenment,
14–15
   in Scotland, 14–15

professionalization, of chemistry, 10, 14, 15
role of societies in, 55
professorships. *see* individual universities
Prout, William, 268
Bridgewater Treatise, 197
provinces, English. *see* England, provincial
public culture, science as, 1–10
public education, Priestley's chemistry in, 93–105
public health, 112, 119–120
pneumatic medicine and, 105, 106
public realm, science in, 1–10
public science, chemistry as, 156
Priestley's role in, 50–90
in Scottish Enlightenment, 11–49
publishing
role in Enlightenment, 14
of scientific periodicals, 252–254

Quarterly Review, 257
Quincy, John, 60

radicalism, 8
Ramsden, Jesse, 139
rarities, debates on, 4
Rathbone Place, 108
Reece, Richard, 240, 241
"Chests of Chemistry" of, 262
regenting, abolition by Scottish universities, 15
Reid, William Hamilton, 185
The Rise and Dissolution of the Infidel Societies of this Metropolis, 185
Reign of Terror, 148
Renshaw, Dr. Daniel Lorimer, 174
rhetoric
Beddoes's, 163
Burke's, 179, 181, 183
Davy's, 9, 10, 189, 194–195, 197, 198, 199–200, 203, 212, 216, 219, 239, 243, 255
of Enlightenment public science, 259
Priestley's, 8, 76, 82, 83, 85, 88, 89, 96, 285
scientific, 3, 7
Richter, J. B., 263–264
Ritter, Johann Wilhelm, 208, 213
rivalry, between doctors and instrument makers, 106
Roberts, Lissa, 140, 147
Robinson, Bryan, 22
Robinson, Eric, 41
Robison, John, 173, 185
as Black's student and literary executor, 39, 41, 44–45, 46, 48

opinions on Cullen of, 25
Proofs of a Conspiracy Against All the Religions and Governments of Europe, 184–185
views on chemistry, 9, 184–185
Rockingham, Lord, 70
Roe, Richard, 82
Roget, Peter Mark, 167, 169, 243, 251–252
Thesaurus of, 169
Romé de l'Isle, Jean Baptiste, 273
Rotheram, Caleb, 96, 99
Rousseau, Jean Jacques, 185
Royal College of Arts, 251
Royal College of Physicians (London), 52, 61
Royal College of Surgeons, 251
Royal Institution (RI) (London), 161, 192, 215–216, 218, 241, 247, 251, 256, 265
Dalton's lectures at, 266
Davy as lecturer at, 9, 187, 188–189, 191–203, 217, 230, 231, 237, 238, 245
Journal of Science and the Arts, 253–254
Journal of the Royal Institution, 208
laboratory at, 189, 219–221, 238, 246
Priestley's lectures at, 285
voltaic pile of, 215–216, 218, 234
Royal Medical Society, 162
Royal Society of Edinburgh, 233
Transactions of the Royal Society of Edinburgh, 253
Royal Society of London, 6, 60, 61, 76, 111, 113, 139, 191, 205, 216, 217, 227, 255, 287
Bakerian Lectures, 189, 203–204, 209–210, 211–212, 214, 215, 217
Beddoes' ostracism by, 158, 162
Copley Medal of, 55, 69, 107, 245
Davy's career at, 9, 237, 266
debates on secrecy of scientific knowledge at, 4, 5
Philosophical Transactions, 5, 55, 69, 77, 107, 119, 124, 204, 208, 252, 253, 264
presidency of, 55, 69, 245, 266
Priestley and, 55, 69, 78
Royal Medal, 266, 268
royalty, 71
as patrons of science, 60
Rumford, Count, 191, 201, 257
Rupp, T. L., 151
Russell, Colin, 243
Russell Institution, Singer's battery at, 246

safety lamp, 245
St. Bartholomew's Hospital (London), 249, 251
St. George's Hospital (London), 250
St. John, James, 149
St. Paul's Coffee House (London), 68
St. Thomas's Hospital (London), 53, 61, 175, 249
Salisbury (England), 97
Salisbury and Winchester Journal, 97
salts, 29
  Cullen's classification of, 26–27
  Cullen's work on, 35–36
  Davy's work on, 227
  quantitative analysis of, 263, 264, 270–271
Sandwich, Earl of, 71
Saunders, William, 161
  as chemical and medical lecturer, 61
  as Cullen's student, 38, 39, 60
  Treatise on the Chemical History . . . of . . . Mineral Waters, 272
Savile, Sir George, 71, 112
Sayers, James, 181, 183, 186
Scandinavia, chemical analysis in, 270–271
Scarborough (England), 60
  spa at, 62
The Sceptic, 173, 179, 204
Schaffer, Simon, 5, 65, 68, 96
Scheele, Carl Wilhelm, 223, 272
Schofield, Robert, 65, 79
Schwediauer, Franz Xavier, 40
Schweppe, J. J., 115
scientific books, reviews of, 72–73
scientific community
  as model of ideal open society, 1
  Priestley's view of, 51
scientific discourse, as type of rhetoric, 3
scientific doctrine, interpretation by audiences, 95
scientific instruments. see instruments
scientific knowledge
  methods of dissemination of, 3–4
  public access to, 1–2
scientific periodicals, 253–254
scientific phenomena, privacy role in observation of, 2–3
scientific societies. see also clubs; societies
  in England, role in chemical progress, 54–55
  in Scotland, 26, 27
Scotland, 207
  Chemical Revolution in, 133–134
  chemistry development in, 7, 11–49, 54
  as intellectual center in eighteenth century, 14

medical education in universities of, 52
Scottish Enlightenment, 6, 7
  chemistry in, 11–49
  English Enlightenment compared to, 51, 52, 58, 59
  social acceptance of chemistry in, 37–49
scurvy, 106, 107, 111, 112, 116
  pneumatic therapy of, 157, 161
secrecy
  of manufacturing processes, 40
  in science, debates on, 4
Sedgwick, Adam, 265
Senebier, Jean, Recherches sur l'Influence de la Lumière Solaire, 126
sexual impropriety, accusations of against pneumatic doctors, 174
Shapin, Steven, 1, 5
Sharples, E., 64
Shaw, Peter, 33, 108–109, 264
  Cullen's attack on, 18
  as public scientific lecturer, 59–60
Sheffield (England), 207
  Mechanics' Institute at, 249
Shelburne, Earl of, 88, 97
  as patron of Priestley, 63–64, 67, 70, 94, 95
Shelley, Mary, Frankenstein, 236
Sherwin, Dr. John, 151
Shrewsbury (England), 59
Shrewsbury Infirmary, 165
Siegfried, Robert, 132
Simond, Louis, 195, 197
Singer, George John, 246
slide rule, Wollaston's, 268
Sloane, Sir Hans, as President of Royal Society of London, 55
Small, William, 70
smell, as indicator of air quality, 125
Smith, Adam, 24
  "Essay on the History of Astronomy," 30–31
  as friend of Cullen, 13, 23
Smith, John, as lecturer at Oxford, 53
Smith, Robert, Elementary Parts of Optics, 75
Smollett, Tobias, 13
social structure, of 18th century England, 57
societies. see also clubs; scientific societies
  in provincial England, 56–58
  role in dissemination of chemical knowledge, 13, 14, 66
  role in scientific professionalization, 55, 56, 58
  sociability in, 69–70, 105
Society for Bettering the Condition of the Poor, 1901

Society for Constitutional Information, 243
Society for Improvement of Arts and Manufactures, 33
Society for Philosophical Experiments and Conversations, 149
Society of Arts (London), 60, 63
    encouragement of applied science by, 56, 58
Society of Gentlemen (Exeter), 186
sociology of science, 2, 10, 66
soda, 222
sodium, Davy's discovery of, 9, 189, 212, 222, 234
soil analysis, 270
    Davy's procedure for, 199
Southey, Robert, 159, 167, 205
spas, in England, 62
specialization
    in chemistry. see under chemistry
    of professors, 15
    in science, 8
Stahl, Georg Ernst, 21, 60, 131
    chemical terms derived from, 19, 24
    phlogiston theory of, 23
Stansfield, Dorothy, 155, 158, 174
steam engine, Watt's patent protection for, 43, 44
Stewart, Larry, 6–7
Stubbe, Henry, views on secrecy of commercial processes, 4
student societies, at Edinburgh University, 39
subscriptions
    to scientific books, 72, 74–75
    to support pneumatic medicine, 162
Suffolk (England), Priestley as a Dissenting preacher in, 63
sulfur, 109, 239, 256
surgeons. see doctors
Surrey Institution, 241
Sweden, 12, 41, 224, 271
Sylvester, Charles, 207, 208, 210
    Elementary Treatise on Chemistry, 214, 239, 249

Table of Chemical Nomenclature, Pearson's translation of, 250
tanning, chemistry applied to, 195, 196
Tatum, John, 248
Taunton (England), 75, 110, 111
"tea tray" laboratories, 263, 279
technology. see also arts; chemistry, applied
    chemical principles applied to, 28–29
temperature measurement, Black's work on, 47

textbooks
    on chemistry, 49, 255–259, 272
    Davy's work reported in, 239, 240
    science, 58–59
Thackray, Arnold, 265, 267
Thenard, Louis-Jacques, 214, 215, 224
    Davy's rivalry with, 214, 219, 222, 233
    iodine discovery by [with Gay-Lussac], 232
thermogen, 213
thermometers, 47, 140, 265
Thomson, John, 207
Thomson, Thomas, 141, 231, 233, 234, 253, 257, 260, 265, 268, 269, 277
    on Dalton's atomic theory, 263–264
    dispute with Chenevix, 275–278
    Elements of Chemistry, 258
    as mineralogist, 273, 275, 276
    System of Chemistry, 232, 240, 249, 258, 264
    views on Cullen's importance in chemistry, 11, 18, 34
Thornton, Robert, 16ʳ
    The Philosophy of Medicine, 160
Tilloch, Alexander, 204, 207, 209, 253
Tobin, William, 167–168
Tonbridge (England), Children's Laboratory at, 245
Tooke, John Horne, 155
Tower of London, 243
Traill, Thomas Stewart, 228
Treatise on Soils and Manures, 199
Trotter, Thomas, 157
tuberculosis, pneumatic therapy of, 157, 161
Tunbridge Wells spa, 62
Turner, Matthew, as lecturer at Warrington Academy, 54, 95
Turner, William, 57
Tuscany, Grand Duke of, 118

Underwood, Mr. 201
United States
    Priestley in, 65, 67, 88, 129, 207, 276
    Wollaston's slide rule and scale use in, 258
universities. see also individual universities
    English, chemical instruction at, 53–54
    Scottish, reforms in, 15
Uppsala University, 271
Ure, Andrew, 233, 241, 268
urine, chemical analysis of, 251

Varley, Samuel, 248
Vaughan, Benjamin, 100, 122
vegetable acid, 26

Venel, G. F., Cullen's attack on, 18
Vigani, John Francis, as medical lecturer at Cambridge, 53
vitriolic acid, 26
vocations, in 18th century science, 55
Volta, Alessandro, 124, 205, 206, 208
  eudiometer of, 126, 135
voltaic pile, 10, 202
  construction by French physicists, 214–215
  Davy's work on, 9, 189, 203–218, 233, 237, 238, 270, 284
  Pepys's, 246
  use in chemical analysis, 212–213, 214, 222–223, 234, 237, 251
  use in mineral analysis, 238
Voltaire, 185

Wakefield (England), 96
Wakley, Thomas, 247
Waldman, Professor, 236
Wales, Prince of, 71
Walker, Adam, 110, 115, 194, 214
  as disseminator of Priestley's discoveries, 96–97, 98, 99, 101, 102, 103
  Philosophical Estimate of the Causes . . . of Unwholesome Air, 112
Walker, Ezekiel, 213
Wall, Martin, as lecturer at Oxford, 53–54
Waller, John, 53
Wallerius, J. G., 271
Warltire, John, 135
  as lecturer at Warrington Academy, 95–96
  as public lecturer, 97, 99–100, 101–102, 103
Warren, John, 110, 111, 112
Warrington Academy, 111, 112, 270
  chemistry teaching at, 54, 95–96
  Priestley at, 54, 63, 78, 79, 82, 95
water
  composition of, 133, 135, 136, 137, 143, 153, 264
  electrolysis of, 205, 206–207, 210, 211–212, 213–214, 216, 247, 250–251
  Lavoisier's experiment on, 133, 135, 153
  Priestley's work on, 136
water impregnation, 93, 109, 112, 113, 114–115
  apparatus for, 113, 114–115, 128
Watson, Richard, 75
  Chemical Essays, 53
  as lecturer at Cambridge, 53

Watson, William, 68
Watt, Gregory [son of James], 167, 170, 188
Watt, James, 56, 57, 76, 116, 162, 174, 263
  collaboration with Beddoes, 157, 158
  Considerations on the Medicinal Use of Factitious Airs [with Beddoes], 157
  as instrument maker, 39, 164
  Joseph Black and, 39–40, 41, 46, 58
  steam-engine patent rights of, 43
  thermometer manufacture by, 47
Watt, James, Jr., 163, 170, 186
Watt, Jessie [daughter of James], 157
Webster, John, 239, 240
Wedgwood, Josiah, 56, 57, 67, 68, 76, 96, 97, 136, 141, 167
Wedgwood, Thomas, 162, 167, 168
Wedgwood pestles and mortars, 199
Weindling, Paul, 254–255, 282–283
Weldon, Walter, 260
  Popular Explanation of Chemistry, 255, 260–261
Werner, Abraham Gottlob, 272–273
  crystallographic system of, 272–278
  Haüy's system compared to, 274–276
  On the External Characters of Minerals, 273
West Indies, Cullen as surgeon in, 16
wet way chemical tests, 270, 281–282, 283
White, William, 119
Whitehurst, John, 70
Wilcke, Johan Carl, 41
Wilkinson, William, 68
Wilson, George, as public scientific lecturer, 59
wine glasses, as chemical apparatus, 262, 263
Withering, William, 54, 58, 67, 70, 110, 116, 160, 271
  as Cullen's student, 38, 39
Wittgenstein, Ludwig, 284
Wollaston, William Hyde, 201, 264, 265
  blowpipe of, 281
  Copley Medal awarded to, 245
  goniometer of, 278
  slide rule and scale of, 268–269, 278
  as supporter of Dalton, 268
  "A Synoptic Scale of Chemical Equivalents," 268
women
  in Enlightenment culture, 76
  exclusion from chemical research, 261
  patients, of pneumatic physicians, 174

women (*cont.*)
  in science lecture audiences, 76, 194, 241, 261
  scientific educators' view of, 194
  as subscribers to scientific books, 76

Yarmouth (England), 75
Yelloly, John, 250, 251
York (England), 119
York Courant, 97
Young, Arthur, 124